Project Management, Planning, and Control

Project Management, Planning, and Control

Managing Engineering, Construction, and Manufacturing Projects to PMI, APM, and BSI Standards

Sixth Edition

Albert Lester

ELSEVIER

AMSTERDAM • BOSTON • HEIDELBERG • LONDON
NEW YORK • OXFORD • PARIS • SAN DIEGO
SAN FRANCISCO • SINGAPORE • SYDNEY • TOKYO
Butterworth-Heinemann is an imprint of Elsevier

Butterworth-Heinemann is an imprint of Elsevier
The Boulevard, Langford Lane, Kidlington, Oxford, OX5 1GB, UK
225 Wyman Street, Waltham, 02451, USA

Notice
No responsibility is assumed by the publisher for any injury and/or damage to persons or property as a matter of products liability, negligence or otherwise, or from any use or operation of any methods, products, instructions or ideas contained in the material herein. Because of rapid advances in the medical sciences, in particular, independent verification of diagnoses and drug dosages should be made

British Library Cataloguing in Publication Data
A catalogue record for this book is available from the British Library

Library of Congress Cataloging-in-Publication Data
A catalog record for this book is available from the Library of Congress

ISBN–13: 978-0-08-098324-0

For information on all Butterworth-Heinemann publications
visit our website at books.elsevier.com

Printed and bound in the United States

06 07 08 09 10   10 9 8 7 6 5 4 3 2 1

Working together
to grow libraries in
developing countries

www.elsevier.com • www.bookaid.org

Contents

Sample exam questions and answers can be found on the companion website for this book:
http://booksite.elsevier.com\9780080983240

Preface

The shortest distance between two points is a straight line.
—Euclid

The longest distance between two points is a shortcut.
—Lester

Project management, like every other aspect of life, is constantly changing and developing, so that for a textbook to remain relevant, it inevitably has to be updated periodically to reflect these changes. In the case of project management training, these changes are mainly ones of emphasis on particular topics as perceived by setters of examination papers or compilers of national or international standards. The soft topics now feature more widely in examination questions and although these are not unique to project management they are, of course, part and parcel of good general management. Whether industry attaches the same importance to some of these emphasised topics is a matter of debate. Nevertheless, a number of changes have been incorporated in this 6th edition to enable the reader to keep abreast of events.

Shortly after the 5th edition was published, The APM added Project Governance to the syllabus of the APMP examination and a new chapter on this important topic has therefore been added. This was specially written by David Shannon, Managing Director of Oxford Project Management Ltd., who is an acknowledged expert on the subject and was the founder of the APM Specific Interest Group on Governance.

A major change in the book is the replacement of the Hornet Windmill computer program by the well-known Primavera P6 project management program. Although Hornet Windmill still is an excellent system, it is unfortunately not marketed anymore by its originators, Claremont Controls. I decided therefore to include a new chapter specially written by Arnaud Morvan from Milestone Ltd., which describes the latest version of Primavera P6 (now part of Oracle), which was used on numerous infrastructure projects such as Transport for London, Network Rail, and Heathrow Airport. Although P6 incorporates planning, performance, cost, earned value, and risk management I must still emphasise that although modern computerised project management programs are now very sophisticated, more user friendly, and have greater functionality, unless a planning network (especially for large or complex projects) is first drafted by hand by the project manager and his team, the full benefits of network analysis, such as the earliest possible completion, may not be realised.

Other changes include incorporating precedence (AoN) diagrams, scaled networks, and bar charts in the chapter on basic network principles, merging the sections on graphical and computer analysis, and combining the section on arithmetical network analysis with float, since the manual calculation of float is really only a matter of arithmetic. The importance of an understanding and the use of float to obtain the shortest completion date with the most efficient use of resources is still not fully appreciated by many planners and this has therefore been highlighted by additional text in this new combined chapter.

On network analysis, the book covers the whole spectrum, from the first principles of CPM through some early techniques (now only of historical interest) to the latest sophisticated computer software.

Although, with the use of modern computers, it is not necessary anymore to number the activities before carrying out an analysis (either manually or by computer), the section on numbering has still been retained for the benefit of readers interested in the historical development of network analysis.

The chapter on cost control and earned value analysis (EVA) has been amended to make it clearer that there is no difference between SMAC, the earned value system developed by Foster Wheeler Power Products in 1978 and first mentioned in the 2nd edition, and EVA (or EVM) as it is now universally called. A number of new developments have taken place over the last few years, and these have also been incorporated.

Apart from the changes to the text of the last (5th) edition, two major topics have been added. The first is a new chapter on agile project management written by Graham Collins, who, with colleagues, has pioneered methods that are used successfully to develop strategic programmes. His teaching at University College London covers the latest approaches to agile project management and his current research and consulting is in collaboration with Thought-Works. The second is a new chapter on BIM (Building Information Modelling) contributed by Clive Robinson, Products Manager from Tekla (UK) Ltd. BIM is now being adopted by many of the large consulting and construction companies, and with strong backing from government departments, will no doubt become the standard methodology for the construction industry in the future.

The glossary has been updated with the latest terminology adopted by the revised BS 6079-1:2010 (Project Management) and the new ISO 21500:2012 (Guidance on Project Management), as have the cross-references to the APM and PMI Bodies of Knowledge. Finally, some of the older publications have been dropped from the revised bibliography, which has been brought up to date by including the latest books on project management.

A. Lester

Foreword to the First Edition

by Geoffrey Trimble, Professor of Construction Management

University of Technology, Loughborough

A key word in the title of this book is 'control'. This word, in the context of management, implies the observation of performance in relation to plan and the swift taking of corrective action when the performance is inadequate. In contrast to many other publications which purport to deal with the subject, the mechanism of control permeates the procedures that Mr. Lester advocates. In some chapters, such as that on Manual and Computer Analysis, it is there by implication. In others, such as that on Cost Control, it is there in specific terms.

The book, in short, deals with real problems and their real solutions. I commend it therefore both to students who seek to understand the subject and to managers who wish to sharpen their performance.

Acknowledgements

The author and publishers acknowledge with thanks all the individuals and organisations whose contributions were vital in the preparation of this book.

Particular acknowledgement is given to the following four contributors:

Oracle and Milestone Ltd. for providing the description of their highly regarded Primavera P6 computer software package.

David Shannon for writing the chapter on Project Governance.

Andrew Bellerby and Clive Robinson of Tekla (UK) Ltd. for contributing the description and procedures for BIM.

Graham Collins from UCL for providing the description of Agile Project Management.

The author would also like to thank the following for their help and cooperation:

The National Economic Development Office for permission to reproduce the relevant section of their report 'Engineering Construction Performance Mechanical & Electrical Engineering Construction, EDC, NEDO December 1976'.

Foster Wheeler Power Products Limited for assistance in preparing the text and manuscripts and permission to utilize the network diagrams of some of their contracts.

Mr. Peter Osborne for assistance in producing some of the computerized examples.

Mr. Tony Benning, my co-author of 'Procurement in the Process Industry', for permission to include certain texts from that book.

British Standards Institution for permission to reproduce extracts from BS 6079-1-10 (Project management life cycle and BS5499-10-2006 (Safety signs). British Standards can be obtained in PDF or hard copy formats from the BSI online shop: "http://www.bsigroup.com/Shop" www.bsigroup.com/Shop or by contacting BSI Customer Services for hardcopies only: Tel: +44 (0)20 8996 9001, Email: "mailto:cservices@bsigroup.com" cservices@bsigroup.com.

A.P. Watt for permission to quote the first verse of Rudyard Kipling's poem, 'The Elephant's Child'.

Daimler Chrysler for permission to use their diagram of the Mercedes-Benz 190 car.

The Automobile Association for the diagram of a typical motor car engine.

WPMC for their agreement to use some of the diagrams in the chapters on risk and quality management.

Jane Walker and University College London for permission to include diagrams in the chapters on project context, leadership, and negotiations.

CHAPTER 1

Project Definition

Chapter Outline

Project Definition

Many people and organizations have defined what a project is, or should be, but probably the most authoritative definition is that given in BS 6079-2:2000 *Project Management Vocabulary*, which states that a project is:

> '*A unique process, consisting of a set of co-ordinated and controlled activities with start and finish dates, undertaken to achieve an objectives conforming to specific requirements, including constraints of time, cost and resources.*'

The next question that can be asked is 'Why does one need project management?' What is the difference between project management and management of any other business or enterprise? Why has project management taken off so dramatically in the last 20 years?

The answer is that project management is essentially management of change, while running a functional or ongoing business is managing a continuum or 'business-as-usual'.

Project management is not applicable to running a factory making sausage pies, but it will be the right system when there is a requirement to relocate the factory, build an extension, or produce a different product requiring new machinery, skills, staff training, and even marketing techniques.

It is immediately apparent therefore that there is a fundamental difference between project management and functional or line management where the purpose of management is to continue the ongoing operation with as little disruption (or change) as possible. This is reflected in the characteristics of the two types of managers. While the project manager thrives on and is *proactive* to change, the line manager is *reactive* to change and hates disruption. In practice this often creates friction and organizational problems when a change has to be introduced.

Projects may be undertaken to generate revenue, such as introducing methods for improving cash flow, or be capital projects that require additional expenditure and resources to introduce a change to the capital base of the organization. It is to this latter type of project that the techniques and methods described in this book can be most easily applied.

Figure 1.1 shows the type of operations that are suitable for a project type of organization and which are best managed as a functional or 'business-as-usual' organization.

Both types of operations have to be managed, but only the ones in column (a) require project management skills.

It must be emphasized that the suitability of an operation being run as a project is independent of size. Project management techniques are equally suitable for building a cathedral or a garden shed. Moving house, a very common project for many people, lends itself as effectively to project management techniques such as tender analysis and network analysis as relocating a major government department from the capital city to another town. There just is no upper or lower limit to projects!

As stated in the definition, a project has a definite starting and finishing point and must meet certain specified objectives.

Broadly these objectives, which are usually defined as part of the business case and set out in the project brief, must meet three fundamental criteria:

1. The project must be completed on time;
2. The project must be accomplished within the budgeted cost; and
3. The project must meet the prescribed quality requirements.

(a) Project organisation	**(b)** Functional or line organisation
Building a house	Manufacturing bricks
Designing a car	Mass-producing cars
Organising a party	Serving the drinks
Setting up a filing system	Doing the filing
Setting up retail cash points	Selling goods & operating tills
Building a process plant	Producing sausages
Introducing a new computer system	Operating credit control procedures

Figure 1.1
Organization comparison

These criteria can be graphically represented by the well-known project triangle (Figure 1.2). Some organizations like to substitute the word 'quality' with 'performance', but the principle is the same – the operational requirements of the project must be met, and met safely.

In certain industries like airlines, railways, and mining, etc., the fourth criterion, safety, is considered to be equally important, if not more so. In these organizations, the triangle can be replaced by a diamond now showing the four important criteria (Figure 1.3).

The order of priority given to any of these criteria is dependent not only on the industry but also on the individual project. For example, in designing and constructing an aircraft, motor car, or railway carriage, safety must be paramount. The end product may cost more than budgeted or it may be late in going into service, and certain quality requirements in terms of comfort may have to be sacrificed, but under no circumstances can safety be compromised. Airplanes, cars, and railways *must* be safe under all operating conditions.

The following (rather obvious) examples show where different priorities on the project triangle (or diamond) apply.

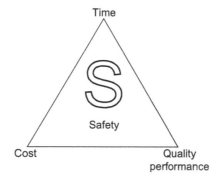

Figure 1.2
Project triangle

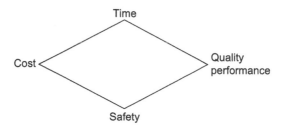

Figure 1.3
Project diamond

Time-Bound Project

A scoreboard for a prestigious tennis tournament must be finished in time for the opening match, even if it costs more than anticipated and the display of some secondary information, such as the speed of the service, has to be abandoned. In other words, cost and performance may have to be sacrificed to meet the unalterable starting date of the tournament.

(In practice, the increased cost may well be a matter of further negotiation and the temporarily delayed display can usually be added later during the non-playing hours.)

Cost-Bound Project

A local authority housing development may have to curtail the number of housing units and may even overrun the original construction programme, but the project cost cannot be exceeded, because the housing grant allocated by central government for this type of development has been frozen at a fixed sum. Another solution to this problem would be to reduce the specification of the internal fittings instead of reducing the number of units.

Performance (Quality)-Bound Project

An armaments manufacturer has been contracted to design and manufacture a new type of rocket launcher to meet the client's performance specification in terms of range, accuracy, and rate of fire. Even if the delivery has to be delayed to carry out more tests and the cost has increased, the specification must be met. Again, if the weapons were required during a war, the specification might be relaxed to get the equipment into the field as quickly as possible.

Safety-Bound Project

Apart from the obvious examples of public transport given previously, safety is a factor that is required by law and enshrined in the Health & Safety at Work Act.

Not only must safe practices be built into every project, but constant monitoring is an essential element of a safety policy. To that extent it could be argued that *all* projects are safety-bound, since, if it became evident after an accident that safety was sacrificed for speed or profitability, some or all of the project stakeholders could find themselves in real trouble, even in jail. This is true for almost every industry, especially agriculture, food/drink production and preparation, pharmaceuticals, chemicals, toy manufacture, aircraft production, motor vehicle manufacture, and, of course, building and construction.

A serious accident that may kill or injure people will not only cause anguish among the relatives, but, while not necessarily terminating the project, could very well destroy the company. For this reason the 'S' symbol when shown in the middle of the project management triangle gives more emphasis of its importance (see Figure 1.2).

While the other three criteria (Cost, Time, and Quality/Performance) can be juggled by the project manager to suit the changing requirements and environment of a project, safety cannot under any circumstances be compromised. As any project manager knows, the duration (Time) may be reduced by increasing resources (Cost), and cost may be saved by sacrificing quality or performance, but any diminution of safety can quickly lead to disaster, death, and even the closure of an organisation. The catastrophic explosions on the Piper Alpha gas platform in the North Sea in July 1988 killed 167 men and cost millions of dollars to Occidental and its insurers, and the explosion at the Buncefield, England, oil depot in 2009 caused massive destruction of its surroundings and huge costs to Total Oil Co. Additionally, the explosion on its Texas City refinery in March 2005, which killed 15 men and injured 170, and the blowout of the Deepwater Horizon drilling rig in the Gulf of Mexico in April 2010, causing 11 fatalities, have seriously damaged the reputation of BP and resulted in a considerable drop in its share price. In the transport industry, the series of railway accidents in 2000 resulted in the winding up of British Rail and subsequently one of its main contractors. More recently, Toyota had to recall millions of cars to rectify an unsafe breaking and control system, after which Mr. Toyoda, the Chairman of the company, publicly stated that Toyota's first priority is safety, the second is quality, and the third is volume (Quantity). These occurrences clearly show that safety must head the list of priorities for any project or organisation.

The priorities of the other three criteria can of course change with the political climate or the commercial needs of the client, even within the life cycle of the project, and therefore the project manager has to constantly evaluate these changes to determine the new priorities. Ideally, all the main criteria should be met (and indeed on many well-run projects, this is the case), but there are times when the project manager, with the agreement of the sponsor or client, has to make difficult decisions to satisfy the best interests of most, if not all, the stakeholders.

However, the examples given above highlight the importance of ensuring a safe operating environment, even at the expense of the other criteria. It is important to note that while a project manager can be reprimanded or dismissed for not meeting any of the three 'corner criteria', the one transgression for which a project manager can actually be jailed is not complying with the provisions of the Health and Safety regulations.

If one were to list the four project management criteria in the order of their importance, the sequence would be safety, performance, time, and cost, which can be remembered using the acronym SAPETICA. The rationale for this order is as follows:

If the project is not safe, it can cost lives and/or destroy the constructor and other stakeholders.

If the performance is not acceptable, the project will have been a waste of time and money.

If the project is not on time, it can still be a success, but may have caused a financial loss. Even if the cost exceeds the budget, the project can still be viable, as extra money can usually be found. (The most famous (or infamous) example is the Sydney Opera House, which was so much over budget that the extra money had to be raised via a New South Wales State lottery but is now celebrated as a great Sydney landmark.)

Project Management

It is obvious that project management is not new. Noah must have managed one of the earliest recorded projects in the Bible – the building of the ark. He may not have completed it to budget, but he certainly had to finish it by a specified time – before the flood — and it must have met his performance criteria, as it successfully accommodated a pair of all the animals.

There are many published definitions of *project management,* (see BS 6079 and ISO 21500), but the following definition covers all the important ingredients:

> *The planning, monitoring, and control of all aspects of a project and the motivation of all those involved in it, in order to achieve the project objectives within agreed criteria of time, cost, and performance.*

While this definition includes the fundamental criteria of time, cost, and performance, the operative word, as far as the management aspect is concerned, is *motivation.* A project will not be successful unless all (or at least most) of the participants are not only competent but also motivated to produce a satisfactory outcome.

To achieve this, a number of methods, procedures, and techniques have been developed, which, together with the general management and people skills, enable the project manager to meet the set criteria of time cost and performance/quality in the most effective way.

Many textbooks divide the skills required in project management into *hard* skills (or topics) and *soft* skills. This division is not exact and some are clearly interdependent. Furthermore it depends on the type of organization, type and size of project, and authority given to a project manager, and which of the listed topics are in his or her remit for a particular project. For example, in many large construction companies, the project manager is not permitted to get involved in industrial (site) disputes as these are more effectively resolved by specialist industrial relations managers who are conversant with the current labour laws, national or local labour agreements, and site conditions.

Project Management, Planning, and Control. http://dx.doi.org/10.1016/B978-0-08-098324-0.00002-0

The hard skills cover such subjects as business case, cost control, change management, project life cycles, work breakdown structures, project organization, network analysis, earned value analysis, risk management, quality management, estimating, tender analysis, and procurement.

The soft topics include health and safety, stakeholder analysis, team building, leadership, communications, information management, negotiation, conflict management, dispute resolutions, value management, configuration management, financial management, marketing and sales, and law.

A quick inspection of the two types of topics shows that the hard subjects are largely only required for managing projects, while the soft ones can be classified as general management and are more or less necessary for any type of business operation whether running a design office, factory, retail outlet, financial services institution, charity, public service organization, national or local government, or virtually any type of commercial undertaking.

A number of organisations, such as APM, PMI, ISO, OGC, and licencees of PRINCE (Project In a Controlled Environment) have recommended and advanced their own methodology for project management, but by and large the differences are on emphasis or sequence of certain topics. For example, PRINCE requires the resources to be determined before the commencement of the time scheduling and the establishment of the completion date, while in the construction industry the completion date or schedule is often stipulated by the customer and the contractor has to provide (or recruit) whatever resources (labour, plant, equipment, or finance) are necessary to meet the specified objectives and complete the project on time.

Project Manager

A *project manager* may be defined as:

"The individual or body with authority, accountability and responsibility for managing a project to achieve specific objectives" (BS 6079-2:2000)

Few organizations will have problems with the above definition, but unfortunately in many instances, while the responsibility and accountability are vested in the project manager, the authority given to him or her is either severely restricted or non-existent. The reasons for this may be a reluctance of a department (usually one responsible for the accounts) to relinquish financial control, or it is perceived that the project manager has not sufficient experience to handle certain tasks such as control of expenditure. There may indeed be good reasons for these restrictions which depend on the size and type of project, the size and type of the organization, and of course the personality and experience of the project manager, but if the project manager is supposed to be in effect the managing director of the project (as one large construction organisation liked to put it), he or she must have control over costs and expenditure, albeit within specified and agreed limits.

Apart from the conventional responsibilities for time, cost, and performance/quality, the project manager must ensure that all the safety requirements and safety procedures are complied with. For this reason the word *safety* has been inserted into the project management triangle to reflect the importance of ensuring the many important health and safety requirements are met. Serious accidents not only have personal tragic consequences, but they also can destroy a project or indeed a business overnight. Lack of attention to safety is just bad business, as any oil company, airline, bus, or railroad company can confirm.

Project Manager's Charter

Because the terms of engagement of a project manager are sometimes difficult to define in a few words, some organizations issue a *project manager's charter*, which sets out the responsibilities and limits of authority of the project manager. This makes it clear to the project manager what his or her areas of accountability are, and if this document is included in the project management plan, all stakeholders will be fully aware of the role the project manager will have in this particular project.

The project manager's charter is project specific and will have to be amended for every manager as well as the type, size, complexity, or importance of a project (see Figure 2.1).

Project Office

On large projects, the project manager will have to be supported either by one or more assistant project managers (one of whom can act as deputy) or a specially created *project office*. The main duties of such a project office is to establish a uniform organisational approach for systems, processes, and procedures, carry out the relevant configuration management functions, disseminate project instructions and other information, and collect, retrieve, or chase information required by the project manager on a regular or ad hoc basis. Such an office can assist greatly in the seamless integration of all the project systems and would also prepare programs, schedules, progress reports, cost analyses, quality reports, and a host of other useful tasks that would otherwise have to be carried out by the project manager himself. In addition the project office can also be required to service the requirements of a programme or portfolio manager, in which case it will probably have its own office manager responsible for the onerous task of satisfying the different and often conflicting priorities set by the various projects managers. (See also Chapter 10.)

PROJECT MANAGER'S CHARTER

1. **Project Manager:**

   Name: _____

   Appointment/Position: _____

   Date of Appointment: _____

2. **Project Title:** _____

3. **Responsibility and Authority given to the Project Manager:**

   The above named Project Manager has been given the authority, responsibility and accountability for _____

4. **Project Goals and Deliverables are:**

   a: _____

   b: _____

   c: _____

5. **The Project will be reviewed.**

   _____

6. **Financial Authority:**

   The Project Manager's delegated financial powers are: _____

   _____

7. **Intramural Resources:**

   The following resources have been/are to be made available:

8. **Trade-offs:**

   a: Cost: _____ %

   b: Time: _____ days/weeks.

   c: Performance: _____

9. **Charter Review: No charter review is expected to take place for the duration of this project unless it becomes clear that the PM cannot fulfil his/her duties or a reassessment of the trade-offs is required.**

10. **Approved:**

    Sponsor/Client/Customer/Programme Manager: _____

    Project Manager: _____

    Line Manager: _____

11. **Distribution:**

    a: Sponsor; b: Programme Manager; c: Line Manager

Figure 2.1
Project manager's charter

CHAPTER 3

Programme and Portfolio Management

Chapter Outline

Programme management may be defined as 'The co-ordinated management of a group of related projects to ensure the best use of resources in delivering the projects to the specified time, cost, and quality/performance criteria'.

A number of organizations and authorities have coined different definitions, but the operative word in any definition is *related*. Unless the various projects are related to a common objective, the collection of projects would be termed a 'portfolio' rather than a 'programme'.

A programme manager could therefore be defined as 'The individual to whom responsibility has been assigned for the overall management of the time, cost, and performance aspects of a group of related projects and the motivation of those involved'.

Again, different organizations have different definitions for the role of the programme manager or portfolio manager. In some companies he or she would be called manager of projects or operations manager or operations director, etc., but it is generally understood that the programme manager's role is to co-ordinate the individual projects that are linked to a common objective. Whatever the definition, it is the programme manager who has the overall picture of the organization's project commitments.

Many organizations carrying out a number of projects have limited resources. It is the responsibility of the programme manager to allocate these resources in the most cost-effective manner, taking into consideration the various project milestones and deadlines as well as the usual cost restrictions. It is the programme manager who may have to obtain further authority to engage any external resources as necessary and decide on their disposition.

As an example, the construction of a large cruise ship would be run by a programme manager who co-ordinates the many (often very large) projects such as the ship's hull, propulsion system and engines, control systems, catering system, interior design, etc. One of the associated projects might even include recruitment and training of the crew.

A manager responsible for diverse projects such as the design, supply, and installation of a computerized supermarket check-out and stock control system, an electronic scoreboard for a cricket ground, or a cheque-handling system for a bank, would be a portfolio manager, because although all the projects require computer systems, they are for different clients at different locations and are independent of each other. Despite this diversity of the projects, the portfolio manager, like the project manager, still has the responsibility to set priorities, maximize the efficient use of the organization's resources, and monitor and control the costs, schedule, and performance of each project.

As with project management, programme management and the way programmes are managed depend primarily on the type of organization carrying out the programme. There are two main types of organizations:

- Client organizations
- Contracting organizations

In a client type of organization, the projects or programmes will probably not be the main source of income and may well constitute or require a major change in the management structure and culture. New resources may have to be found and managers involved in the normal running of the business may have to be consulted, educated, and finally convinced of the virtues, not only of the project itself, but also of the way it has to be managed.

The programme manager in such an organization has to ensure that the project fits into the corporate strategy and meets the organization's objectives. He or she has to ensure that established project management procedures, starting with the business case through implementation, and ending with disposal, are correctly employed. In other words, the full life cycle systems using all the 'soft' techniques to create a project environment have to be in place in an organization that may well be set up for 'business-as-usual', employing only well-established line management techniques. In addition the programme manager has to monitor all projects to ensure that they meet the strategic objectives of the organization as well as fulfilling the more obvious requirements of being performed safely, minimizing and controlling risks, and meeting the cost, time, and performance criteria for every project.

Programme management can, however, mean more than co-ordinating a number of related projects. The prioritization of the projects themselves, not just the required resources, can be a function of programme management. It is the programme manager who decides which project, or which type of project, is the best investment and which one is the most cost-effective one to start. It may even be advantageous to merge two or more small projects into one larger project, if they have sufficient synergy or if certain resources or facilities can be shared.

Another function of programme management is to monitor the performance of the projects that are part of the programme and check that the expected deliverables have produced the specified benefits, whether to the parent organization or the client. This could take several days or months depending on the project, but unless it is possible to measure these benefits, it is not possible to assess the success of the project or indeed say whether the whole exercise was worthwhile. It can be seen therefore that it is just as important for the programme manager to set up the monitoring and close-out reporting system for the end of a project as the planning and control systems for the start.

In a contracting organization, such a culture change will either not be necessary, as the organization will already be set up on a project basis, or the change to a project-oriented company will be easier because the delivery of projects is after all the 'raison d'être' of the organization. Programme management in a contracting organization is therefore more the co-ordination of the related or overlapping projects covering such topics as resource management, cost management, and procurement and ensuring conformity with standard company systems and procedures. The cost, time, and performance/quality criteria therefore relate more to the obligations of the contractor (apart from performance) than those of the client.

The life cycles of projects in a contracting organization usually start after the feasibility study has been carried out and finishes when the project is handed over to the client for the operational phase. There are clearly instances when these life cycle terminal points occur earlier or later, but a contractor is rarely concerned with whether the strategic or business objectives of the client have been met.

Portfolio Management

Portfolio management, which can be regarded as a subset of corporate management, is very similar to programme management, but the projects in the programme manager's portfolio, though not necessarily related, are still required to meet an organization's objectives. Furthermore, portfolios (unlike projects or programmes) do not necessarily have a defined start and finish date. Indeed portfolios can be regarded as a rolling set of programmes that are monitored in a continuous life cycle from the strategic planning stage to the delivery of the programme. In a large organization a portfolio manager may be in charge of several programme managers, while in a smaller company, he or she may be in direct control of a number of project managers.

Companies do not have unlimited resources, so the portfolio manager has to prioritize the deployment of these resources for competing projects, each of which has to be assessed in terms of:

(a) Profitability and cost/benefit
(b) Return on investment
(c) Cash flow
(d) Risks
(e) Prestige
(f) Importance of the client
(g) Company strategy and objectives

Portfolio management therefore involves the identification of these project attributes and the subsequent analysis, prioritization, balancing, monitoring, and reporting of progress of each project, or in the case of large organizations, each programme. As each project develops, different pressures and resource requirements will occur, often as a result of contractual changes or the need to rectify errors or omissions. Unforeseen environmental issues may require immediate remedial action to comply with health and safety requirements, and there is always the danger of unexpected resignations of key members of one of the project teams.

A portfolio manager must therefore possess the ability to reassign resources, both human and material (such as office equipment, construction plant, and bulk materials), in an effective and economical manner, often in emergency or other stressful situations and always taking into account the cost/benefit calculations, the performance and sustainability criteria, and the overall strategic objectives of the organization.

Project Context (Project Environment)

Projects are influenced by a multitude of factors which can be external or internal to the organization responsible for its management and execution. The important thing for the project manager is to recognize what these factors are and how they impact on the project during the various phases from inception to final handover, or even disposal.

These external or internal influences are known as the *project context* or *project environment*. The external factors making up this environment are the client or customer, various external consultants, contractors, suppliers, competitors, politicians, national and local government agencies, public utilities, pressure groups, the end users, and even the general public. Internal influences include the organization's management, the project team, internal departments, (technical and financial), and possibly the shareholders.

Figure 4.1 illustrates the project surrounded by its external environment.

All these influences are neatly encapsulated by the acronym PESTLE, which stands for

- Political
- Economic
- Social
- Technical
- Legal
- Environmental

A detailed discussion of these areas of influence is given below.

Political

Here, two types of politics have to be considered.

Project Management, Planning, and Control. http://dx.doi.org/10.1016/B978-0-08-098324-0.00004-4

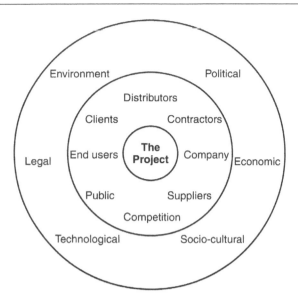

Figure 4.1
The project environment

First, there are the internal politics that inevitably occur in all organizations whether governmental, commercial, industrial, or academic and which manifest themselves in the opinions and attitudes of the different stakeholders in these organizations. The relationships to the project by these stakeholders can vary from the very supportive to the downright antagonistic, but depending on their field of influence, they must be considered and managed. Even within an apparently cohesive project, team jealousies and personal vested interests can have a disruptive influence the project manager has to recognize and diffuse.

The fact that a project relies on clients, consultants, contractors (with their numerous subcontractors), material and service suppliers, statutory authorities, and, of course, the end user, all of which may have their own agenda and preferences, gives some idea of the potential political problems that may occur.

Second, there are the external politics, over which neither the sponsor nor the project manager may have much, if any, control. Any project that has international ramifications is potentially subject to disruption due to the national or international political situation. In the middle of a project, the government may change and impose additional import, export, or exchange restrictions, impose penal working conditions, or even cancel contracts altogether. For overseas construction contracts in countries with inherently unstable economies or governments, sudden coups or revolutions may require the whole construction team to be evacuated at short notice. Such a situation should have been envisaged, evaluated, and planned for as part of the political risk assessment when the project was first considered.

Even on a less dramatic level, the political interplay between national and local government, lobbyists, and pressure groups has to be taken into consideration as can be appreciated when the project consists of a road by-pass, reservoir, power station, or airport extension.

Economic

Here again there are two levels of influence: internal or micro-economic, and external or macro-economic.

The internal economics relate to the viability of the project and the soundness of the business case. Unless there is a net gain, whether financial or non-financial, such as required by prestige, environmental, social service, or national security considerations, there is no point in even considering embarking on a project. It is vital therefore that financial models and proven accountancy techniques are applied during the evaluation phase to ensure the economic viability of the project. These tests must be applied at regular intervals throughout the life of a project to check that with the inevitable changes that may be required, it is still worthwhile to proceed. The decision to abort the whole project at any stage after the design stage is clearly not taken lightly, but once the economic argument has been lost, it may in the end be the better option. A typical example is the case of an oil-fired power station that had to be mothballed over halfway through construction, when the price of fuel oil rose above the level at which power generation was no longer economic. It is not uncommon for projects to be shelved when the cost of financing the work has to be increased and the resulting interest payments exceed the foreseeable revenues.

The external economics, often related to the political climate, can have a serious influence on the project. Higher interest rates or exchange rates, and additional taxes on labour, materials, or the end product, can seriously affect the viability of the project. A manufacturer may abandon the construction of a factory in its home country and transfer the project abroad if just one of these factors changes enough to make such a move economically desirable. Again, changes to fiscal and interest movements must be constantly monitored so that representations can be made to government or the project curtailed. Other factors that can affect a project are tariff barriers, interstate taxes, temporary embargoes, shipping restrictions such as only being permitted to use conference line vessels, and special licenses.

Social (or Sociological)

Many projects and indeed most construction projects inevitably affect the community in whose area they are carried out. It is vital therefore to inform the residents in the affected areas as early as possible of the intent, purpose, and benefits to the organization and community of the project.

This may require a public relations campaign to be initiated, which includes meetings, exploratory discussions, consultations at various levels, and possible trade-offs. This is particularly important when public funding from central or local government is involved or when public spaces and access facilities are affected. A typical example of a trade-off is when a developer wishes to build a shopping centre, the local authority may demand that it include a recreation area or leisure park for free use by the public.

Some projects cannot even be started without first being subjected to a public enquiry, environmental impact assessment, route surveys, or lengthy planning procedures. There are always pressure groups who have a special interest in a particular project, and it is vital that they are given the opportunity to state their case while at the same time informing them of the positive and often less desirable implications. The ability to listen to their points of view and give sympathetic attention to their grievances is essential, but as it is almost impossible to satisfy all the parties, compromises may be necessary. The last thing a project manager wants are constant demonstrations and disruptions while the project is being carried out.

On another level, the whole object of the project may be to enhance the environment and facilities of the community, in which case the involvement of local organizations can be very helpful in focusing on areas which give the maximum benefit and avoiding pitfalls which only people with local knowledge are aware of. A useful method to ensure local involvement is to set up advisory committees or even invite a local representative to be part of the project management team.

Technical

It goes without saying that unless the project is technically sound, it will end in failure. Whether the project involves rolling out a new financial service product or building a power station, the technology must be in place or be developed as the work proceeds. The mechanisms by which these technical requirements are implemented have to be firmed up at a very early stage after a rigorous risk assessment of all the realistically available options. Each option may then be subjected to a separate feasibility study and investment appraisal. Alternatives to be considered may include:

- Should in-house or external design, manufacture, or installation be used?
- Should existing facilities be used or should new ones be acquired?
- Should one's own management team be used or should specialist project managers be appointed?
- Should existing components (or documents) be incorporated?
- What is the anticipated life of the end product (deliverable) and how soon must it be updated?

- Are materials available on a long-term basis and what alternatives can be substituted?
- What is the nature and size of the market and can this market be expanded?

These and many more technical questions have to be asked and assessed before a decision can be made to proceed with the project. The financial implications of these factors can then be fed into the overall investment appraisal, which includes the commercial and financing and environmental considerations.

Legal

One of the fundamental requirements of a contract, and by implication a project, is that it is legal. In other words, if it is illegal in a certain country to build a brewery, little protection can be expected from the law.

The relationships between the contracting parties must be confirmed in a legally binding contract that complies with the laws (and preferably customs) of the participating organizations. The documents themselves have to be legally acceptable and equitable, and unfair and unreasonable clauses must be eliminated.

Where suppliers of materials, equipment, or services are based in countries other than the main contracting parties, the laws of those countries have to be complied with in order to minimize future problems regarding deliveries and payments.

In the event of disputes, the law under which the contract is administered and adjudicated must be written into the contract together with the location of the court for litigation.

Generally, project managers are strongly advised always to take legal advice from specialists in contract law and especially, where applicable, in international law.

The project context includes the established conditions of contract and other standard forms and documents used by industry, and can also include all the legal, political, and commercial requirements stipulated by international bodies as well as national and local governments in their project management procedures and procurement practices.

Environmental

Some of the environmental aspects of a project have already been alluded to under 'Social', from which it became apparent that environmental impact assessments are highly desirable where they are not already mandatory.

The location of the project clearly has an enormous influence on the cost and completion time. The same type of plant or factory can be constructed in the UK, in the Sahara desert, in China, or even on an offshore platform, but the problems, costs, and construction times can be very different. The following considerations must therefore be taken into account when deciding to carry out a project in a particular area of the world:

- Temperature (daytime and night time) in different seasons
- Rainy seasons (monsoon)
- Tornado or typhoon seasons
- Access by road, rail, water, or air
- Ground conditions and earthquake zones
- Possible ground contamination
- Nearby rivers and lakes
- Is the project onshore or offshore?
- Tides and storm conditions
- Nearby quarries for raw materials
- Does the project involve the use of radioactive materials?

Most countries now have strict legislation to prevent or restrict emissions of polluting substances whether solids, liquids, or gases. In addition noise restrictions may apply at various times and cultural or religious laws may prohibit work at specified times or on special days in the year.

The following list is a very small sample of over 15000 web pages covering European Economic Community (EEC) directives and gives some idea of the regulations that may have to be followed when carrying out a project.

EC Directive		
	85/337/EEC	Environmental impact assessment
	97/11/EEC	Assessment of effects on certain public & private projects
	92/43/EEC	Chapter 4 Environment
	86/278/EEC	Protection of the environment
	90/313/EEC	Sustainable development
	90/679/EEC	Substances hazardous to health
	79/409/EEC	Conservation of natural habitats
	96/82/EEC	COMAH (Control of major accident hazards)
	91/156/EEC	Control of pollution
	87/217/EEC	Air pollution
	89/427/EEC	Air pollution
	80/779/EEC	Air quality limit values
	75/442/EEC	Ozone depleting substances
	89/427/EEC	Quality limit of sulphur dioxide
	80/1268/EEC	Fuel & CO_2 emissions
	91/698/EEC	Hazardous waste
	78/659/EEC	Quality standard of water
	80/68/EEC	Ground water directive
	80/778/EEC	Spring waters
	89/336/EEC	Noise emissions

Business Case

Chapter Outline

Before embarking on a project, it is clearly necessary to show that there will be a benefit either in terms of money or service or both. The document that sets out the main advantages and parameters of the project is called the *business case* and is (or should be) produced by either the client or the sponsor of the project who in effect becomes the owner of the document.

A business case in effect outlines the 'why' and 'what' of the project as well as making the financial case by including the investment appraisal.

As with all documents, a clear procedure for developing the business case is highly desirable, and the following headings give some indication of the subjects to be included:

1. Why is the project required?
2. What are we trying to achieve?
3. What are the deliverables?
4. What is the anticipated cost?
5. How long will it take to complete?
6. What quality standards must be achieved?
7. What are the performance criteria?
8. What are key performance indicators (KPI)?
9. What are the main risks?
10. What are the success criteria?
11. Who are the main stakeholders?

In addition any known information such as location, key personnel, resource requirements, etc., should be included so that the recipients, usually a board of directors, are in a position to accept or reject the case for carrying out the project.

The Project Sponsor

It is clear that the business case has to be prepared before the project can be started. Indeed, the business case is the first document to be submitted to the directorate of an organization to

  21

enable this body to discuss the purpose and virtues of the project before making any financial commitments. It follows therefore that the person responsible for producing the business case cannot normally be the project manager but must be someone who has a direct interest in the project going ahead. This person, who is often a director of the client's organization with a special brief to oversee the project, is the *project sponsor* (sometimes also known as the project champion).

The role of the project sponsor is far greater than being the initiator or champion of the project. Even after the project has started the sponsor's role is to:

1. Monitor the performance of the project manager
2. Constantly ensure that the project's objectives and main criteria are met
3. Ensure that the project is run effectively as well as efficiently
4. Assess the need and viability of variations and agree to their implementation
5. Assist in smoothing out difficulties with other stakeholders
6. Support the project by ensuring sufficient resources (especially financial) are available
7. Act as business leader and top-level advocate to the company board
8. Ensure that the perceived benefits of the project are realized

Depending on the value, size, and complexity of the project, the sponsor is a key player who, as a leader and mentor, can greatly assist the project manager to meet all the project's objectives and key performance indicators.

Requirements Management

As has been explained previously, the main two components of a business case are 'what' is required and 'why' it is required. *Requirements management* is concerned with the 'what'.

Clients, end users, and indeed most stakeholders have their own requirements on what they expect from the project even if the main objectives have been agreed. Requirements management is concerned with the eliciting, capturing, collating, assessing, analysing, testing, prioritizing, organizing, and documenting of all these different requirements. Many of these may of course be the common needs of a number of stakeholders and will therefore be high on the priority list, but it is the project manager who is responsible for deciding on the viability or desirability of a particular requirement and to agree with the stakeholder on whether it should or could be incorporated, taking into account the cost, time, and performance factors associated with the requirement. Once agreed, these requirements become the benchmark against which the success of the project is measured.

Ideally all the requirements should have been incorporated as clear deliverables in the objectives enshrined in the business case and confirmed by the project manager in the project management plan. It is always possible, however, that one or more stakeholders may wish to change these requirements either just before or even after the project scope has been agreed and finalized. The effect of such a change of requirement will have to be carefully examined by the project manager, who must take into account any cost implication, effects on the project programme, changes to the procedures and processes needed to incorporate the new requirement, and the environmental impact in its widest sense.

In such a situation, the project manager must immediately advise all the relevant stakeholders of the additional cost, time, and performance implications and obtain their approval before amending the objectives, scope, and cost of the project.

If the change of requirements is requested after the official start of the project, that is, after the cost and time criteria have been agreed, the new requirements will be subject to the normal project (or contract) change procedure and configuration management described elsewhere.

To log and control the requirement documents during the life of the project, a simple 'reporting matrix', as shown below, will be helpful.

No.	Document (requirement)	Prepared by	Information from	Sent or copied to	Issued date

Testing and periodic reviews of the various requirements will establish their viability and ultimate effect on the outcome of the project. The following are some of the major characteristics that should be examined as part of this testing process:

- Feasibility, operability, and time constraints
- Functionality, performance, and quality requirements and reliability
- Compliance with health and safety regulations and local by-laws
- Buildability, delivery (transportability), storage, and security
- Environmental and sociological impact
- Labour, staffing, outsourcing, and training requirements

There may be occasions where the project manager is approached by a stakeholder, or even the client, to incorporate a 'minor' requirement 'as a favour'. The dangers of agreeing to such a request without following the normal change management procedures are self-apparent. A small request can soon escalate into a large change once all the ramifications and spin-off effects have become apparent as this leads to the all too common 'scope creep'. All changes to requirements, however small, must be treated as official and handled accordingly. It may of course be politically expedient not to charge a client for any additional requirement, but this is a commercial decision taken by senior management for reasons of creating goodwill, obtaining possible future contracts, or succumbing to political pressure.

Investment Appraisal

Chapter Outline

The investment appraisal, which is part of the business case, will, if properly structured, improve the decision-making process regarding the desirability or viability of the project. It should have examined all the realistic options before making a firm recommendation for the proposed case. The investment appraisal must also include a cost/benefit analysis and take into account all the relevant factors such as:

- Capital costs, operating costs, and overhead costs
- Support and training costs
- Dismantling and disposal costs
- Expected residual value (if any)
- Any cost savings that the project will bring
- Any benefits that cannot be expressed in monetary terms

To enable some of the options to be compared, the payback, return on capital, net present value, and anticipated profit must be calculated. In other words, the project viability must be established.

Project Viability

Return on Investment (ROI)

The simplest way to ascertain whether the investment in a project is viable is to calculate the *return on investment* (ROI).

If a project investment is £10 000, and gives a return of £2000 per year over 7 years,

$$\text{the average return/year} = \frac{(7 \times £2000) - £10000}{7}$$

$$= \frac{£4000}{7} = £571.4$$

25

The return on the investment, usually given as a percentage, is the average return over the period considered × 100, divided by the original investment, i.e.,

$$\text{return on investment } \% = \frac{\text{average return} \times 100}{\text{investment}}$$

$$= \frac{£571.4 \times 100}{£10000} = 5.71 \%$$

This calculation does not, however, take into account the cash flow of the investment, which in a real situation may vary year by year.

Net Present Value

As the value of money varies with time due to the interest it could earn if invested in a bank or other institution, the actual cash flow must be taken into account to obtain a realistic measure of the profitability of the investment.

If £100 were invested in a bank earning an interest of 5%:

The value in 1 year would be £100 × 1.05   = £105
The value in 2 years would be £100 × 1.05 × 1.05   = £110.25
The value in 3 years would be £100 × 1.05 × 1.05 × 1.05 = £115.76

It can be seen therefore that, today, to obtain £115.76 in 3 years it would cost £100. In other words, the present value of £115.76 is £100.

Another way of finding the *present value* (PV) of £115.76 is to divide it by 1.05 × 1.05 × 1.05 or 1.157, for

$$\frac{115.76}{1.05 \times 1.05 \times 1.05} = \frac{115.76}{1.157} = £100.$$

If instead of dividing the £115.76 by 1.157, it is multiplied by the inverse of 1.157, one obtains the same answer, since

$$£115.76 \times \frac{1}{1.157} = £115.76 \times 0.8638 = £100.$$

The 0.8638 is called the *discount factor* or present value factor and can be quickly found from discount factor tables, a sample of which is given in Figure 6.1.

It will be noticed from these tables that 0.8638.5 is the PV factor for a 5% return after 3 years. The PV factor for a 5% return after 2 years is 0.9070 or

$$\frac{1}{1.05 \times 1.05} = \frac{1}{1.1025} = 0.9070$$

In the above example the income (5%) was the same every year. In most projects, however, the projected annual net cash flow (income minus expenditure) will vary year by year, and to

Table A Present value of £1

Years Hence	1%	2%	4%	6%	8%	10%	12%	14%	15%	16%	18%	20%	22%	24%	25%	26%	28%	30%	35%	40%	45%	50%
1	0.990	0.980	0.962	0.943	0.926	0.909	0.893	0.877	0.870	0.862	0.847	0.833	0.820	0.806	0.800	0.794	0.781	0.769	0.741	0.714	0.690	0.667
2	0.980	0.961	0.925	0.890	0.857	0.826	0.797	0.769	0.756	0.743	0.718	0.694	0.672	0.650	0.640	0.630	0.610	0.592	0.549	0.510	0.476	0.444
3	0.971	0.942	0.889	0.840	0.794	0.751	0.712	0.675	0.658	0.641	0.609	0.579	0.551	0.524	0.512	0.500	0.477	0.455	0.406	0.364	0.328	0.296
4	0.961	0.924	0.855	0.792	0.735	0.683	0.636	0.592	0.572	0.552	0.516	0.482	0.451	0.423	0.410	0.397	0.373	0.350	0.301	0.260	0.226	0.198
5	0.951	0.906	0.822	0.747	0.681	0.621	0.567	0.519	0.497	0.476	0.437	0.402	0.370	0.341	0.328	0.315	0.291	0.269	0.223	0.186	0.156	0.132
6	0.942	0.888	0.790	0.705	0.630	0.564	0.507	0.456	0.432	0.410	0.370	0.335	0.303	0.275	0.262	0.250	0.227	0.207	0.165	0.133	0.108	0.088
7	0.933	0.871	0.760	0.665	0.583	0.513	0.452	0.400	0.376	0.354	0.314	0.279	0.249	0.222	0.210	0.198	0.178	0.159	0.122	0.095	0.074	0.059
8	0.923	0.853	0.731	0.627	0.540	0.467	0.404	0.351	0.327	0.305	0.266	0.233	0.204	0.179	0.168	0.157	0.139	0.123	0.091	0.068	0.051	0.039
9	0.914	0.837	0.703	0.592	0.500	0.424	0.361	0.308	0.284	0.263	0.225	0.194	0.167	0.144	0.134	0.125	0.108	0.094	0.067	0.048	0.035	0.026
10	0.905	0.820	0.676	0.558	0.463	0.386	0.322	0.270	0.247	0.227	0.191	0.162	0.137	0.116	0.107	0.099	0.085	0.073	0.050	0.035	0.024	0.017
11	0.896	0.804	0.650	0.527	0.429	0.350	0.287	0.237	0.215	0.195	0.162	0.135	0.112	0.094	0.086	0.079	0.066	0.056	0.037	0.025	0.017	0.012
12	0.887	0.788	0.625	0.497	0.397	0.319	0.257	0.208	0.187	0.168	0.137	0.112	0.092	0.076	0.069	0.062	0.052	0.043	0.027	0.018	0.012	0.008
13	0.879	0.773	0.601	0.469	0.368	0.290	0.229	0.182	0.163	0.145	0.116	0.093	0.075	0.061	0.055	0.050	0.040	0.033	0.020	0.013	0.008	0.005
14	0.870	0.758	0.577	0.442	0.340	0.263	0.205	0.160	0.141	0.125	0.099	0.078	0.062	0.049	0.044	0.039	0.032	0.025	0.015	0.009	0.006	0.003
15	0.861	0.743	0.555	0.417	0.315	0.239	0.183	0.140	0.123	0.108	0.084	0.065	0.051	0.040	0.035	0.031	0.025	0.020	0.011	0.005	0.004	0.002
16	0.853	0.728	0.534	0.394	0.292	0.218	0.163	0.123	0.107	0.093	0.071	0.054	0.042	0.032	0.028	0.025	0.019	0.015	0.008	0.005	0.003	0.002
17	0.844	0.714	0.513	0.371	0.270	0.198	0.146	0.108	0.093	0.080	0.060	0.045	0.034	0.026	0.023	0.020	0.015	0.012	0.006	0.003	0.002	0.001
18	0.836	0.700	0.494	0.350	0.250	0.180	0.130	0.095	0.081	0.069	0.051	0.038	0.028	0.021	0.018	0.016	0.012	0.009	0.005	0.002	0.001	0.001
19	0.828	0.686	0.475	0.331	0.232	0.164	0.116	0.083	0.070	0.060	0.043	0.031	0.023	0.017	0.014	0.012	0.009	0.007	0.003	0.002	0.001	
20	0.820	0.673	0.456	0.312	0.215	0.149	0.104	0.073	0.061	0.051	0.037	0.026	0.019	0.014	0.012	0.010	0.007	0.005	0.002	0.001	0.001	
21	0.811	0.660	0.439	0.294	0.199	0.135	0.093	0.064	0.053	0.044	0.031	0.022	0.015	0.011	0.009	0.008	0.006	0.004	0.002	0.001	0.001	
22	0.803	0.647	0.422	0.278	0.184	0.123	0.083	0.056	0.046	0.038	0.026	0.018	0.013	0.009	0.007	0.006	0.004	0.003	0.001	0.001		
23	0.795	0.634	0.406	0.262	0.170	0.112	0.074	0.049	0.040	0.033	0.022	0.015	0.010	0.007	0.006	0.005	0.003	0.002	0.001			
24	0.788	0.622	0.390	0.247	0.158	0.102	0.066	0.043	0.035	0.028	0.019	0.013	0.008	0.006	0.005	0.004	0.003	0.002	0.001			
25	0.780	0.610	0.375	0.233	0.146	0.092	0.059	0.038	0.030	0.024	0.016	0.010	0.007	0.005	0.004	0.003	0.002	0.001	0.001			
26	0.772	0.598	0.361	0.220	0.135	0.084	0.053	0.033	0.026	0.021	0.014	0.009	0.006	0.004	0.003	0.002	0.002	0.001				
27	0.764	0.586	0.347	0.207	0.125	0.076	0.047	0.029	0.023	0.018	0.011	0.007	0.005	0.003	0.002	0.002	0.001	0.001				
28	0.757	0.574	0.333	0.196	0.116	0.069	0.042	0.026	0.020	0.016	0.010	0.006	0.004	0.002	0.002	0.002	0.001	0.001				
29	0.749	0.563	0.321	0.185	0.107	0.063	0.037	0.022	0.017	0.014	0.008	0.005	0.003	0.002	0.002	0.001	0.001	0.001				
30	0.742	0.552	0.308	0.174	0.099	0.057	0.033	0.020	0.015	0.012	0.007	0.004	0.003	0.002	0.001	0.001	0.001					
40	0.672	0.453	0.208	0.097	0.046	0.022	0.011	0.005	0.004	0.003	0.001	0.001										
50	0.608	0.372	0.141	0.054	0.021	0.009	0.005	0.004	0.001	0.001												

Figure 6.1
Discount factors

obtain a realistic assessment of the *net present value* (NPV) of an investment, the net cash flow must be discounted separately for every year of the projected life.

The following example will make this clear.

Year	Income £	Discount rate	Discount factor	NPV £
1	10000	5%	$1/1.05 = 0.9523$	$10000 \times 0.9523 = 9523.8$
2	11000	5%	$1/1.05^2 = 0.9070$	$10000 \times 0.9070 = 9070.3$
3	12000	5%	$1/1.05^3 = 0.8638$	$12000 \times 0.8638 = 10365.6$
4	12000	5%	$1/1.05^4 = 0.8227$	$12000 \times 0.8227 = 9872.4$
Total	45000			39739.1

One of the main reasons for finding the NPV is to be able to compare the viability of competing projects or different repayment modes. Again an example will demonstrate the point.

A company decides to invest £12000 for a project which is expected to give a total return of £24000 over 6 years. The discount rate is 8%.

There are two options of receiving the yearly income.

1. £6000 for years 1 and 2 = £12000
   £4000 for years 2 and 3 = £8000
   £2000 for years 5 and 6 = £4000
   Total   £24000

2. £5000 for years 1, 2, 3, and 4 = £20000
   £2000 for years 5 and 6   = £4000
                                     £24000

The DCF method will quickly establish the most profitable option to take as will be shown in the following table.

Year	Discount factor	Cash flow A £	NPV A £	Cash flow B £	NPV B £
1	$1/1.08 = 0.9259$	6000	5555.40	5000	4629.50
2	$1/1.08^2 = 0.8573$	6000	5143.80	5000	4286.50
3	$1/1.08^3 = 0.7938$	4000	3175.20	5000	3969.00
4	$1/1.08^4 = 0.7350$	4000	2940.00	5000	3675.00
5	$1/1.08^5 = 0.6806$	2000	1361.20	2000	1361.20
6	$1/1.08^6 = 0.6302$	2000	1260.40	2000	1260.40
Total		24000	19437.00	24000	19181.50

Clearly, A gives the better return, and after deducting the original investment of £12000, the net discounted return for A = £7437.00 and for B = £7181.50.

The mathematical formula for calculating the NPV is as follows:

If NPV = Net Present Value

r = the interest rate

n = number of years the project yields a return

B1, B2, B3, etc. = the annual net benefits for years 1, 2, and 3, etc.

NPV for year 1 = $B1/(1 + r)$

for year 2 = $B1/(1 + r) + B2/(1 + r)^2$

for year 3 = $B1/(1 + r) + B2/(1 + r)^2 + B3/(1 + r)^3$ and so on

If the annual net benefit is the same for each year for n years, the formula becomes

$$NPV = B/(1 + r)^n .$$

As explained previously, the discount rate can vary year by year, so the rate relevant to the year for which it applies must be used when reading off the discount factor table.

Two other financial calculations need to be carried out to enable a realistic decision to be taken as to the viability of the project.

Payback

Payback is the period of time it takes to recover the capital outlay of the project, having taken into account all the operating and overhead costs during this period. Usually this is based on the undiscounted cash flow. A knowledge of the payback is particularly important when the capital must be recouped as quickly as possible as would be the case in short-term projects or projects whose end products have a limited appeal due to changes in fashion, competitive pressures, or alternative products. Payback is easily calculated by summating all the net incomes until the total equals the original investment (e.g., if the original investment is £600 000, and the net income is £75000 per year for the next 10 years, the payback is £600000/£75000 = 8 years).

Internal Rate of Return (IRR)

It has already been shown that the higher the discount rate (usually the cost of borrowing) of a project, the lower the net present value (NPV). There must therefore come a point at which the discount rate is such that the NPV becomes zero. At this point the project ceases to be viable and the discount rate is the *internal rate of return* (IRR). In other words, it is the discount rate at which the NPV is 0.

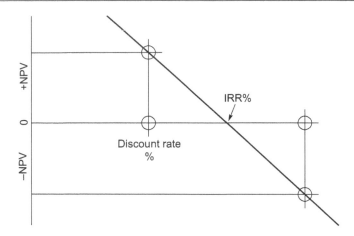

Figure 6.2
Internal rate of return (IRR) graph

While it is possible to calculate the IRR by trial and error, the easiest method is to draw a graph as shown in Figure 6.2.

The horizontal axis is calibrated to give the discount rates from 0 to any chosen value, say 20%. The vertical axis represents the NPVs, which are + above the horizontal axis and – below.

By choosing two discount rates (one low and one high), two NPVs can be calculated for the same envisaged net cash flow. These NPVs (preferably one +ve and one –ve) are then plotted on the graph and joined by a straight line. Where this line cuts the horizontal axis, i.e., where the NPV is zero, the IRR can be read off.

The basic formulae for the financial calculations are given below.

Investment appraisal definitions

NPV (net present value)	=	summation of PVs – original investment
Net income	=	incoming moneys – outgoing moneys
Payback period	=	no. of years it takes for net income to equal original investment
Profit	=	total net income – original investment
Average return/annum	=	$\dfrac{\text{total net income}}{\text{no. of years}}$
Return on investment (%)	=	$\dfrac{\text{average return} \times 100}{\text{investment}}$
	=	$\dfrac{\text{net income} \times 100}{\text{no. of years} \times \text{investment}}$
IRR (internal rate of return)	=	% discount rate for NPV = 0

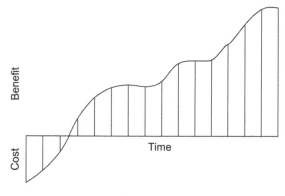

Figure 6.3
Cost/benefit profile

Cost/Benefit Analysis

Once the cost of the project has been determined, an analysis has to be carried out which compares these costs with the perceived benefits. The first cost/benefit analysis should be carried out as part of the business case investment appraisal, but in practice such an analysis should really be undertaken at the end of every phase of the life cycle to ensure that the project is still viable. The phase interfaces give management the opportunity to proceed with or, alternatively, abort the project if there is an unacceptable escalation in costs or a diminution of the benefits due to changes in market conditions such as a reduction in demand caused by political, economic, climatic, demographic, or a host of other reasons.

It is relatively easy to carry out a cost/benefit analysis where there is a tangible deliverable producing a predictable revenue stream. Provided there is an acceptable NPV, the project can usually go ahead. However, where the deliverables are intangible, such as better service, greater customer satisfaction, lower staff turnover, higher staff morale, etc., there may be considerable difficulty in quantifying the benefits. It will be necessary in such cases to run a series of tests and reviews and assess the results of interviews and staff reports.

Similarly while the cost of redundancy payments can be easily calculated, the benefits in terms of lower staff costs over a number of years must be partially offset by lower production volume or poorer customer service. Where the benefits can only be realized over a number of years, a benefit profile curve as shown in Figure 6.3 should be produced, making due allowance for the NPV of the savings.

The following lists some of the benefits that have to be considered, from which it will be apparent that some will be very difficult to quantify in monetary terms:

* Financial
* Statutory
* Economy

- Risk reduction
- Productivity
- Reliability
- Staff morale
- Cost reduction
- Safety
- Flexibility
- Quality
- Delivery
- Social

Stakeholder Management

Chapter Outline

Almost any person or organization with an interest in a project can be termed a *stakeholder*.

The type and interest of a stakeholder are of great importance to a project manager since they enable him or her to use these to the greatest benefit of the project. The process of listing, classifying, and assessing the influence of these stakeholders is termed *stakeholder analysis*.

Stakeholders can be divided into two main groups:

1. Direct (or primary) stakeholders
2. Indirect (or secondary) stakeholders

Direct Stakeholders

This group is made up, as the name implies, of all those directly associated or involved in the planning, administration, or execution of the project. These include the client, project sponsor, project manager, members of the project team, technical and financial services providers, internal or external consultants, material and equipment suppliers, site personnel, contractors and subcontractors, as well as end users. In other words, people or organizations directly involved in all or some of the various phases of the project.

Indirect Stakeholders

This group covers all those indirectly associated with the project such as internal managers of the organization and support staff not directly involved in the project, including the HR department, accounts department, secretariat, senior management levels not directly responsible for the project, and last but not least the families of the project manager and team members.

A subsection of indirect stakeholders are those representing the regulatory authorities such as national and local government, public utilities, licensing and inspecting organizations, technical institutions, professional bodies, and personal interest groups such as stockholders, labour unions, and pressure groups.

Project Management, Planning, and Control. http://dx.doi.org/10.1016/B978-0-08-098324-0.00007-X

Each of these groups can contain

- *Positive stakeholders* who support the aims and objectives of the project, and
- *Negative stakeholders* who do not support the project and do not wish it to proceed.

Direct stakeholders mainly consist of positive stakeholders as they are the ones concerned with the design and implementation of the project with the object of completing it within the specified parameters of time, cost, and quality/performance. They therefore include the sponsor, the project manager, and the project design and construction/installation teams. This group could also have negative stakeholders such as employees of the end user, who would prefer to retain the existing facility because the new installation might result in relocation or even redundancy.

The indirect group contains probably the greatest number of potential negative stakeholders. These could include environmental pressure groups, trade (labour) unions, local residents' associations, and even politicians (usually in opposition) who object to the project on principle or on environmental grounds.

Local residents' associations can be either positive or negative. For example, when it has been decided to build a by-pass road around a town, the residents in the town may well be in favour to reduce traffic congestion in the town centre, while residents in the outer villages whose environment will be degraded by additional noise and pollution will undoubtedly protest and will try to stop the road being constructed. It is these pressure groups who cause the greatest problems to the project manager.

In some situations, statutory/regulatory authorities or even government agencies who have the power to issue or withhold permits, access, wayleaves, or other consents can be considered as negative stakeholders.

Figure 7.1 shows some of the types of people or organizations in the different groups and subgroups.

Positive stakeholders				Negative stakeholders	
Direct		Indirect		Indirect	
Internal	External	Internal	External	Internal	External
Sponsor	Client	Management	Stockholders	Disgruntled employees	Disgruntled end user
Project manager	Contractors Suppliers Consultants	Accounts dept HR dept Tech. depts Families	Banks Insurers Utilities Local authorities Government agencies		Pressure groups Unions Press (media) Competitors Politicians Residents' associations
Project team Project office					

Figure 7.1
Stakeholder groups

Stakeholder	Interest	Influence impact	Probability	Action to maximize support	Reaction to minimize disruption

Figure 7.2
Stakeholder analysis

Although most negative stakeholders are clearly disruptive and tend to hamper progress, often in ingenious ways, they must nevertheless be given due consideration and afforded the opportunity to state their case. Whether it is possible to change their attitude by debate or argument depends on the strength of their convictions and the persuasiveness of the project supporters.

Diplomacy and tact are essential when negotiating with potentially disruptive organizations, and it is highly advisable to enlist experts to participate in the discussion process. Most large organizations employ labour and public relation experts as well as lawyers well versed in methods for dealing with difficult stakeholders. Their services can be of enormous help to the project manager.

It can be seen therefore that for the project manager to be able to take advantage of the positive contributions of stakeholders and counter the negative ones most effectively, a detailed analysis must be carried out setting out the interests of each positive and negative stakeholder, the impact of these interests on the project, the probability of occurrence, particularly in the case of action by negative stakeholders and the actions, or reactions, to be taken.

Figure 7.2 shows how this information can best be presented for analysis.

The Stakeholder column should contain the name of the organization and the main person or contact involved.

The Interest column states whether it is + or – and whether it is financial, technical, environmental, organizational, commercial, political, etc.

The Influence/impact column sets out the possible effect of stakeholder interference, which may be helpful or disruptive. This influence could affect the cost, time, or performance criteria of the project. Clearly stakeholders with financial muscle must be of particular interest.

The Probability column can only be completed following a cursory risk analysis based on experience and other techniques such as brainstorming, Delphi, and historical surveys.

The Action column relates to positive stakeholders and lists the best ways to generate support, such as maintaining good personal relations, invitations to certain meetings, updated information, etc.

The Reaction column sets out the tactics to assuage unfounded fears, kill malicious rumours, and minimize physical disruption.

The key to all these procedures is a good communication and intelligence-gathering system.

CHAPTER 8

Project Success Criteria

Chapter Outline
Key Performance Indicators   38

One of the topics in the project management plan is *project success criteria*. These are the most important attributes and objectives that must be met to enable the project to be termed a success. The most familiar success criteria are completion on time, keeping the project costs within budget, and meeting the performance and quality requirements set out in the specification. However, there are additional criteria that in some industries are equally or even more important. These can be safety, sustainability, reliability, legacy (long-term performance), and meeting the desired business benefit. It can be argued that all these are enshrined in performance, but there is undoubtedly a difference of emphasis between industries, organisations, and public perceptions. With the realisation that climate change has a significant impact on the environment and our future lives, sustainability in the form of conservation of energy and natural resources and the control of carbon emissions have all become performance criteria in their own right, especially as these will be subject to stricter and stricter government legislation across the whole industrial spectrum. In this context, sustainability is of course linked with legacy, as future generations will thrive or suffer depending on our success in meeting these important criteria.

It is always possible that during the life of a project, problems arise that demand that certain changes have to be made which may involve compromises and trade-offs to keep the project either on programme or within the cost boundaries. The extent to which these compromises are acceptable or permissible depends on their scope and nature and requires the approval of the project manager and possibly also the sponsor and client. However, where such an envisaged change will affect one of the project success criteria, a compromise of the affected success criterion may not be acceptable under any circumstance.

For example, if one of the project success criteria is that the project finishes by or before a certain date, then there can be no compromise of the date, but the cost may increase or quality may be sacrificed.

Success criteria can of course be subjective and depend often on the point of view of the observer. Judged by the conventional criteria of a well-managed project, i.e., cost, time, and

Project Management, Planning, and Control. http://dx.doi.org/10.1016/B978-0-08-098324-0.00008-1

performance, the Sydney Opera House failed in all three, as it was vastly over budget, very late in completion, and is considered to be too small for grand opera. Despite this, most people consider it to be a great piece of architecture and a wonderful landmark for the city of Sydney.

While it is not difficult to set the success criteria, they can only be achieved if a number of success factors are met. The most important of these are given below:

- Clear objectives and project brief agreed with client
- Good project definition
- Good planning and scheduling methods
- Accurate time control and feedback system
- Rigorous performance monitoring and control systems
- Rigorous change control (variations) procedures
- Adequate resource availability (finance, labour, plant, materials)
- Full top management and sponsor support
- Competent project management
- Tight financial control
- Comprehensive quality control procedures
- Motivated and well-integrated team
- Competent design
- Good contractual documentation
- Good internal and external communications
- Good client relationship
- Well-designed reporting system to management and client
- Political stability

* Awareness of environmental issues and related legislation

This list is not exhaustive, but if only one of the functions or systems listed is not performed adequately, the project may well end in failure.

Key Performance Indicators

A *key performance indicator* (KPI) is a major criterion against which a particular part of the project can be measured. A KPI can be a milestone that must be met, a predetermined design, delivery, installation, production, testing, erection, or commissioning stage, a payment date (in or out), or any other important stage in a project. In process plants, KPIs can include the contractual performance obligations such as output or throughput, pressure, temperature, or other quality requirements. Even when the project has been commissioned and handed over, KPIs relating to performance over defined time spans (reliability and repeatability) are often still part of the contractual requirements. Some KPIs cannot be measured or proven until the project or the operations following project completion have been running for a number of years, but these, which could also include performance and sustainability criteria, should nevertheless be considered and incorporated at both the planning and execution stages.

Organization Structures

Chapter Outline

To manage a project, a company or authority has to set up a project organization, which can supply the resources for the project and service it during its life cycle. There are three main types of project organizations:

1. Functional
2. Matrix
3. Project or taskforce

Functional Organization

This type of organization consists of specialist or functional departments, each with their own departmental manager responsible to one or more directors. Such an organization is ideal for routine operations where there is little variation of the end product. Functional organizations are usually found where items are mass produced, whether they are motor cars or sausages. Each department is expert at its function and the interrelationship between them is well established. In this sense a functional organization is not a project-type organization at all and is only included because when small, individual, one-off projects have to be carried out, they may be given to a particular department to manage. For projects of any reasonable size or complexity, it will be necessary to set up one of the other two types of organizations.

Matrix Organization

This is probably the most common type of project organization, since it utilizes an existing functional organization to provide the human resources without disrupting the day-to-day operation of the department.

The personnel allocated to a particular project are responsible to a project manager for meeting the three basic project criteria: time, cost, and quality. The departmental manager is, however, still responsible for their 'pay and rations' and their compliance with the department's standards and procedures, including technical competence and conformity to company quality standards. The members of this project team will still be working at their desks in their department, but will be booking their time to the project. Where the project does not warrant a full-time contribution, only those hours actually expended on the project will be allocated to it.

The advantages of a matrix organization are:

1. Resources are employed efficiently, since staff can switch to different projects if held up on any one of them;
2. The expertise built up by the department is utilized and the latest state-of-the-art techniques are immediately incorporated;
3. Special facilities do not have to be provided and disrupting staff movements are avoided;
4. The career prospects of team members are left intact;
5. The organization can respond quickly to changes of scope; and
6. The project manager does not have to concern himself with staff problems.

The disadvantages are:

1. There may be a conflict of priorities between different projects;
2. There may be split loyalties between the project manager and the departmental manager due to the dual reporting requirements;
3. Communications between team members can be affected if the locations of the departments are far apart; and
4. Executive management may have to spend more time to ensure a fair balance of power between the project manager and the department manager.

Matrix organizations can sometimes be categorized as strong or weak, depending on the degree of dominance or authority of the project manager or department managers respectively. This can of course create friction as both sides will try to assert themselves.

However, all the above problems can be resolved if senior management ensures (and indeed insists) that there is a good working relationship between the project manager and the department heads. At times both sides may have to compromise on the interests of the project and the organization as a whole.

Project Organization (Taskforce)

From a project manager's point of view this is the ideal type of project organization, since with such a setup he or she has complete control over every aspect of the project. The project team will usually be located in one area, which can be a room for a small project or a complete building for a very large one.

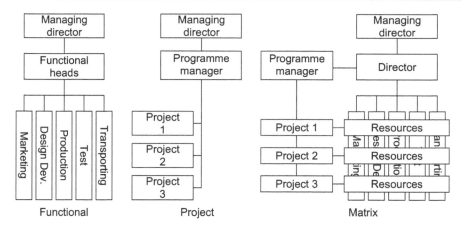

Figure 9.1
Types of organization

Lines of communication are short and the interaction of the disciplines reduces the risk of errors and misunderstandings. Not only are the planning and technical functions part of the team but also the project cost control and project accounting staff. This places an enormous burden and responsibility on the project manager, who will have to delegate much of the day-to-day management to special project coordinators whose prime function is to ensure a good communication flow and timely receipt of reports and feedback information from external sources.

On large projects with budgets often greater than £0.5 billion, the project manager's responsibilities are akin to those of a managing director of a medium-size company. He or she is not only concerned with the technical and commercial aspects of the project but must also deal with the staff, financial, and political issues, which are often more difficult to delegate.

There is no doubt that for large projects a taskforce type of project organization is essential, but as with so many areas of business, the key to success lies with the personality of the project manager and his or her ability to inspire the project team to regard themselves as personal stakeholders in the project.

One of the main differences between the two true project organizations (matrix and taskforce) and the functional organization is the method of financial accounting. For the project manager to retain proper cost control during the life of the project, it is vital that a system of project accounting is instituted, whereby all incomes and expenditures, including a previously agreed overhead allocation and profit margin, are booked to the project as if it were a separate self-standing organization. The only possible exceptions are certain corporate financial transactions such as interest payments on loans taken out by the host organization and interest receipts on deposits from a positive cash flow.

Figure 9.1 shows a diagrammatic representation of the three basic project management organizations, functional, project (or taskforce), and matrix.

CHAPTER 10

Organization Roles

Every project has a number of key people, who have important roles to play. Each of these role players has specific responsibilities, which, if carried out in their prescribed manner, will ensure a successful project. The following list gives the some examples of these organizational roles:

- Project sponsor
- Portfolio manager
- Programme manager
- Project manager
- Project planning manager or project planner
- Cost manager or cost control manager
- Risk manager
- Procurement manager
- Configuration manager
- Quality manager
- Project board or steering committee
- Project support office

In practice, the need for some of these roles depends on the size and complexity of the project and the organizational structure of the company or authority carrying out the project. For example, a small organization carrying out only two or three small projects a year may not require, or indeed be able to afford, many of these roles. Having received the commission from the client, a good project manager supported by a planner, a procurement manager, and established well-documented quality control and risk procedures may well be able to deliver the specified requirements.

The detailed descriptions of the above roles are given in the relevant sections dealing with each topic, but the last two roles warrant some further discussion.

A *project board* or steering committee is sometimes set up for large projects to act as a supervising authority, and sometimes as a champion, during the life cycle of the project. Its job is to ensure that the interests of the sponsoring organization (or client) are protected and that the project is run and delivered to meet the requirements of these organizations. The project board appoints the project manager, and the project manager must report to the project board on a regular basis and obtain from it the authority to proceed after certain predetermined stages have been reached.

45

National or local government projects following the PRINCE methodology frequently have a project board, as the project may affect a number of different departments who are all stakeholders to a greater or lesser extent. The board is thus usually constituted from senior managers (or directors) of the departments most closely involved in the project. A project board may also be desirable where the client consists of a number of companies who are temporary partners in a special consortium set up to deliver the project. A typical example is a consortium of a number of oil producers who each have a share in a refinery, drilling platform, or pipeline either during the construction or operating phase, or both.

Ideally, the board should only be consulted or be required to make decisions on major issues, as unnecessary interference with the normal running of a project will undermine the authority of the project manager. However, there may be instances where fundamental disagreements or differences of interest between functional departments or stakeholders need to be discussed and resolved, and it is in such situations that the project board can play a useful role.

The project support office briefly described in Chapter 2 is only required on large projects, where it is desirable to service all the project administration and project technology by a central department or office, supervised by a service manager. The functions most usually carried out by the project support office are the preparation and updating of the project schedule (programme), collection and processing of time sheets, progress reports, project costs and quality reports, operating configuration management, and controlling the dissemination of specifications, drawings, schedules, and other data.

The project support office is in fact the secretariat of the project and its size and constitution will therefore depend on the size of the project, its technical complexity, and its administrative procedures and reporting requirements. The last mentioned can be very onerous where the client organization is a consortium of companies, each of which requires its own reporting procedures, cost reports, and timetable for submission.

Ideally the accounting system for the project should be self-contained and independent of the corporate accounting system. Clearly a monthly cost report will have to be submitted to the organization's accounts department, but project accounting will give speedier and more accurate cost information to the project manager, which will enable him or her to take appropriate action before the costs spiral out of control. The project support office can play a vital role in this accounting function provided that the office staff includes the project accountants. These accountants control not only the direct costs such as labour, materials, equipment, and plant but also those indirect costs and expenses related to the project.

CHAPTER 11

Project Life Cycles

Most, if not all, projects go through a life cycle that varies with the size and complexity of the project. On medium to large projects the life cycle will generally follow the following pattern:

1	Concept	Basic ideas, business case, statement of requirements, scope;
2	Feasibility	Tests for technical, commercial, and financial viability, technical studies, investment appraisal, DCF, etc.;
3	Evaluation	Application for funds, stating risks, options, TCQ criteria;
4	Authorization	Approvals, permits, conditions, project strategy;
5	Implementation	Development design, procurement, fabrication, installation, commissioning;
6	Completion	Performance tests, handover to client, post-project appraisal;
7	Operation	Revenue earning period, production, maintenance;
8	Termination	Close-down, decommissioning, disposal.

Items 7 and 8 are not usually included in a project life cycle where the project ends with the issue of an acceptance certificate after the performance tests have been successfully completed. Where these two phases are included, as, for example, with defence projects, the term *extended project life cycle* is often used.

The project life cycle of an IT project may be slightly different as the following list shows:

1	Feasibility	Definition, cost benefits, acceptance criteria, time, cost estimates;
2	Evaluation	Definitions of requirements, performance criteria, processes;
3	Function	Functional and operational requirements, interfaces, system design;
4	Authorization	Approvals, permits, firming up procedures;
5	Design and build	Detail design, system integration, screen building, documentation;
6	Implementation	Integration and acceptance testing, installation, training;
7	Operation	Data loading, support setup, handover.

Running through the period of the life cycle are control systems and decision stages at which the position of the project is reviewed. The interfaces of the phases of the life cycle form convenient milestones for progress payments and reporting progress to top management, who can then make the decision to abort or provide further funding. In some cases the interfaces of the phases overlap, as in the case of certain design and construct contracts, where construction starts before the design is finished. This is known as concurrent engineering and is often employed to reduce the overall project programme.

As the word *cycle* implies, the phases may have to be amended in terms of content, cost, and duration as new information is fed back to the project manager and sponsor. Projects are essentially dynamic organizations that are not only specifically created to effect change, but are also themselves subject to change.

On some projects it may be convenient to appoint a different project manager at a change of phase. This is often done where the first four stages are handled by the development or sales department, who then hands the project over to the operations department for the various stages of the implementation and completion phases.

When the decommissioning and disposal is included, it is known as an extended life cycle, since these two stages could occur many years after commissioning and could well be carried out by a different organization.

While all institutions associated with project management stress the importance of the project cycle, both BSI and ISO preferred describing what operations should be carried out during the various phases, rather than giving the phases specific names.

Different organizations tend to have different descriptions and sequences of the phases, and Figure 11.1 shows two typical life cycles prepared by two different organizations. The first example, as given by APM in their latest Body of Knowledge, is a very simple generic life cycle consisting of only six basic phases. The second life cycle shown, as formulated by the

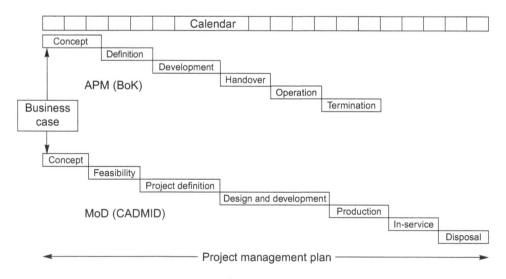

Figure 11.1
Examples of project life cycles

Ministry of Defence (MoD), clearly shows the phases required for a typical weapons system, where concept, feasibility, and project definition are the responsibility of the MoD, design, development, and production are carried out by the manufacturer, and in-service and disposal are the phases when the weapon is in the hands of the armed forces.

The diagram also shows a calendar scale over the top. While this is not strictly necessary, it can be seen that if the lengths of the bars representing the phases are drawn proportional to the time taken by the phases, such a presentation can be used as a high-level reporting document, showing which phases are complete or partially complete in relation to the original schedule.

The important point to note is that each organization should develop its own life cycle diagram to meet its particular needs. Where the life cycle covers all the phases from cradle to grave as it were, it is often called a *programme life cycle*, since it spans over the full programme of the deliverable. The term *project life cycle* is then restricted to those phases that constitute a project within the programme (e.g., the design, development, and manufacturing periods).

Figure 11.2 shows how decision points or milestones (sometimes called trigger points or gates) relate to the phases of a life cycle. At each gate, a check should be carried out to ensure that the project is still viable, that it is still on schedule, that costs are still within budget, that sufficient resources are available for the next phase, and that the perceived risks can be managed.

Figure 11.3 shows how the life cycle of the MoD project shown in Figure 11.1 could be split into the *project life cycle*, i.e., the phases under the control of the project team (conception to

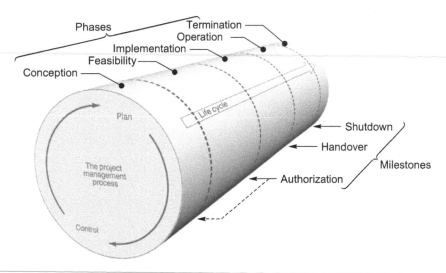

Figure 11.2
Project management life cycle

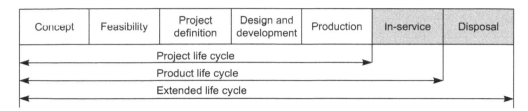

Figure 11.3
Life cycle of MoD project

production), the *product life cycle*, the phases of interest to the sponsor, which now includes the in-service performance, and lastly the *extended life cycle*, which includes disposal. From the point of view of the contractor, the *project life cycle* may only include design and development and production. It can be seen therefore that there are no hard-and-fast rules as to where the demarcation points are, as each organization will define its own phases and life cycles to suit its method of working.

Work Breakdown Structures

Chapter Outline

An examination of the project life cycle diagram will show that each phase can be regarded as a project in its own right, although each will be of very different size and complexity. For example, when a company is considering developing a new oil field, the feasibility study phase could be of considerable size although the main project would cover the design, development, and production phases. To be able to control such a project, the phase must be broken down further into stages or tasks, which in turn can be broken down further into subtasks or work packages until one can be satisfied that an acceptable control structure has been achieved.

The choice of tasks incorporated in the *work breakdown structure* (WBS) is best made by the project team drawing on their combined experience or engaging in a brainstorming session.

Once the main tasks have been decided upon, they can in turn be broken down into subtasks or work packages, which should be coded to fit in with the project cost coding system. This will greatly assist in identifying the whole string of relationships from overall operational areas down to individual tasks. For this reason the WBS is the logical starting point for subsequent planning networks. Another advantage is that a cost allocation can be given to each task in the WBS and, if required, a risk factor can be added. This will assist in building up the total project cost and creates a risk register for a subsequent, more rigorous risk assessment.

The object of all this is to be able to control the project by allocating resources (human, material, and financial) and giving time constraints to each task. It is always easier to control a series of small entities that make up a whole than to control the whole enterprise as one operation. What history has proven to be successful for armies, which are divided into divisions, regiments, battalions, companies, and platoons, or corporations, which have area organizations, manufacturing units, and sales territories, is also true for projects, whether they are large, small, sophisticated, or straightforward.

The tasks will clearly vary enormously with the type of project in both size and content, but by representing their relationships diagrammatically, a clear graphical picture can be created.

51

This, when distributed to other members of the project team, becomes a very useful tool for disseminating information as well as a reporting medium to all stakeholders. As the completion of the main tasks are in effect the major project milestones, the WBS is an ideal instrument for reporting progress upwards to senior management, and for this reason it is essential that the status of each work package or task is regularly updated.

As the WBS is produced in the very early stages of a project, it will probably not reflect all the tasks that will eventually be required. Indeed the very act of draughting the WBS often throws up the missing items or work units, which can then be formed into more convenient tasks. As these tasks are decomposed further, they may be given new names such as unit or work package. It is then relatively easy for management to allocate task owners to each task or group of tasks, who have the responsibility for delivering each task to the normal project criteria of cost, time, and quality/performance.

The abbreviation WBS is a generic term for a hierarchy of stages of a project. However, some methodologies like PRINCE call such a hierarchical diagram a *product breakdown* structure (PBS). The difference is basically what part of speech is being used to describe the stages. If the words used are *nouns*, it is, strictly speaking, a PBS, because we are dealing with products or things. If, on the other hand, we are describing the work that has to be performed on the nouns and use *verbs*, we call it a WBS. Frequently, a diagram starts as a PBS for the first three or four levels and then becomes a WBS as more detail is introduced.

Despite this unfortunate lack of uniformity of nomenclature in the project management fraternity, the principles of subdividing the project into manageable components are the same.

It must be pointed out, however, that the WBS is *not* a programme, although it looks like a precedence diagram. The interrelationships shown by the link lines do not necessarily imply a time dependence or indeed any sequential operation.

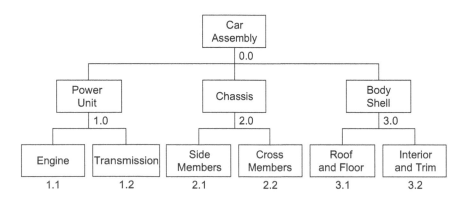

Figure 12.1
PBS

The corresponding WBS shown in Figure 12.2 uses verbs, and the descriptions of the packages or tasks then become: *assemble* car, *build* power unit, *weld* chassis, *press* body shell, etc.

The degree to which the WBS needs to be broken down before a planning network can be drawn will have to be decided by the project manager, but there is no reason why a whole family of networks cannot be produced to reflect each level of the WBS.

Once the WBS (or PBS) has been drawn, a bottom-up cost estimate can be produced starting at the lowest branch of the family tree. In this method, each work package is costed and arranged in such a way that the total cost of the packages on any branch must add up to the cost of the package of the parent package on the branch above. If the parent package has a cost value of its own, this must clearly be added before the next stage of the process. This is shown in Figure 12.3, which not only explains the bottom-up estimating process, but also shows how the packages can be coded to produce a project cost coding system that can be carried through to network analysis and earned value analysis.

The resulting diagram is now a *cost breakdown structure* (CBS).

It can be seen that a WBS is a powerful tool that can show clearly and graphically who is responsible for a task, how much it should cost, and how it relates to the other tasks in the project. It was stated earlier that the WBS is not a programme, but once it has been accepted as a correct representation of the project tasks, it will become a good base for drawing up the network diagram. The interrelationships of the tasks will have to be shown more accurately, and the only additional items of information to be added are the durations.

An alternative to the *bottom-up* cost allocation is the *top-down* cost allocation. In this method, the cost of the total project (or subproject) has been determined and is allocated to the top package of the WBS (or PBS) diagram. The work packages below are then forced to accept

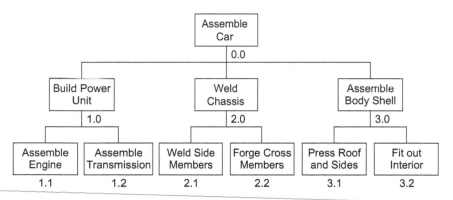

Figure 12.2
WBS

the appropriate costs so that the total cost of each branch cannot exceed the total cost of the package above. Such a top-down approach is shown in Figure 12.4.

In practice both methods may have to be used. For example, the estimator of a project may use the bottom-up method on a WBS or PBS diagram to calculate the cost. When this is given to the project manager, he or she may break this total down into the different departments of an organization and allocate a proportion to each, making sure that the sum total does not exceed the estimated cost. Once names have been added to the work packages of a WBS or PBS it becomes an organization breakdown structure (OBS).

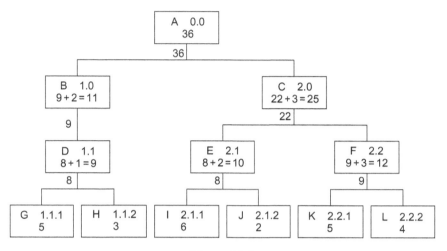

Figure 12.3
CBS

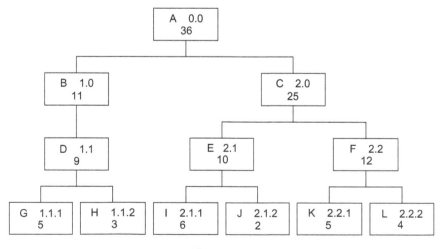

Figure 12.4
Top-down cost allocation

It did not take long for this similarity to be appreciated, so that another name for such an organization diagram became *organization breakdown*. This is the family tree of the organization in the same way that the WBS is in effect the family tree of the project. It is in fact more akin to a family tree or organization chart (organogram).

Figure 12.5 shows a typical OBS for a manufacturing project such as the assembly of a prototype motor car. It can be seen that the OBS is not identical in layout to the WBS, as one manager or task owner can be responsible for more than one task.

The OBS shown is typical of a matrix-type project organization where the operations manager is in charge of the actual operating departments for 'pay and rations', but each departmental head (or his or her designated project leader) also has a reporting line to the project manager. If required, the OBS can be expanded into a responsibility matrix to show the responsibility and authority of each member of the organization or project team.

The quality assurance (QA) manager reports directly to the director to ensure independence from the operating and projects departments. He or she will, however, assist all operating departments with producing the quality plans and give ongoing advice on QA requirements and procedures as well as pointing out any shortcomings he or she may discover.

Although the WBS may have been built up by the project team, based on their collective experience or by brainstorming, there is always the risk that a stage or task has been

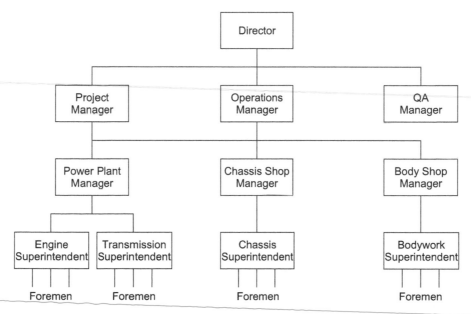

Figure 12.5
OBS

Table 12.1

Project Risks			
Organization	**Environment**	**Technical**	**Financial**
Management	Legislation	Technology	Financing
Resources	Political	Contracts	Exchange rates
Planning	Pressure groups	Design	Escalation
Labour	Local customs	Manufacture	Financial stability of
Health and safety	Weather	Construction	(a) project
Claims	Emissions	Commissioning	(b) client
Policy	Security	Testing	(c) suppliers

forgotten. An early review then opens up an excellent opportunity to refine the WBS and carry out a risk identification for each task, which can be the beginning of a risk register. At a later date a more rigorous risk analysis can then be carried out. The WBS does in effect give everyone a better understanding of the risk assessment procedure.

Indeed a further type of breakdown structure is the *risk breakdown structure*. Here the main risks are allocated to the WBS or PBS in either financial or risk rating terms, giving a good overview of the project risks.

In another type of risk breakdown structure the main areas of risk are shown in the first level of the risk breakdown structure, and the possible risk headings are listed below. (See Table 16.1 in Chapter 16, Risk Management.)

Responsibility Matrix

By combining the WBS with the OBS it is possible to create a *responsibility matrix*. Using the car assembly example given in Figures 12.1 and 12.5, the matrix is drawn by writing the WBS work areas vertically and the OBS personnel horizontally, as shown in Figure 12.6.

By placing a suitable designatory letter into the intersecting boxes, the level of responsibility for any work area can be recorded on the matrix.

A = Receiving monthly reports

B = Receiving weekly reports

C = Daily supervision

The vertical list giving the WBS stages could be replaced by the PBS stages. The horizontal list showing the different departmental managers could instead have the names of the departments or even consultants, contractors, subcontractors, and suppliers of materials or services.

	Director	Project Manager	Operations Manager	QA Manager	Power Plant Manager	Chassis Shop Manager	Body Shop Manager	Engine Superintendent	Transmission Superintendent	Chassis Superintendent	Bodywork Superintendent
Car Assembly	A	B	A	B							
Power Unit	A	B	A	B	C						
Chassis	A	B	A	B		C					
Body Shell	A	B	A	B			C				
Engine		B	A	B	C			C			
Transmission		B	A	B	C				C		
Side Members		B	A	B		C				C	
Cross Members		B	A	B		C				C	
Roof and Floor		B	A	B			C				C
Interior and Trim		B	A	B			C				C

Figure 12.6

Responsibility matrix

Estimating

Chapter Outline

Estimating is an essential part of project management, since it becomes the baseline for subsequent cost control. If the estimate for a project is too low, a company may well lose money in the execution of the work. If the estimate is too high, the company may well lose the contract due to overpricing.

As explained in the section on work breakdown structures, there are two basic methods of estimating: top down and bottom up. However, unfortunately in only a few situations are the costs available in a form for simply slotting into the work package boxes. It is necessary therefore to produce realistic estimates of each package and indeed the entire project before a meaningful cost estimate can be carried out. In most estimates that require any reasonable degree of accuracy the method used must be bottom up. This principle is used in bills of quantities, which literally start at the bottom of the construction process, the ground clearance, and foundations and work up through the building sequence to the final stages such as painting and decorating.

Estimating the cost of a project requires a structured approach, but whatever method is used, the first thing is to decide the level of accuracy required. This depends on the status of the project and the information available. There are four main estimating methods in use, varying from the very approximate to the very accurate. These are:

1. Subjective (degree of accuracy +/– 20% to 40%)
2. Parametric (degree of accuracy +/– 10% to 20%)
3. Comparative (degree of accuracy +/– 10%)
4. Analytical (degree of accuracy +/– 5%)

Subjective

At the proposal stage, a contractor may well be able to give only a 'ballpark figure' to give a client or sponsor an approximate 'feel' of the possible costs. The estimating method used in this case would either be *subjective or approximate parametric*. In either case the degree of accuracy would largely depend on the experience of the estimator. When using the subjective method, the estimator relies on his or her experience of similar projects to give a cost indication based largely on 'hunch'. Geographical and political factors as well as the more obvious labour and material content must be taken into account. Such an approximate method of estimating is often given the disparaging name of 'guesstimating'.

Parametric

The *parametric* method would be used at the budget preparation stage, but relies on good historical data-based past jobs or experience. By using well-known empirical formulae or ratios in which costs can be related to specific characteristics of known sections or areas of the project, it is possible to produce a good estimate on which firm decisions can be based. Clearly such estimates need to be qualified to enable external factors to be separately assessed. For example, an architect will be able to give a parametric estimate of a new house once he or she is given the cube (height × length × depth) of the proposed building and the standard of construction or finish. The estimate will be in £/cubic metre of structure. Similarly, office blocks are often estimated in £/square metre of floor space. The qualifications would be the location, ground conditions, costs of the land, etc. Another example of a parametric estimate is when a structural steel fabricator gives the price of fabrication in £/tonne of steel, depending on whether the steel sections are heavy beams and columns or light latticework. In both cases the estimate may or may not include the cost of the steel itself.

Comparative (By Analogy)

As an alternative to the parametric method, the *comparative* method of estimating can be used for the preparation of the budget. When a new project is very similar to another project recently completed, a quick comparison can be made of the salient features. This method is based on the costs of a simplified schedule of major components that were used on previous similar jobs. It may even be possible to use the costs of a similar-sized complete project of which one has had direct (and preferably recent) experience. Due allowance must clearly be made for the inevitable minor differences, inflation, and other possible cost escalations. An example of such a comparative estimate is the installation of a new computer system in a building when an almost identical (and proven) system was installed six months earlier in another building. It must be stressed that such an estimate does not require a detailed breakdown.

Analytical

Once the project has been sanctioned a working budget estimate will be necessary against which the cost of the project will be controlled. This will normally require an *analytical* estimate or bill of quantities. This type of estimate may also be required where a contractor has to submit a fixed price tender, since once the contract is signed there can be no price adjustment except by inflation factors or client-authorised variations.

As the name implies, this is the most accurate estimating method, but it requires the project to be broken down into sections, subsections, and finally individual components. Each component must then be given a cost value (and preferably also a cost code) including both the material and labour content. The values, which are sometimes referred to as 'norms', are usually extracted from a database or company archives and must be individually updated or factored to reflect the present-day political and environmental situation.

Examples of analytical estimates are the norms used by the petrochemical industry where a value exists for the installation of piping depending on pipe diameter, wall thickness, material composition, height from ground level, and whether flanged or welded. The norm is given as a cost/linear metre, which is then multiplied by the meterage including an allowance for waste. Contingencies, overheads, and profit are then added to the total sum.

Quantity surveyors will cost a building or structure by measuring the architect's drawings and applying a cost to every square metre of wall or roof, every door and window, and such systems as heating, plumbing, electrics, etc. Such estimates are known as bills of quantities and together with a schedule of rates for costing variations form the basis of most building and civil engineering contracts. The accuracy of such estimates are better than plus or minus 5%, depending on the qualifications accompanying the estimate. The rates used in bills of quantities (when produced by a contractor) are usually inclusive of labour, materials, plant, overheads, and anticipated profit, but when produced by an independent quantity surveyor the last two items may have to be added by the contractor.

Unfortunately such composite rates are not ideal for planning purposes as the time factor only relates to the labour content. To overcome this problem, the UK Building Research Station in 1970 developed a new type of bill of quantities called 'operational bills' in which the labour was shown separately from the other components, thus making it compatible with critical path planning techniques. However, these new methods were never really accepted by industry and especially not by the quantity surveying profession.

To assist the estimator a number of estimating books have been published which give in great detail the materials and labour costs of nearly every operation or trade used in the building trade. These costs are given separately for labour based on the number of man-hours required and the materials cost per the appropriate unit of measurement such as the metre length, square metre, or cubic metre. Most of these books also give composite rates including materials, labour, overhead, and profit. As rates for materials and labour change every year due to inflation or other factors, these books will have to be republished yearly to reflect the current rates. It is important, however, to remember that these books are only guides and require the given rates to be factorized depending on site conditions, geographical location, and any other factor the estimator may consider to be significant.

The percentage variation at all stages should always be covered by an adequate contingency allowance that must be added to the final estimate to cover for possible, probable, and unknown risks, which could be technical, political, environmental, administrative, etc., depending on the results of a more formal risk analysis. The further addition of overheads and profit gives the price (i.e., what the customer is being asked to pay).

It must be emphasized that such detailed estimating is not restricted to the construction (building or civil engineering) industry. Every project, given sufficient time, can be broken down into its labour, material, plant, and overhead content and costed very accurately.

Sometimes an estimate produced by the estimator is drastically changed by senior management to reflect market conditions, the volume of work currently in the company, or the strength of the perceived competition. However, from a control point of view, such changes to the final price should be ignored, which are in any case normally restricted to the overhead and profit portion and are outside the control of the project manager. Where such a price adjustment was downward, every effort should be made to recover these 'losses' by practising value management throughout the period of the project.

Computer systems and software preparation, which are considerably more difficult to estimate than construction work due to their fundamentally innovative and untried processes, may be estimated using:

1. Function point analysis, where the number of software functions such as inputs, outputs, files, interfaces, etc., are counted, weighted, and adjusted for complexity and importance. Each function is then given a cost value and aggregated to find the overall cost.
2. Lines of code to be used in the program. A cost value can be ascribed to each line.
3. Plain man-hour estimates based on experience of previous or similar work, taking into account such new factors as inflation, the new environment, and the client organization.

While it is important to produce the best possible estimate at every stage, the degree of accuracy will vary with the phase of the project, as shown in Figure 13.1. As the project develops and additional or more accurate information becomes available, it is inevitable that the estimate becomes more accurate. This is sometimes known as rolling wave estimating, and while these revised costs should be used for the next estimating stage, once the actual final budget stage has been reached and the price has been accepted by the client, any further cost refinements will only be useful for updating the monthly cost estimate, which may affect the profit or loss without changing the price or control budget as used in earned value methods.

When estimating the man-hours related to the activities in a network programme, it may be difficult to persuade certain people to commit themselves to giving a firm man-hour estimate.

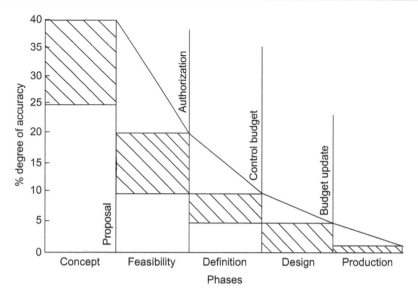

Figure 13.1
Phase/accuracy curve

In such cases, just in order to elicit a realistic response, it may be beneficial to employ the 'three time estimate' approach, $t = (a + 4m + b)/6$, as described in Chapter 20. In this formula, t is the expected or most likely time, a is the most optimistic time, b is the most pessimistic time, and m is the most probable time.

In most cases m, the most probable time, is sufficient for the estimate, as the numerical difference between this and the result obtained by rigorously applying the formula is in most cases very small.

Project Management Plan

Chapter Outline

As soon as the project manager has received his or her brief or project instructions, he or she must produce a document that distils what is generally a vast amount of information into a concise, informative, and well-organized form that can be distributed to all members of the project team and indeed all the stakeholders in the project. This document is called a project management plan (PMP). It is also sometimes just called a project plan, or in some organizations a coordination procedure.

The PMP is one of the key documents required by the project manager and his or her team. It lists the phases and encapsulates all the main parameters, standards, and requirements of the project in terms of time, cost, and quality/performance by setting out the '*Why*', '*What*', '*When*', '*Who*', '*Where*', and '*How*' of the project. In some organizations the PMP also includes the '*How much*', that is, the cost of the project. There may, however, be good commercial reasons for restricting this information to key members of the project team.

The contents of a PMP vary depending on the type of project. While it can run to several volumes for a large petrochemical project, it need not be more than a slim binder for a small, unsophisticated project.

There are, however, a number of areas and aspects that should always feature in such a document. These are set out very clearly in Table 1 of BS 6079-1-2002. With the permission of the British Standards Institution, the main headings of what is termed the *model project plan* are given below, but augmented and rearranged in the sections given above.

General

1. Foreword
2. Contents, distribution, and amendment record
3. Introduction
   3.1 Project diary
   3.2 Project history

Project Management, Planning, and Control. http://dx.doi.org/10.1016/B978-0-08-098324-0.00014-7
65

The Why

4. Project aims and objectives
   4.1 Business case

The What

5. General description
   5.1 Scope
   5.2 Project requirement
   5.3 Project security and privacy
   5.4 Project management philosophy
   5.5 Management reporting system

The When

6. Programme management
   6.1 Programme method
   6.2 Program software
   6.3 Project life cycle
   6.4 Key dates
   6.5 Milestones and milestone slip chart
   6.6 Bar chart and network if available

The Who

7. Project organization
8. Project resource management
9. Project team organization
   9.1 Project staff directory
   9.2 Organizational chart
   9.3 Terms of reference (TOR)
      (a) for staff
      (b) for the project manager
      (c) for the committees and working group

The Where

10. Delivery requirements
    10.1 Site requirements and conditions
    10.2 Shipping requirements
    10.3 Major restrictions

The How

11. Project approvals required and authorization limits
12. Project harmonization
13. Project implementation strategy
    13.1 Implementation plans
    13.2 System integration
    13.3 Completed project work
14. Acceptance procedure
15. Procurement strategy
    15.1 Cultural and environmental restraints
    15.2 Political restraints
16. Contract management
17. Communications management
18. Configuration management
    18.1 Configuration control requirements
    18.2 Configuration management system
19. Financial management
20. Risk management
    20.1 Major perceived risks
21. Technical management
22. Tests and evaluations
    22.1 Warranties and guarantees
23. Reliability management (see also BS 5760: Part 1)
    23.1 Availability, reliability, and maintainability (ARM)
    23.2 Quality management
24. Health and safety management
25. Environmental issues
26. Integrated logistic support (ILS) management
27. Close-out procedure

The numbering of the main headings should be standardized for all projects in the organization. In this way the reader will quickly learn to associate a clause number with a subject. This will not only enable him or her to find the required information quickly but will also help the project manager when he or she has to write the PMP. The numbering system will in effect serve as a convenient checklist. If a particular item or heading is not required, it is best simply to enter 'not applicable' (or NA), leaving the standardized numbering system intact.

Apart from giving all the essential information about the project between two covers, for quick reference, the PMP serves another very useful function. In many organizations the scope, technical, and contractual terms of the project are agreed on in the initial stages by the proposals or sales department. It is only when the project becomes a reality that the project manager is appointed. By having to assimilate all these data and write such a PMP (usually within two weeks of the handover meeting), the project manager will inevitably obtain a thorough understanding of the project requirements as he or she digests the often voluminous documentation agreed on with the client or sponsor.

Clearly not every project requires the exact breakdown given in this list, and each organization can augment or expand this list to suit the project. If there are any subsequent changes, it is essential that the PMP is amended as soon as changes become apparent so that every member of the project team is immediately aware of the latest revision. These changes must be numbered on the amendment record at the front of the PMP and annotated on the relevant page and clause with the same amendment number or letter.

The contents of the project management plan are neatly summarized in the first verse of the little poem from *The Elephant's Child* by Rudyard Kipling:

> I keep six honest serving-men
> (They taught me all I knew);
> Their names are What and Why and When,
> And How and Where and Who.

Methods and Procedures

Methods and procedures are the very framework of project management and are necessary throughout the life cycle of the project. All the relevant procedures and processes are set out in the project management plan, where they are customized to suit the particular project.

Methods and procedures should be standardized within an organization to ensure that project managers do not employ their own pet methods or 'reinvent the wheel'.

All the organization's standard methods and procedures as well as some of the major systems and processes should be enshrined in a company project manual. This should then be read and signed by every project manager who will then be familiar with the company systems and thus avoid wasteful and costly duplication. The main contents of such a manual are the methods and procedures covering:

- Company policy and mission statement
- Company organization and organization chart
- Accountability and responsibilities
- Estimating

- Risk analysis
- Cost control
- Planning and network analysis
- Earned value management
- Resource management
- Change management (change control)
- Configuration management
- Procurement (bid preparation, purchasing, expediting, inspection, shipping)
- Contract management and documentation
- Quality management and control
- Value management and value engineering
- Issue management
- Design standards
- Information management and document distribution
- Communication
- Health and Safety
- Conflict management
- Close-out requirements and reviews

It will be seen that this list is very comprehensive but in every case a large proportion of the documentation required can be standardized. There are always situations where a particular method or procedure has to be tailored to suit the circumstances or where a system has to be simplified, but the standards set out in the manual form a baseline that acts as a guide for any necessary modification.

Certain UK government departments, a number of local authorities, and other public bodies have adopted a project management methodology called PRINCE 2 (an acronym for Projects In a Controlled Environment). This was developed by the Central Computer and Telecommunications Agency (CCTA) for IT and government contracts but has not found favour in the construction industry due to a number of differences in approach to reporting procedures, management responsibilities, and assessing durations with respect to resources.

Risk Management

Every day we take risks. If we cross the street we risk being run over. If we go down the stairs, we risk missing a step and tumbling down. Taking risks is such a common occurrence that we tend to ignore it. Indeed, life would be unbearable if we constantly worried whether we should or should not carry out a certain task or take an action, because the risk is, or is not, acceptable.

With projects, however, this luxury of ignoring the risks cannot be permitted. By their very nature, because projects are inherently unique and often incorporate new techniques and procedures, they are risk prone, and risk has to be considered right from the start. It then has to be subjected to a disciplined regular review and investigative procedure known as risk management.

Before applying risk management procedures, many organizations produce a *risk management plan*. This is a document produced at the start of the project that sets out the strategic requirements for risk assessment and the whole risk management procedure. In certain situations the risk management plan should be produced at the estimating or contract tender stage to ensure that adequate provisions are made in the cost build-up of the tender document.

The project management plan (PMP) should include a résumé of the risk management plan, which will first of all define the scope and areas to which risk management applies, particularly the risk types to be investigated. It will also specify which techniques will be used for risk identification and assessment, whether SWOT (strengths, weaknesses, opportunities, and threats) analysis is required and which risks (if any) require a more rigorous quantitative analysis such as Monte Carlo simulation methods.

Project Management, Planning, and Control. http://dx.doi.org/10.1016/B978-0-08-098324-0.00015-9

The risk management plan will set out the type, content, and frequency of reports, the roles of risk owners, and the definition of the impact and probability criteria in qualitative and/or quantitative terms covering cost, time, and quality/performance.

The main contents of a risk management plan are as follows:

- *General introduction* explaining the need for the risk management process;
- *Project description*. Only required if it is a stand-alone document and not part of the PMP;
- *Types of risks*. Political, technical, financial, environmental, security, safety, programme, etc.;
- *Risk processes*. Qualitative and/or quantitative methods, maximum number of risks to be listed;
- *Tools and techniques*. Risk identification methods, size of P-I matrix, computer analysis, etc.;
- *Risk reports*. Updating periods of risk register, exception reports, change reports, etc.;
- *Attachments*. Important project requirements, dangers, exceptional problems, etc.

The risk management plan of an organization should follow a standard pattern in order to increase its familiarity (rather like standard conditions of contract), but each project will require a bespoke version to cover its specific requirements and anticipated risks.

Risk management consists of the following five stages, which, if followed religiously, will enable one to obtain a better understanding of those project risks that could jeopardize the cost, time, quality, and safety criteria of the project. The first three stages are often referred to as *qualitative analysis* and are by far the most important stages of the process.

Stage 1: Risk Awareness

This is the stage at which the project team begins to appreciate that there are risks to be considered. The risks may be pointed out by an outsider, or the team may be able to draw on their own collective experience. The important point is that once this attitude of mind has been achieved, i.e., that the project, or certain facets of it, are at risk, it leads very quickly to …

Stage 2: Risk Identification

This is essentially a team effort at which the scope of the project, as set out in the specification, contract and WBS (if drawn), is examined and each aspect investigated for a possible risk.

To get the investigation going, the team may have a brainstorming session and use a prompt list (based on specific aspects such as legal or technical problems) or a checklist compiled from risk issues from similar previous projects. It may also be possible to obtain expert opinion or carry out interviews with outside parties. The end product is a long list of activities that may be affected by one or a number of adverse situations or unexpected occurrences. The risks that generally have to be considered may be conveniently split into four main areas as shown in Table 15.1.

Table 15.1

Project risks			
Organization	*Environment*	*Technical*	*Financial*
Management	Legislation	Technology	Financing
Resources	Political	Contracts	Exchange rates
Planning	Pressure groups	Design	Escalation
Labour	Local customs	Manufacture	Financial stability of
Health and safety	Weather	Construction	(a) project
Claims	Emissions	Commissioning	(b) client
Policy	Security	Testing	(c) suppliers

Any applicable risk in each area can then be examined by a further screening process as shown by the samples given below:

- Technical   New technology or materials. Test failures;
- Environmental   Unforeseen weather conditions. Traffic restrictions;
- Operational   New systems and procedures. Training needs;
- Cultural   Established customs and beliefs. Religious holidays;
- Financial   Freeze on capital. Bankruptcy of stakeholder. Currency fluctuations;
- Legal   Local laws. Lack of clarity of contract;
- Commercial   Change in market conditions or customers;
- Resource   Shortage of staff, operatives, plant, or materials;
- Economic   Slow-down of economy. Change in commodity prices;
- Political   Change of central or local government or government policy;
- Security   Safety. Theft. Vandalism. Civil disturbance.

Some risks could be categorised in more than one area or section, such as civil unrest, which could be a political as well as a security problem.

The following gives the advantages and disadvantages:

Brainstorming
Advantages: Wide range of possible risks suggested for consideration
                     Involves a number of stakeholders
Disadvantages: Time consuming
                     Requires firm control by the facilitator

Prompt list
Advantages: Gives benefit of past problems
                     Saves time by focusing on real possibilities

Disadvantages	Restricts suggestions to past experience
	Past problems may not be applicable

Checklist

Advantages:	Similar to prompt list: Company standard
Disadvantages	Similar to prompt list

Work breakdown structure

Advantages:	Focused on specific project risks;
	Quick and economical.
Disadvantages:	May limit scope of possible risks.

Delphi technique

Advantages:	Offers wide experience of experts;
	Can be wide ranging.
Disadvantages:	Time consuming if experts are far away;
	Expensive if experts have to be paid;
	Advice may not be specific enough.

Asking experts

Advantages:	As Delphi.
Disadvantages:	As Delphi.

At this stage it may be possible to identify who is the best person to manage each risk. This person becomes the *risk owner*.

To reduce the number of risks being seriously considered from what could well be a very long list, some form of screening will be necessary. Only those risks that pass certain criteria need be examined more closely, which leads to the next stage …

Stage 3: Risk Assessment

This is the qualitative stage at which the two main attributes of a risk, *probability* and *impact*, are examined.

The *probability* of a risk becoming a reality has to be assessed using experience and/or statistical data such as historical weather charts or close-out reports from previous projects. Each risk can then be given a probability rating of HIGH, MEDIUM, or LOW.

In a similar way, by taking into account all the available statistical data, past project histories, and expert opinion, the *impact* or effect on the project can be rated as SEVERE, MEDIUM, or LOW.

A simple matrix can now be drawn up which identifies whether a risk should be taken any further. Such a matrix is shown in Figure 15.1.

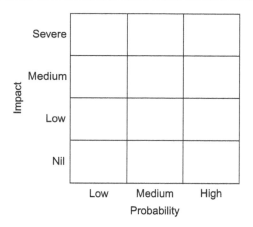

Figure 15.1
Probability versus impact table. Such a table could be used for each risk worthy of further assessment, and to assess, for example, all major risks to a project or programme.

		Risk Summary Chart		
Risk no.	Description	Probability rating	Impact rating	Risk owner

Figure 15.2
Risk Summary Chart

Each risk can now be given a *risk number*, so that it is now possible to draw up a simple chart that lists all the risks considered so far. This chart will show the risk number, a short description, the risk category, the probability rating, the impact rating (in terms of high, medium, or low), and the *risk owner* who is charged with monitoring and managing the risk during the life of the project.

Figure 15.2 shows the layout of such a chart.

A *quantitative analysis* can now follow. This is known as …

Stage 4: Risk Evaluation

It is now possible to give comparative values, often on a scale 1 to 10, to the probability and impact of each risk and by drawing up a matrix of the risks, an order of importance or priority can be established. By multiplying the *impact rating* by the *probability rating*, the *exposure rating* is obtained. This is a convenient indicator that may be used to reduce the list to only the top dozen that require serious attention, but an eye should nevertheless be kept on even the minor ones, some of which may suddenly become serious if unforeseen circumstances arise.

An example of such a matrix is shown in Figure 15.3. Clearly the higher the value, the greater the risk and the more attention it must receive to manage it.

Another way to quantify both the impact and probability is to number the ratings as shown in Figure 15.4 from 1 for very low to 5 for very high. By multiplying the appropriate numbers in the boxes, a numerical (or quantitative) exposure rating is obtained, which gives a measure of seriousness and hence importance for further investigation.

For example, if the impact is rated 3 (i.e., medium) and the probability 5 (very high), the exposure rating is $3 \times 5 = 15$.

Further sophistication in evaluating risks is possible by using some of the computer software developed specifically to determine the probability of occurrence. These programs use sampling techniques like 'Monte Carlo simulations' that carry out hundreds of iterative sampling calculations to obtain a probability distribution of the outcome.

One application of the Monte Carlo simulation is determining the probability to meet a specific milestone (like the completion date) by giving three time estimates to every activity. The program will then carry out a great number of iterations resulting in a frequency/time histogram and a cumulative 'S' curve from which the probability of meeting the milestone can be read off (see Figure 15.5).

Exposure table							
			Probability				
	Rating		Very low	Low	Medium	High	Very high
		Value	0.1	0.2	0.5	0.7	0.9
Impact	Very high	0.8					
	High	0.5					
	Medium	0.2					
	Low	0.1					
	Very low	0.05					

Figure 15.3
Exposure table

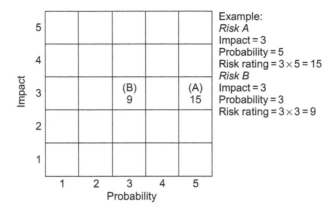

Figure 15.4
5×5 matrix

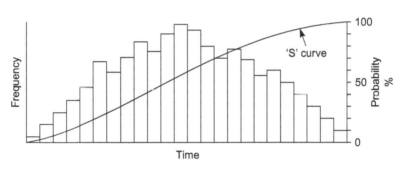

Figure 15.5
Frequency/time histogram

At the same time a *Tornado* diagram can be produced, which shows the sensitivity of each activity as far as it affects the project completion (see Figure 15.6).

Other techniques such as sensitivity diagrams, influence diagrams, and decision trees have all been developed in an attempt to make risk analysis more accurate or more reliable. It must be remembered, however, that any answer is only as good as the initial assumptions and input data, and the project manager must give serious consideration as to the cost effectiveness of these methods for his/her particular project.

Stage 5: Risk Management

Having listed and evaluated the risks and established a table of priorities, the next stage is to decide how to manage the risks; in other words, what to do about them and who should be

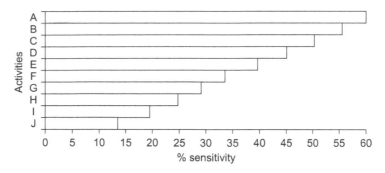

Figure 15.6
Tornado diagram

responsible for managing them. For this purpose it is advisable to appoint a *risk owner* for every risk that has to be monitored and controlled. A risk owner may, of course, be responsible for a number or even all of the risks. There are a number of options available to the project manager when faced with a set of risks. These are:

- Avoidance
- Reduction
- Sharing
- Transfer
- Deference
- Mitigation
- Contingency
- Insurance
- Acceptance

These options are perhaps most easily explained by a simple example.

The owner of a semi-detached house decides to replace part of his roof with solar panels to save on his hot water heating bill. The risks in carrying out this work are as follows:

Risk 1   The installer may fall off the roof;
Risk 2   The roof may leak after completion;
Risk 3   The panels may break after installation;
Risk 4   Birds may befoul the panels;
Risk 5   The electronic controls may not work;
Risk 6   The heat recovered may not be sufficient to heat the water on a cold day;
Risk 7   It may not be possible to recover the cost if the house is sold within 2–3 years;
Risk 8   The cost of the work will probably never pay for itself;
Risk 9   The cost may escalate due to unforeseen structural problems.

These risks can all be managed by applying one or several of the above options:

Risk 1	Transfer	Employ a builder who is covered by insurance;
Risk 2	Transfer	Insist on a two-year guarantee for the work (at least two season cycles);
Risk 3	Insurance	Add the panel replacement to the house insurance policy;
Risk 4	Mitigation	Provide access for cleaning (this may increase the cost);
Risk 5	Reduction	Ensure a control unit is used which has been proven for a number of years;
Risk 6	Contingency	Provide for an electric immersion heater for cold spells;
Risk 7	Deference	Wait 3 years before selling the house;
Risk 8	Acceptance	This is a risk one must accept if the work goes ahead, or
Risk 8	Avoidance	Don't go ahead with the work;
Risk 9	Sharing	Persuade the neighbour in the adjoining house to install a similar system at the same time.

Monitoring

To keep control of the risks, a *risk register* should be produced that lists all the risks and their method of management. Such a list is shown in Figure 15.7. Where risk owners have been appointed, these will be identified on the register. The risks must be constantly monitored and at preset periods, the register must be reassessed and if necessary amended to reflect the latest position. Clearly as the project proceeds, the risks reduce in number, so that the contingency sums allocated to cover the risk of the completed activities can be allocated to other sections of the budget. These must be recorded in the register under the heading of *risk closure*. However, sometimes new rules emerge that must be taken into account.

The summary of the risk management procedure is then as follows:

1. Risk awareness;
2. Risk identification (checklists, prompt lists, brainstorming);
3. Risk owner identification;
4. Qualitative assessment;
5. Quantification of probability;
6. Quantification of impact (severity);
7. Exposure rating;
8. Mitigation;
9. Contingency provision;
10. Risk register;
11. Software usage (if any);
12. Monitoring and reporting.

Project:.................................. Key: H, high: M, medium: L, low					Prepared by:...................................			Reference:.................... Date:............................		
Type of risk	Description of risk	Probability			Impact			Risk reduction strategy	Contingency plans	Risk owner
		H	M	L	Perf.	Cost	Time			

Figure 15.7
Risk register (risk log)

To aid the process of risk management, a number of software tools have been developed. The most commonly used ones are *@Risk, Predict, Pandora*, and *Plantrac Marshal*, but no doubt new ones will be developed in the future.

Example of Effective Risk Management

One of the most striking and beneficial advantages of risk analysis is associated with the temporary jetties system known as Mulberry Harbour, which was towed across the Channel to support Allied landings in Normandy on D-Day, 6 June 1944.

The two jetties (A) at Gold beach for the British army and (B) at Omaha for the American army consisted of floating roadways, pontoons, and caissons, protected by breakwaters, which were pre-fabricated at numerous sites across Britain and towed across the Channel where the caissons were sunk onto the seabed. All site construction of the roadways was carried out by British and American military engineers. The harbours were completed 12 days later on 18 June, and on 19 June an enormous storm blew up, which by 22 June destroyed the American and badly damaged the British harbour.

Fortunately someone carried out a risk analysis that identified the possibility of such a disaster. As a result, the provision had been made for a large quantity of spare units to be ready at British ports to be towed out as replacements. By using these spare sections and cannibalising the wrecked American harbour, the British harbour could be repaired and until it was abandoned following the capture of Cherbourg 8 months later, more than 7000 tons of vital military supplies per day were delivered to the allied armies in France.

Positive Risk or Opportunity

Although most risks are generally regarded as negative or undesirable, and indeed most mitigation strategies have been devised to reduce the impact or probability of negative risk, there is paradoxically also such a thing as positive risk, or opportunistic risk. This is basically the risk that any entrepreneur or investor takes when he/she invests in a new enterprise. A simple case of 'Nothing ventured, nothing gained.' A case may also arise where a perceived negative risk becomes a positive risk or opportunity. For example, in an attempt to reduce the risk of skidding, a car manufacturer may invent an anti-skid device that can be marketed independently at a profit. If there had been no risk, there would have been no need for the antidote.

Local authorities tend to use the term "positive risk taking" when referring to the pro-active approach of providing services and facilities to certain sections of the community (usually challenged by a disability). However, apart from discussing and agreeing on their risks with clients, the process is in effect a comprehensive programme of ascertaining, prioritizing, and mitigating perceived risks before they occur.

Quality Management

Chapter Outline

'Quality is remembered long after the price is forgotten'.
Gucci

Quality (or performance) forms the third corner of the time–cost–quality triangle, which is the basis of project management.

Project Management, Planning, and Control. http://dx.doi.org/10.1016/B978-0-08-098324-0.00016-0

A project may be completed on time and within the set budget, but if it does not meet the specified quality or performance criteria it will at best attract criticism and at worst be considered a failure. Striking a balance between meeting the three essential criteria of time, cost, and quality is one of the most onerous tasks of a project manager, but in practice usually one will be paramount. Where quality is synonymous with safety, as with aircraft or nuclear design, there is no question which point of the project management triangle is the most important. However, even if the choice is not so obvious, a failure in quality can be expensive and dangerous and can destroy an organization's reputation far quicker than it took to build up.

Quality management is therefore an essential part of project management, and as with any other attribute, it does not just happen without a systematic approach. To ensure a quality product it has to be defined, planned, designed, specified, manufactured, constructed (or erected), and commissioned to an agreed set of standards that involve every department of the organization from top management to dispatch.

It is not possible to build quality into a product. If a product meets the specified performance criteria for a specified minimum time, it can be said to be a quality product. Whether the cost of achieving these criteria is high or low is immaterial, but to ensure that the criteria are met will almost certainly require additional expenditure. If these costs are then added to the normal production costs, a quality assured product will normally cost more than an equivalent one that has not gone through a quality control process.

Quality is an attitude of mind, and to be most effective, every level of an organization should be involved and be committed to achieving the required performance standards by setting and operating procedures and systems which ensure this. It should permeate right through an organization from the board of directors down to the operatives on the shop floor.

Ideally everyone should be responsible for ensuring that his or her work meets the quality standards set down by management. To ensure that these standards are met, quality assurance requires checks and audits to be carried out on a regular basis.

However, producing a product that has not undergone a series of quality checks and tests and therefore not met customers' expectations could be very much more expensive, since there will be more returns of faulty goods and fewer returns of customers. In other words, quality assurance is good business. It is far better to get it right first time, every time, than to have a second attempt or carry out a repair.

To enable this consistency of performance to be obtained (and guaranteed), the quality assurance, control, review, and audit procedures have to be carried out in an organised manner and the following functions and actions implemented:

1. The quality standards have been defined;
2. The quality requirements have been disseminated;

3. The correct equipment has been set up;
4. The staff and operatives have been trained;
5. The materials have been tested and checked for conformity;
6. Adequate control points have been set up;
7. The designated components have been checked at predetermined stages and intervals;
8. A feedback and rectification process has been set up;
9. Regular quality audits and reviews are carried out;
10. All these steps, which make up quality control, are enshrined in the quality manual together with the quality policy, quality plan, and quality programme.

History

The first quality standards were produced in the USA for the military as MIL-STD and subsequently used by NATO.

In the 1970s, the MOD issued the Defence Standard series 05 to 20 (Def Std) based on the American MIL-Q-9858A, but it was then superseded by 15 parts of the Allied Quality Assurance Publication (AQAP).

Defence contractors and other large companies adopted the MOD system until in 1979 BSI produced the BS 5750 series of Quality Systems. These were updated in 1987 and then became an international standard (ISO), the ISO 9000:1987 series, which also covers the European standard E 29000 series.

To understand the subject, it is vital that the definitions of the various quality functions are understood. These are summarized in the following list and explained more fully in the subsequent sections.

Quality Management Definitions

Quality

The totality of features and characteristics of a product, service, or facility that bear on its ability to satisfy a given need.

Quality Policy

The overall quality intentions and direction of an organization as regards quality, as formally expressed by top management.

Quality Management

That aspect of overall quality functions that determines and implements the quality policy.

Quality Assurance (QA)

All the planned and systematic actions necessary to provide confidence that an item, service, or facility will meet the defined requirements.

Quality Systems (Quality Management Systems) (QMS)

The organizational structures, responsibilities, procedures, processes, and resources for implementing quality management.

Quality Control (QC)

Those quality assurance actions that provide a means of control and measure the characteristics of an item, process, or facility to established requirements.

Quality Manual

A set of documents that communicates the organization's quality policy, procedures, and requirements.

Quality Programme

A contract (project)-specific document that defines the quality requirements, responsibilities, procedures, and actions to be applied at various stages of the contract.

Quality Plan

A contract (project)-specific document defining the actions and processes to be undertaken together with the hold points for reviews and inspections. It also defines the control document, applicable standards, inspection methods, and inspection authority. This authority may be internal and/or may include the client's inspectors or an independent/statutory inspection authority.

Quality Audit

A periodic check that the quality procedures set out in the quality plan have been carried out.

Quality Reviews

Periodic reviews of the quality standards, procedures, and processes to ensure their applicability to current requirements.

Total Quality Management (TQM)

The company-wide approach to quality beyond the prescriptive requirements of a quality management standard such as ISO 9001.

Explanation of the Definitions

Quality Policy

The quality policy has to be set by top management and issued to the whole organization, so that everyone is aware what the aims of management regarding quality are. The quality policy might be to produce a component that lasts a specific period of time under normal use, withstands a set number of reversing cycles before cracking, withstands a defined load or pressure, or, on the opposite scale, lasts only a limited number of years so that a later model can be produced to replace it. A firm of house builders might have a quality policy to build all their houses to the highest standards in only the most desirable locations, or the top management of a car manufacturer might dictate to their design engineers to design a car using components that will not fail for at least five years. There is clearly a cost and marketing implication in any quality policy that must be taken into account.

Quality Management

Quality management can be divided into two main areas: quality assurance (QA) and quality control (QC). All the quality functions, such as the procedures, methods, techniques, programmes, plans, controls, reviews, and audits, make up the science of quality management.

It also includes all the necessary documentation and its distribution, the implementation of the procedures, and the training and appointment of quality managers, testers, checkers, auditors, and other staff involved in quality management.

Quality Assurance

Quality assurance (QA) is the process that ensures that adequate quality systems, processes, and procedures are in place. It is the term given to a set of documents that give evidence of how and when the different quality procedures and systems are actually being implemented. These documents give proof that quality systems are in place and adequate controls have been set up to ensure compliance with the quality policy. To satisfy him- or herself that the quality of a product he or she needs is to the required standard, the buyer may well ask all tenderers or suppliers to produce their quality assurance documents with their quotations or tenders.

Guidelines for quality management and quality assurance standards are published by various national and international institutions including the British Standards Institution, which publishes the following quality standards:

ISO 9000	Quality Management and Quality Assurance Standards
ISO 9001	Quality Systems – Model for Quality Assurance in design and development, production, installation and servicing
ISO 9002	Quality Systems – Model for Quality Assurance in Production and Installation
ISO 9003	Quality Systems – Model for Quality Assurance in Inspection and Testing
ISO 9004	Guide to Quality Management and Quality Systems Elements
ISO 10006	Guidelines for quality in project management
ISO 10007	Guidelines for configuration management

Quality Systems

Quality systems or quality management systems, as they are often called, are the structured procedures that in effect enable quality control to be realized. The systems required include the levels of responsibility for quality control, such as hierarchical diagrams (family trees) showing who is responsible to whom and for which part of the quality spectrum they are accountable for as well as the procedures for recruiting and training staff and operatives. Other systems cover the different quality procedures and processes that may be common for all as well as all components or specifics for particular ones.

Quality systems also include the procurement, installation, operation, and maintenance of equipment for carrying out quality checks. These cover such equipment as measuring tools, testing bays, non-destructive testing equipment for radiography (X-rays), magnetic particle scans, ultrasonic inspection, and all the different techniques being developed for testing purposes.

Documentation plays an important role in ensuring that the tests and checks have been carried out as planned and the results accurately recorded and forwarded to the specified authority. Suitable action plans for recovering from deviations of set criteria will also form part of the quality systems. The sequence of generating the related quality documentation is shown in Figure 16.1.

Quality Control

The means to control and measure characteristics of a component and the methods employed for monitoring and measuring a process or facility are part of quality control. Control covers the actions to be taken by the different staff and operators employed in the quality environment

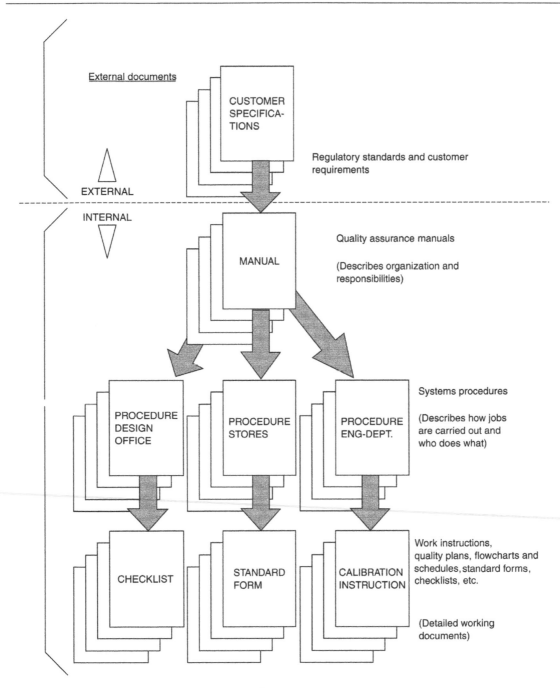

Figure 16.1
Quality-related documentation

and the availability of the necessary tools to enable this work to be done. Again, the provision of the right documentation to the operatives and their correct, accurate, and timely completion has to be controlled. This control covers the design, material specification, manufacture, assembly, and distribution stages. The performance criteria are often set by the feedback from market research and customer requirements and confirmed by top management.

Quality Manual

The quality manual is in effect the 'bible' of quality management. It is primarily a communication document which, between its covers, contains the organization's quality policy, the different quality procedures, the quality systems to be used, and the list of personnel involved in implementing the quality policy. The manual will also contain the various test certificates required for certain operatives such as welders, the types of tests to be carried out on different materials and components, and the sourcing trails required for specified materials.

Quality Programme

This is a document written specifically for the project in hand and contains all the requirements for that project. The different levels and stages for quality checks or tests will be listed together with the names of the staff and operatives required at each stage. Included also will be the reporting procedures and the names or organizations authorised to approve or reject results and instruct what remedial action (if required) has to be taken, especially when concessions for non-compliance have been requested.

Quality Plan

This is also a contract-specific document that can vary in content and size greatly from company to company. As a general rule it defines in great detail:

1. The processes to be employed;
2. The hold points of each production process;
3. The tests to be carried out for different materials and components, including:
    * dimensional checks and weight checks
    * material tests (physical and chemical)
    * non-destructive tests (radiography, ultrasonic, magnetic particle, etc.)
    * pressure tests
    * leak tests
    * electrical tests (voltage, current, resistance, continuity, etc.)
    * qualification and capability tests for operatives;
4. The control documents including reports and concession requests.
5. The standards to be applied for the different components;

6. The method of inspection;
7. The percentage of items or processes (such as welds) to be checked;
8. The inspection authorities, whether internal, external, or statutory; and
9. The acceptance criteria for the tests and checks.

Most organizations have their own standards and test procedures but may additionally be required to comply with the client's quality standards. A sample quality plan of components of a boiler is shown in Figure 16.2.

The quality plan is part of the project management plan and because of its size is usually attached as an appendix.

Quality Audit

To ensure that the various procedures are implemented correctly, regular quality audits must be carried out across the whole spectrum of the quality systems. These audits vary in scope and depth and can be carried out by internal members of staff or external authorities. Where an organization is officially registered as being in compliance with a specific quality standard such as ISO 9000, an annual audit by an independent inspection authority may be carried out to ensure that the standards are still being met.

Quality Reviews

As manufacturing or distribution processes change, a periodic review must be carried out to ensure that the quality procedures are still relevant and applicable in light of the changed conditions. Statutory standards may also have been updated, and these reviews check that the latest versions have been incorporated.

As part of the reviewing process, existing, proposed, or new suppliers and contractors have to confirm their compliance with quality assurance procedures. Figure 16.3 shows the type of letter that should be sent periodically to all vendors.

All the above procedures are sometimes described as the 'tools' of quality management to which can be added the following techniques:

- Failure mode analysis (cause and effect analysis)
- Pareto analysis
- Trend analysis

Failure Mode Analysis (Cause and Effect Analysis)

This technique involves selecting certain (usually critical) items and identifying all the possible modes of failure that could occur during its life cycle. The probability, causes, and

CLIENT : X Y Z	QUALITY PLAN	Sheet : 3 of 3
CONTRACT No : 2-31-07797	FOR	Rev : 0
CODE : DS. 1113 :	BOILERS	Issued by : *[signature]*
		Approved by: *[signature]* 5/6/85

LEGEND
D = Domestic Inspection
C = Certificate Required
T = Internal Record
R = Random Insp. (i.e. R10 = 10%)
E = External Surveillance
▮ = Mandatory Inspection

SURVEY AUTHORITY
B = British Engine
P = P.S.A.

Description of Part	OPERATION	Material Identification	Inspection of Set Up	Inspection of Marking Off/ Drilling and Machining	Tube Manipulation	Visual Inspection of Welding	Radiographic Examination	Magnetic Particle Examination	Hydraulic Test	Release for Despatch				
REFERENCE DOCUMENT		BR. 600/108M BS. 1113	BS. 1113	F.W.P.P.	F.W.P.P.	F.W.P.P.	BS. 1113	BS. 1113	BS. 1113	F.W.P.P.				
SUPERHEATER AND PANELS		1	2	3	4	5	6	7	8	9				
Headers and End Caps	D	C	I	I										
	E	B												
Header Circ. Welds	D					I	C							
	E						B							
Stubs/Panel Tubes	D	C		I10										
	E													
Stubs to Header/ Panel to Hdr.Welds	D		I10			I		C10						
	E							B						
Tubes	D	C	I10			I10								
	E													
Tube Butt Welds	D					I	C10							
	E						B							
Attachments	D	C	I5											
	E													
Attachment Welds	D					I								
	E													
Complete Superheater	D							C	C					
	E							B						
Completed Panels	D							C	C					
	E							B						

Figure 16.2
Quality assurance approval

Quality assurance approval

1121/DAR/QA
Our ref: Your ref:

DATE

F.A.O: <u>Quality Assurance Manager</u>

Dear Sirs,

<u>QUALITY ASSURANCE APPROVAL</u>

In order to meet the increasing Quality Assurance demands of our Clients, we are revising our Approved Vendor Lists. Should you wish to either remain on, or be added to these lists it will be necessary for you to complete the attached document and return it to us without delay.

It is of the utmost importance that the document is fully completed and gives all relevant information asked for regarding your existing Q.A. Approvals including the following details:-

<u>Sub-Contract Quality Assurance - Form A</u>

1)   The level approved at.

2)   The organisation or body who have awarded the approval.

3)   Certificate Number

4)   The date approved.

5)   The period of validity of the approval.

6)   Commodity and materials approved.

The completed form should be returned to:-

(state address here)

Marked for the attention of Mr. John Brown - Procurement Manager

Figure 16.3
Confirmation of compliance request

impact of such a failure are then assessed and the necessary controls and rectification processes put in place. Clearly, as with risk analysis, the earlier in the project this process is carried out, the more opportunity there is to anticipate a problem and, if necessary, change the design to 'engineer' it out.

The following example illustrates how this technique can be applied to find the main causes affecting the operability of a domestic vacuum cleaner.

The first step is to list all the main causes that are generally experienced when using such a machine. These causes (or quality shortcomings), which may require a brainstorming session to generate them, are:

* Electrical
* Physical (weight and size)
* Mechanical (brush wear)
* Suction (dust collection)

The second step is drawing a *cause and effect* diagram as shown in Figure 16.4, which is also known as an *Ishikawa* or *fishbone* diagram, from which it is possible to see clearly how these causes affect the operation of the vacuum cleaner.

The third step requires all the sub-causes (or reasons) of a main cause to be written against the tributary lines (or fishbones) of each cause. For example, the sub-causes of electrical failure are the lead being too short, thus pulling the plug out of the socket, or hauling the cleaner by the lead and causing a break in the cable.

The last step involves an assessment of the number of times over a measured period each cause has resulted in a failure. However, it is highly advantageous to concentrate on those causes that are responsible for the most complaints, and when this has been completed and assessed by applying the next technique, Pareto analysis, appropriate action can be taken to resolve any problem or rectify any error.

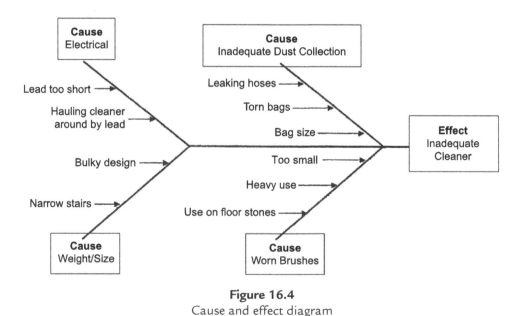

Figure 16.4
Cause and effect diagram

Pareto Analysis

In the nineteenth century, Vilfredo Pareto discovered that in Italy 90% of income was earned by 10% of the population. Further study showed that this distribution was also true for many other situations from political power to industrial problems. He therefore formulated Pareto's law, which states that 'In any series of elements to be controlled, a small fraction in terms of the number of elements always account for a large fraction in terms of effect'.

In the case of the vacuum cleaner, this is clearly shown in the Pareto chart in Figure 16.5, which plots the impact Y in terms of percentage of problems encountered against the number of causes X identified. The survey of faults shows that of the four main causes examined, inadequate dust collection is responsible for 76% (nearly 80%) of the failures or complaints. This is why Pareto's law is sometimes called the 80/20 rule.

The percentage figure can be calculated by tabulating the causes and the number of times they resulted in a failure over a given period, say 1 year, and then converting these into a percentage of the total number of failures. This is shown in Figure 16.6. Clearly such ratios

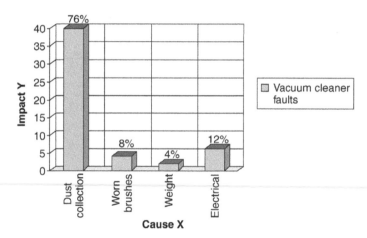

Figure 16.5
Pareto chart

Cause	No. of failures	% of failures
Electrical	6	12
Physical	2	4
Mechanical	4	8
Suction	38	76
Total	50	100

Figure 16.6
Failure table

are only approximate and can vary widely, but in general only a relatively small number of causes are responsible for the most serious effects. Anyone who is involved in club committee activities will know that there are always a few keen members who have the greatest influence.

Trend Analysis

Part of the quality control process is to issue regular reports that log non-conformance, accepted or non-accepted variances, delays, cost overruns, or other problems. If these reports are reviewed on a regular basis, trends may be discerned which, if considered to be adverse, can then be addressed by taking appropriate corrective action. At the same time the opportunity can be taken to check whether there has been a deficiency in the procedures or documents and whether other components could be affected. Most importantly, the cause and source of a failure has to be identified, which may require a review of all the suppliers and subcontractors involved.

CHAPTER 17

Change Management

Chapter Outline
Document Control   100
Issue Management   100

There are very few projects that do not change in some way during their life cycle. Equally there are very few changes that do not affect in some way either (or all) the time, cost, or quality aspects of the project. For this reason it is important that all changes are recorded, evaluated, and managed to ensure that the effects are appreciated by the originator of the change, and the party carrying out the change is suitably reimbursed where the change is a genuine extra to the original specification or brief.

In cases where a formal contract exists between the client and the contractor, an equally formal procedure of dealing with changes (or variations) is essential to ensure that:

1. No unnecessary changes are introduced;
2. The changes are only issued by an authorized person;
3. The changes are evaluated in terms of cost, time, and performance;
4. The originator is made aware of these implications before the change is put into operation (In practice this may not always be possible if the extra work has to be carried out urgently for safety or security reasons. In such a case the evaluation and report of the effect must be produced as soon as possible.); and
5. The contractor is compensated for the extra costs and given extra time to complete the contract.

Unfortunately clients do not always appreciate what effect even a minor change can have on a contract. For example, a client might think that eliminating an item of equipment such as a small pump a few weeks into the contract would reduce the cost. He or she might well find, however, that the changes in the design documentation, data sheets, drawings, bid requests, etc., will actually cost more than the capital value of the pump, so that the overall cost of the project will increase! The watchwords must therefore be: *is the change really necessary?*

In practice as soon as a change or variation has been requested either verbally or by a change order, it must be confirmed back to the originator with a statement to the effect that the cost and time implications will be advised as soon as possible.

A change of contract scope notice must then be issued to all departments who may be affected to enable them to assess the cost, time, and quality implications of the change.

A copy of such a document is shown in Figure 17.1, which should contain the following information:

- Project or contract no.
- Change of scope no.
- Issue date
- Name of originator of change
- Method of transmission (letter, fax, telephone, e-mail, etc.)
- Description of change
- Date of receipt of change order or instruction

When all the affected departments have inserted their cost and time estimates, the form is sent to the originator for permission to proceed or for advice of the implications if the work has had to be started before the form could be completed. The method of handling variations will probably have been set out in the contract documentation but it is important to follow the agreed procedures, especially if there are time limitations for submitting the claims at a later stage.

As soon as a change has been agreed, the cost and time variations must be added to the budget and programme respectively to give the revised target values against which costs and progress will be monitored. However, while all variations have to be recorded and processed in the same way, the *project budget* can only be changed (increased or decreased) when the variation has been requested by the client. When the change was generated internally, for example, by one of the design departments due to a discovered error, omission, or necessary improvement, it is not possible to increase the budget (and hence the price) unless the client has agreed to this. The extra cost must clearly still be recorded and monitored but will only appear as an increase (or decrease) in the *actual* cost column of the cost report. The result will be a reduction or increase of the profit, depending on whether the change required more or fewer resources.

The accurate and timely recording and managing of changes could make the difference between a project making a profit or losing money.

Change management must not be confused with *management of change*, which is the art of changing the culture or systems of an organization and managing the human reactions. Such a change can have far-reaching repercussions on the lives and attitudes of all the members of the organization, from the board level to the operatives on the shop floor. The way such changes are handled and the psychological approaches used to minimize stress and resistance are outside the scope of this book.

FOSTER WHEELER POWER PRODUCTS LTD. HAMSTEAD ROAD LONDON NW1 7QN	**ADVICE OF CHANCE OF CONTRACT SCOPE** DEPARTMENT: ENGINEERING No 82

To: Contract Management Department

☑ Please note that the scope of the subject contract has been altered due to the change(s) detailed below. To: Contract Management Department

☐ The following is a statement of the manhours and expenses incurred due to Contract Variation Notice

reference_____ dated _____17 Dec. 1982_____

ICI BILLINGHAM
2-32-07059

Distribution:
Project Manager
Estimating Department
Management Services
Deaprtmental Manager
Manager Engineering (see note 4 below)
File
   N. Smith
   J. Harris

By internal mail

BRIEF DESCRIPTION OF CHANGE AND EFFECT ON DEPARTMENTAL WORK

The provision of an 'Air to Igniters' control valve.
Scope of work includes purchasing and adding to drawings.
The clients preferred-specified vendor for control valves is Fisher Controls.

Manhour requirements are as follows:-
Dept 1104-63 manhours (1104 Split)
Dept 1102- 8 manhours Req. 60
              Drg. 2
Dept 1105-38 manhours   MH. 1
                63

CHANGE NOTIFIED BY

☐ Minutes of Meeting with client

Date of Meeting :_____
Subject of Meeting :_____
Minute Number :_____

☐ Client's telex

Date of Telex :_____
Reference :_____
Signed by :_____

☒ Client's letter

Date of Letter :____10.12.1992____
Reference :____SGP 3641____
Signed by :____B. Francis____

☐ Client's request by telephone

Date of Call :_____
Name of Contact :_____

☐ Client's Variation Order

V.O.Ref. :_____
Date of V.O. :_____

MANHOURS AND COSTS INCURRED IN

Department No. | 1104, 1102, 1105

☐ Increase   ☐ Decrease

Engineering	69	Manhours
Design/drghtg	37	Manhours
Tech clerks	3	Manhours
TOTAL	109	Manhours
COSTS	£T.B.A.	Manhours

Remarks

NOTES

1. The 'change notified by' section need not be completed if form is used to advise manhours and costs only.

2. This form to be completed IMMEDIATELY ON RECIEPT of definite instructions.

3. Manhours MUST BE REALISTIC. Make FULL ALLOWANCE for all additional and re-cycle work. Take into account 'chain reaction' affect throughout department.

4. Submit copy of this form to Manager Engineering if manhours involved exceed 250.

Initiated by N. Smith	Date 17.12.82
Checked by MWN	Date 22.12.82
Approved by	Date

Figure 17.1
Change of contract scope form

Document Control

Invariably a change to even the smallest part of a project requires the amendment of one or more documents. These may be programmes, specifications, drawings, instructions, and, of course, financial records. The amendment of each document is in itself a change, and it is vital that the latest version of the document is issued to *all* the original recipients. In order to ensure that this takes place, a document control, or version control procedure, must be part of the project management plan.

In practice a document control procedure may be either a single page of A4 or several pages of a spreadsheet as part of the computerized project management system. The format should, however, feature the following columns:

- Document number
- Document title
- Originator of document
- Original issue date
- Issue code (general or restricted)
- Name of originator (or department) of revision
- Revision (or version) number
- Date of revision (version)

The sheet should also include a list of recipients.

A separate sheet records the date the revised document is sent to each recipient and the date of acknowledgement of receipt.

Where changes have been made to one or more pages of a multi-page document, such as a project management plan, it is only necessary to issue the revised pages under a page revision number. This requires a discrete version control sheet for this document with each clause listed and its revision and date of issue recorded.

Issue Management

An issue is a threat that could affect the objectives or operations of a project. Unlike a risk, which is an uncertain event, an issue is a reality that may have already occurred or is about to occur.

The difference between an issue and a problem is that an issue is a change or potential change of external circumstances outside the control of the project manager. A problem, on the other hand, is a day-to-day adverse event the project manager has to deal with as a matter of routine. An issue therefore has to be brought to the attention of a higher authority such as the project board, steering group, or programme manager, since it may well require additional resources, either human, financial, or physical (material), which require sanctioning by senior management.

As with other types of changes, a register of issues, often called an issue log, must be set up and kept up to date. This not only records the type, source, and date of the issue to be addressed, but also shows to whom it was circulated, the effect in terms of cost, time, and performance it has on the project, and how it was resolved. If the issue is large or complex, it may, as with particular risks, be necessary to appoint an issue owner (or resolution owner) who will be responsible for implementing the agreed actions to resolve the matter.

The way an issue is resolved depends clearly on the size and complexity. Again, as with change requests, other departments that may be affected have to be advised and consulted, and in serious situations, special meetings may have to be convened to discuss the matter in-depth and sufficient detail to enable proposals for a realistic solution to be tabled. Care must be taken not to escalate inevitable problems that arise during the course of a project to the status of an issue, as this would take up valuable management time of the members of a steering group or senior management. As is so often the case with project management, judgement and experience are the key attributes for handling the threats and vicissitudes of a project.

CHAPTER 18

Configuration Management

Although in the confined project management context configuration management is often assumed to be synonymous with version control of documentation or software, it is of course very much more far reaching in the total project environment. Developed originally in the aerospace industry, it has been created to ensure that changes and modifications to physical components, software, systems, and documentation are recorded and identified in such a way that replacements, spares, and assembly documentation has conformed to the version in service. It also has been developed to ensure that the design standards and characteristics were reflected accurately in the finished product.

It can be seen that when projects involve complex systems as in the aerospace, defence, or petrochemical industry, configuration management is of the utmost importance as the very nature of these industries involves development work and numerous modifications not only from the original concept or design but also during the whole life cycle of the product.

Keeping track of all these changes to specifications, drawings, support documentation, and manufacturing processes is the essence of configuration management, which can be split into the following five main stages:

1. *Configuration management and planning.* This covers the necessary standards, procedures, support facilities, resources, and training and sets out the scope, definitions, reviews, milestones, and audit dates.
2. *Configuration identification.* This encompasses the logistics and systems and procedures. It also defines the criteria for selection in each of the project phases.
3. *Configuration change management.* This deals with the proposed changes and their investigation before acceptance. At this stage changes are compared with the configuration baseline including defining when formal departure points have been reached.
4. *Configuration status accounting.* This records and logs the accepted (registered) changes and notifications as well as providing traceability of all baselines.
5. *Configuration audit.* This ensures that all the previous stages have been correctly applied and incorporated in the organization. The output of this stage is the audit report.

In all these stages resources and facilities must always be considered and arrangements must be made to feed back comments to the management stage.

Project Management, Planning, and Control. http://dx.doi.org/10.1016/B978-0-08-098324-0.00018-4

| | | Master Record Index | | | |
| | | Documents | | | |
Document Title	Reference Number	Issue	Date	Responsibility	Distribution
Business Case	Rqmt SR 123	Draft A	14/6/86	Mr Sponsor	PM, Line Mgmt
		Draft B	24/7/86		
		Issue 1	30/7/86		
		Issue 2	30/9/86		
Project Mgmt Plan	PMP/MLS/34	Draft A	28/7/86	Ms MLS PM	All Stakeholders
		Issue 1	30/9/86		
WBS	WBS/PD1	Draft A	30/7/86	Mr MLS Deputy PM	IPMT (Project Team)
		Issue 1	2/8/86		
Risk Mgmt Plan, etc.	RMP/MLS/1				

Figure 18.1
Master record index

Essentially the process of identification, evaluation, and implementation of changes requires accurate monitoring and recording and subsequent dissemination of documentation to the interested parties. This is controlled by a master record index (MRI). An example of such an MRI for controlling documents is shown in Figure 18.1.

On large, complex, and especially multinational projects, where the design and manufacture are carried out in different countries, great effort is required to ensure that product configuration is adequately monitored and controlled. To this end a *configuration control committee* is appointed to head up special *interface control groups* and *configuration control boards* that investigate and, where accepted, approve all proposed changes.

Basic Network Principles

It is true to say that whenever a process requires a large number of separate but integrated operations, a critical path network can be used to advantage. This does not mean, of course, that other methods are not successful or that the critical path method (CPM) is a substitute for these methods – indeed, in many cases network analysis can be used in conjunction with traditional techniques – but if correctly applied CPM will give a clearer picture of the complete programme than other systems evolved to date.

Every time we do anything, we string together, knowingly or unknowingly, a series of activities that make up the operation we are performing. Again, if we so desire, we can break down each individual activity into further components until we end up with the movement of an electron around a nucleus. Clearly, it is ludicrous to go to such a limit but we can call a halt to this successive breakdown at any stage to suit our requirements. The degree of the breakdown depends on the operation we are performing or intend to perform.

In the UK it was the construction industry that first realized the potential of network analysis, and most of, if not all, the large construction, civil engineering, and building firms now use CPM regularly for their larger contracts. However, a contract does not have to be large before

CPM can be usefully employed. If any process can be split into twenty or more operations or 'activities', a network will show their interrelationship in a clear and logical manner so that it may be possible to plan and rearrange these interrelationships to produce either a shorter or a cheaper project, or both.

Network Analysis

Network analysis, as the name implies, consists of two basic operations:

1. Drawing the network and estimating the individual activity times
2. Analysing these times in order to find the critical activities and the amount of float in the non-critical ones

The Network

Basically the network is a flow diagram showing the sequence of operations of a process. Each individual operation is known as an activity and each meeting point or transfer stage between one activity and another is an event or node. If the activities are represented by straight lines and the events by circles, it is very simple to draw their relationships graphically, and the resulting diagram is known as the network. In order to show whether an activity has to be performed before or after its neighbour, arrowheads are placed on the straight lines, but it must be explained that the length or orientation of these lines is quite arbitrary. This format of network is called activity on arrow (AoA), as the activity description is written over the arrow.

It can be seen, therefore, that each activity has two nodes or events; one at the beginning and one at the end (Figure 19.1). Thus events 1 and 2 in the figure show the start and finish of activity A. The arrowhead indicates that 1 comes before 2, i.e., the operation flows towards 2.

We can now describe the activity in two ways:

1. By its activity title (in this case, A)
2. By its starting and finishing event nodes 1–2

For analysis purposes (except when using a computer), the second method must be used.

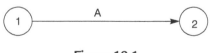

Figure 19.1

Basic Rules

Before proceeding further it may be prudent at this stage to list some very simple but basic rules for network presentation, which must be adhered to rigidly:

1. Where the starting node of an activity is also the finishing node of one or more other activities, it means that *all* the activities with this finishing node must be completed before the activity starting from that node can be commenced. For example, in Figure 19.2, 1–3(A) and 2–3(B) must be completed before 3–4(C) can be started.
2. Each activity must have a different set of starting and finishing node numbers. This poses a problem when two activities start and finish at the same event node, and means that the example shown in Figure 19.3 is incorrect. In order to apply this rule, therefore, an artificial or 'dummy' activity is introduced into the network (Figure 19.4). This 'dummy' has a duration of zero time and thus does not affect the logic or overall time of the project. It can be seen that activity A still starts at 1 and takes 7 units of time before being completed at event 3. Activity B also still takes 7 units of time before being completed at 3, but it starts at node 2. The activity between 1 and 2 is a timeless dummy.

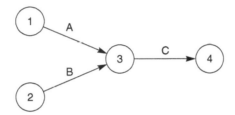

Figure 19.2

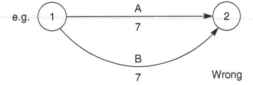

Figure 19.3

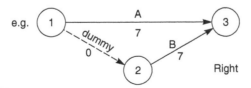

Figure 19.4

3. When two chains of activities are interrelated, this can be shown by joining the two chains either by a linking activity or a 'dummy' (Figure 19.5). The dummy's function is to show that all the activities preceding it, i.e., 1–2(A) and 2–3(B) shown in Figure 20.5, must be completed before activity 7–8(F) can be started. Needless to say, activities 5–6(D), 6–7(E), and 2–6(G) must also be completed before 7–8(F) can be started.

4. Each activity (except the last) must run into another activity. Failure to do so creates a loose end or 'dangle' (Figure 19.6). Dangles create premature 'ends' of a part of a project, so that the relationship between this end and the actual final completion node cannot be seen. Hence the loose ends must be joined to the final node (in this case, node 6 in Figure 19.7) to enable the analysis to be completed.

5. No chain of activities must be permitted to form a loop, i.e., such a sequence that the last activity in the chain has an influence on the first. Clearly, such a loop makes nonsense of any logic since, if one considers activities 2–3(B), 3–4(C), 4–5(E), and 5–2(F) in Figure 19.8, one finds that B, C, and E must precede F, yet F must be completed before B can start. Such a situation cannot occur in nature and defies analysis.

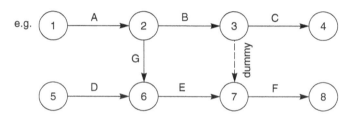

Figure 19.5

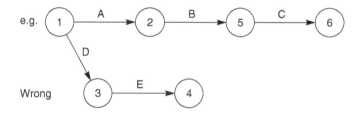

Figure 19.6

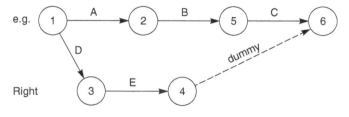

Figure 19.7

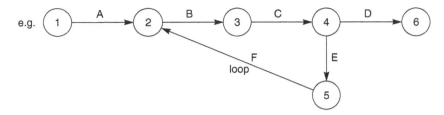

Figure 19.8

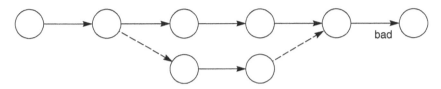

Figure 19.9

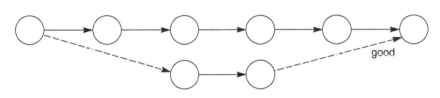

Figure 19.10

Apart from strictly following the basic rules 1 to 5 set out above, the following points are worth remembering to obtain the maximum benefit from network techniques.

1. Maximize the number of activities that can be carried out in parallel. This obviously (resources permitting) cuts down the overall programme time.
2. Beware of imposing unnecessary restraints on any activity. If a restraint is convenient rather than imperative, it should best be omitted. The use of resource restraints is a trap to be particularly avoided since additional resources can often be mustered – even if at additional cost. The exception is when using 'critical chain' project management methods.
3. Start activities as *early* as possible and connect them to the rest of the network as *late* as possible (Figures 19.9 and 19.10). This avoids unnecessary restraints and gives maximum float.
4. Resist the temptation to use a conveniently close node point as a 'staging post' for a dummy activity used as a restraint. Such a break in a restraint could impose an additional unnecessary restraint on the succeeding activity. In Figure 19.11 the intent is to restrain activity E by B and D and activity G by D. However, because the dummy from B uses

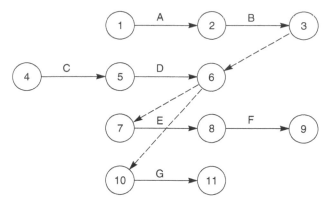

Figure 19.11

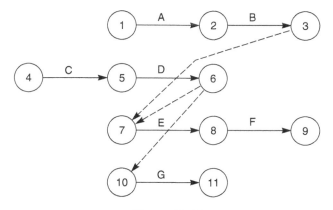

Figure 19.12

node 6 as a staging post, activity G is also restrained by B. The correct network is shown in Figure 19.12. It must be remembered that the restraint on G may have to be added at a later stage, so that the effect of B in Figure 19.11 may well be overlooked.

5. When drawing ladder networks beware of the danger of trying to economize on dummy activities as described later (Figures 19.24 and 19.25).

Durations

Having drawn the network in accordance with the logical sequence of the particular project requirements, the next step is to ascertain the duration or time of each activity. These may be estimated in the light of experience, in the same manner that any programme times are usually ascertained, e.g., using standard industry or company norms, but it must be remembered that the shorter the durations, the more accurate they usually are.

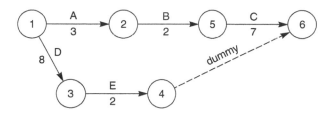

Figure 19.13

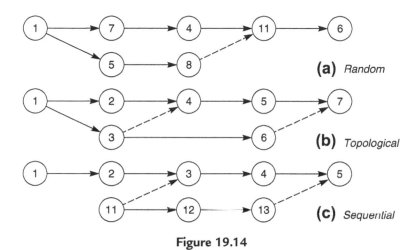

Figure 19.14

The durations are then written against each activity in any convenient time unit, but this must, of course, be the same for every activity. For example, referring to Figure 19.13, if activities 1–2(A), 2–5(B), and 5–6(C) took 3, 2, and 7 days, respectively, one would show this by merely writing these times under the activity.

Numbering

The next stage of network preparation is numbering the events or nodes. Depending on the method of analysis, the following systems shown in Figure 19.14 can be used.

Random

This method, as the name implies, follows no pattern and merely requires each node number to be different. All computers (if used) can, of course, accept this numbering system, but there is always the danger that a number may be repeated.

Topological

This method was necessary for the original early computer programs, which demanded that the starting node of an activity be smaller than the finishing node of that activity. If this law is applied throughout the network, the node numbers will increase in value as the project moves towards the final activity. It had some value for beginners using network analysis since loops are automatically avoided. However, it is very time consuming and requires constant back-checking to ensure that no activity has been missed. The real drawback is that if an activity is added or changed, the whole network has to be renumbered from that point onwards. Clearly, this is an unacceptable restriction in practice and, with the development of modern computer programs, can now be consigned to the history books of project planning.

Sequential

This is a random system from an analysis point of view, but the numbers are chosen in blocks so that certain types of activities can be identified by the nodes. The system therefore clarifies activities and facilitates recognition. The method is quick and easy to use, and should always be used whatever method of analysis is employed. Sequential numbering is usually employed when the network is banded (see Chapter 20). It is useful in such circumstances to start the node numbers in each band with the same prefix number, i.e., the nodes in band 1 would be numbered 101, 102, 103, etc., while the nodes in band 2 are numbered 201, 202, 203, etc. Figure 20.1 would lend itself to this type of numbering.

Coordinates

This method of activity identification can only be used if the network is drawn on a gridded background. In practice, thin lines are first drawn on the *back* of the translucent sheet of drawing paper to form a grid. This grid is then given coordinates or map references with letters for the vertical coordinate and numbers for the horizontal (Figure 19.15).

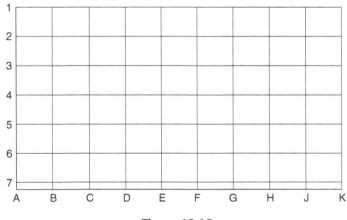

Figure 19.15
Grid

The reason for drawing the lines on the back of the paper is, of course, to leave the grid intact when the activities are changed or erased. A fully drawn grid may be confusing to some people, so it may be preferable to draw a grid showing the intersections only (Figure 19.16).

When activities are drawn, they are confined in length to the distance between two intersections. The node is drawn on the actual intersection so that the coordinates of the intersection become the node number. The number may be written in or the node left blank, as the analyst prefers.

As an alternative to writing the grid letters on the nodes, it may be advantageous to write the letters *between* the nodes as in Figure 23.3. This is more fully described in Chapter 23.

Figure 19.17 shows a section of a network drawn on a gridded background representing the early stages of a design project. As can be seen, there is no need to fill in the nodes, although, for clarity, activities A1–B1, B1–B2, A3–B3, A3–B4, and A5–C5 have had the node numbers

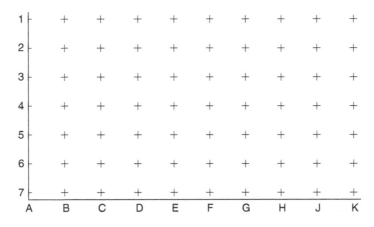

Figure 19.16
Grid (intersections only)

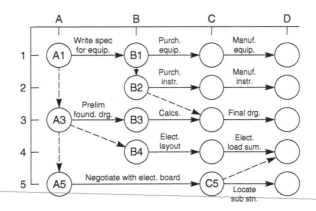

Figure 19.17

added. The node numbers for 'electrical layout' would be B4–C4, and the map reference principle helps to find the activity on the network when discussing the programme on the telephone or quoting it in e-mail.

There is no need to restrict an activity to the distance between two adjacent intersections of coordinates. For example, A5–C5 takes up two spaces. Similarly, any space can also be used as a dummy and there is no restriction on the length or direction of dummies. It is, however, preferable to restrict activities to horizontal lines for ease of writing and subsequent identification.

When required, additional activities can always be inserted in an emergency by using suffix letters. For example, if activity 'preliminary foundation drawings' A3–B3 had to be preceded by, say, 'obtain loads', the network could be redrawn as shown in Figure 19.18.

Identifying or finding activities quickly on a network can be of great benefit, and the above method has considerable advantages over other numbering systems. The use of coordinates is particularly useful in minimizing the risk of duplicating node numbers in a large network. Since each node is, as it were, prenumbered by its coordinates, the possibility of double numbering is virtually eliminated.

Unfortunately, on the earlier computer programs, if the planner entered any number twice, the results could be disastrous, since the machine will, in many instances, interpret the error as a logical sequence. The following example shows how this is possible. The intended sequence is shown in Figure 19.19. If the planner by mistake enters a number 11 instead of 15 for the

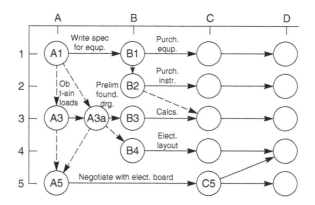

Figure 19.18

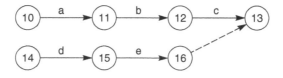

Figure 19.19

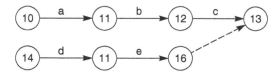

Figure 19.20

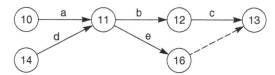

Figure 19.21

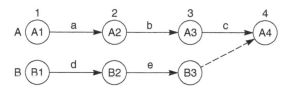

Figure 19.22

last event of activity d, the sequence will, in effect, be as shown in Figure 19.20, but the computer will interpret the error as in Figure 19.21. Clearly, this will give a wrong analysis. If this little network had been drawn on a grid with coordinates as node numbers, it would have appeared as in Figure 19.22. Since the planner knows that all activities on line B must start with a B, the chance of the error occurring is considerably reduced.

Hammocks

When a number of activities are in series, they can be summarized into one activity encompassing them all. Such a summary activity is called a *hammock*. It is assumed that only the first activity is dependent on another activity outside the hammock and only the last activity affects another activity outside the hammock.

On bar charts, hammocks are frequently shown as summary bars above the constituent activities and can therefore simplify the reporting document for a higher management who are generally not concerned with too much detail. For example, in Figure 19.22, activities A1 to A4 could be written as one hammock activity since only A1 and A4 are affected by work outside this activity string.

Ladders

When a string of activities repeats itself, the set of strings can be represented by a configuration known as a ladder. For a string consisting of, say, four activities relating to two stages of excavation, the configuration is shown in Figure 19.23.

This pattern indicates that, for example, hand trim of Stage II can only be done if

1. Hand trim of Stage I is complete
2. Machine excavation of Stage II is complete

This, of course, is what it should be.

However, if the work were to be divided into three stages, the ladder could, on the face of it, be drawn as shown in Figure 19.24.

Again, in Stage II all the operations are shown logically in the correct sequence, but closer examination of Stage III operations will throw up a number of logic errors the inexperienced planner may miss.

What we are trying to show in the network is that Stage III hand trim cannot be performed until Stage III machine excavation is complete and Stage II hand trim is complete. However, what the diagram says is that, in addition to these restraints, Stage III hand trim cannot be performed until Stage I level bottom is also complete.

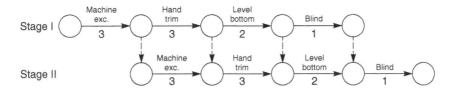

Figure 19.23

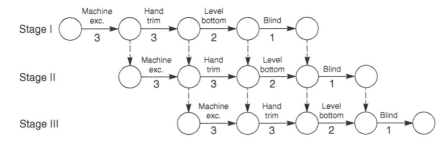

Figure 19.24

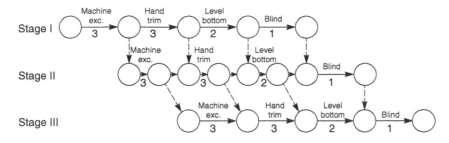

Figure 19.25

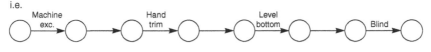

Figure 19.26

Clearly, this is an unnecessary restraint and cannot be tolerated. The correct way of drawing a ladder therefore when more than two stages are involved is as in Figure 19.25.

We must, in fact, introduce a dummy activity in Stage II (and any intermediate stages) between the starting and completion node of every activity except the last. In this way, the Stage III activities will not be restrained by Stage I activities except by those of the same type.

An examination of Figure 19.26 shows a new dummy between the activities in Stage II, i.e.

This concept led to the development of a new type of network presentation called the 'Lester' diagram, which is described more fully in Chapter 23. This has considerable advantages over the conventional arrow diagram and the precedence diagram, also described later.

Precedence or Activity on Node (AoN) Diagrams

Some planners prefer to show the interrelationship of activities by using the node as the activity box and interlinking them by lines. The durations are written in the activity box or node and are therefore called activity on node (AoN) diagrams. This has the advantage that separate dummy activities are eliminated. In a sense, each connecting line is, of course, a dummy because it is timeless. The network produced in this manner is also called variously a precedence diagram or a circle and link diagram. Precedence diagrams have a number of advantages over arrow (AoA) diagrams in that

1. No dummies are necessary;
2. They may be easier to understand by people familiar with flowsheets;

3. Activities are identified by one number instead of two so that a new activity can be inserted between two existing activities without changing the identifying node numbers of the existing activities; and
4. Overlapping activities can be shown very easily without the need for the extra dummies shown in Figure 19.25.

Analysis and float calculation (see Chapter 21) is identical to the methods employed for arrow diagrams and, if the box is large enough, the earliest and latest start and finishing times can be written.

A typical precedence network is shown in Figure 19.27 where the letters in the box represent the description or activity numbers. Durations are shown above-centre and the earliest and latest starting and finish times are given in the corners of the box, as explained in the key diagram. The top line of the activity box gives the earliest start (ES), duration (D), and earliest finish (EF).

Therefore:

$$EF = ES + D$$

The bottom line gives the latest start and the latest finish. Therefore:

$$LS = LF - D$$

The centre box is used to show the total float.

ES is, of course, the *highest* EF of the previous activities leading into it, i.e., the ES of activity E is 8, taken from the EF of activity B.

LF is the *lowest* LS of the previous activity *working backwards*, i.e., the LF of A is 3, taken from the LS of activity B.

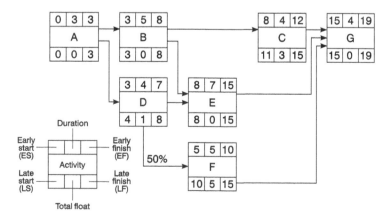

Figure 19.27
AoN diagram

The ES of activity F is 5 because it can start after activity D is 50% complete, i.e.,

ES of activity D is 3.
Duration of activity D is 4.
Therefore 50% of duration is 2.
Therefore ES of activity F is 3 + 2 = 5.

Sometimes it is advantageous to add a percentage line on the bottom of the activity box to show the stage of completion before the next activity can start (Figure 19.28). Each vertical line represents 10% completion. Apart from showing when the next activity starts, the percentage line can also be used to indicate the percentage completion of the activity as a statement of progress once work has started, as in Figure 19.29.

There are two other advantages of the precedence diagram over the arrow diagram.

1. The risk of making the logic errors is virtually eliminated. This is because each activity is separated by a link, so that the unintended dependency from another activity is just not possible.

    This is made clear by referring to Figure 19.30, which is the precedence representation of Figure 19.25.

    As can be seen, there is no way for an activity like 'Level bottom' in Stage I to affect activity 'Hand trim' in Stage III, as is the case in Figure 19.30.

2. In a precedence diagram all the important information of an activity is shown in a neat box.

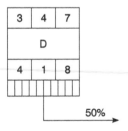

Figure 19.28

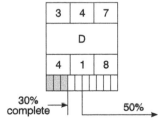

Figure 19.29
Progress indication

A close inspection of the precedence diagram (Figure 19.31) shows that in order to calculate the total float, it is necessary to carry out the forward and backward pass. Once this has been done, the total float of any activity is simply the difference between the latest finishing time (LF) obtained from the backward pass and the earliest finishing time (EF) obtained from the forward pass.

On the other hand, the free float can be calculated from the forward pass only, because it is simply the difference of the earliest start (ES) of a subsequent activity and the earliest finishing time (EF) of the activity in question.

This is clearly shown in Figure 19.31.

Despite the above-mentioned advantages, which are especially appreciated by people familiar with flow diagrams as used in manufacturing industries, many prefer the arrow diagram because it resembles more closely a bar chart. Although the arrows are not drawn to scale, they do represent a forward-moving operation and, by thickening up the actual line in approximately the same proportion as the reported progress, a 'feel' for the state of the job is immediately apparent.

One major disadvantage of precedence diagrams is the practical one of the size of the box. The box has to be large enough to show the activity title, duration, and earliest and latest times, so

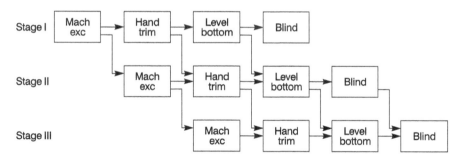

Figure 19.30
Logic to precedence diagram

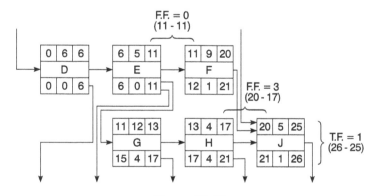

Figure 19.31
Total and free float calculation

that the space taken up on a sheet of paper reduces the network size. By contrast, an arrow diagram is very economical, since the arrow is a natural line over which a title can be written and the node need be no larger than a few millimetres in diameter – if the coordinate method is used.

The difference (or similarity) between an arrow diagram and a precedence network is most easily seen by comparing the two methods in the following example. Figure 19.32 shows a project programme in AoA format and Figure 19.33 the same programme as a precedence diagram, or AoN format. The difference in area of paper required by the two methods is obvious (see also Chapter 33).

Figure 19.33 shows the precedence version of Figure 19.32.

In practice, the only information necessary when drafting the original network is the activity title, the duration, and, of course, the interrelationships of the activities. A precedence diagram can therefore be modified by drawing ellipses just big enough to contain the activity title and duration, leaving the computer (if used) to supply the other information at a later stage. The important thing is to establish an acceptable logic before the end date and the activity floats are computed. In explaining the principles of network diagrams in textbooks (and in examinations), letters are often used as activity titles, but in practice when building up a network, the real descriptions have to be used.

An example of such a diagram is shown in Figure 19.34. Care must be taken not to cross the nodes with the links and to insert the arrowheads to ensure the correct relationship.

One problem of a precedence diagram is that when large networks are being developed by a project team, the drafting of the boxes takes up a lot of time and paper space and the insertion

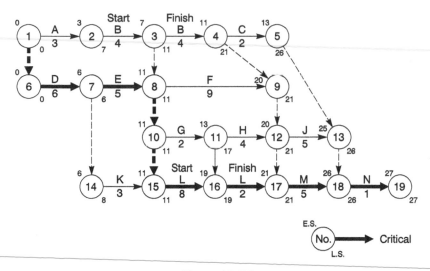

Figure 19.32
Arrow (AoA) network

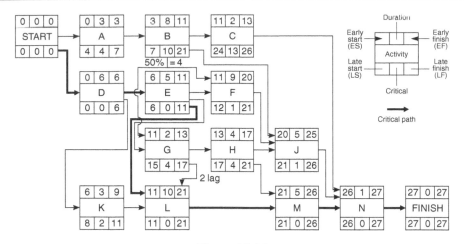

Figure 19.33
Precedence (AoN) network

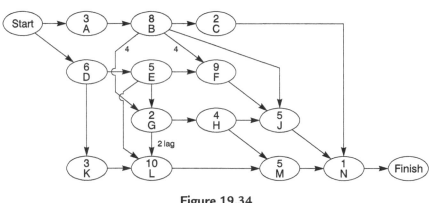

Figure 19.34
Logic draft

of links (or dummy activities) becomes a nightmare, because it is confusing to cross the boxes, which are in effect nodes. It is necessary therefore to restrict the links to run horizontally or vertically between the boxes, which can lead to congestion of the lines, making the tracing of links very difficult.

When a large precedence network is drawn by a computer, the problem becomes even greater, because the link lines can sometimes be so close together that they will appear as one thick black line. This makes it impossible to determine the beginning or end of a link, thus nullifying the whole purpose of a network, i.e., to show the interrelationship and dependencies of the activities. (See Figure 19.35.)

For small networks with few dependencies, precedence diagrams are no problem, but for networks with 200–400 activities per page, it is a different matter. The planner must not feel restricted by the drafting limitations to develop an acceptable logic, and the tendency by some

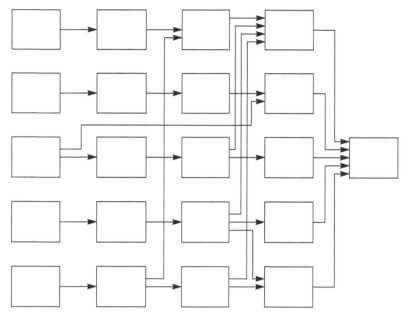

Figure 19.35
Computer-generated AoN diagram

irresponsible software companies to advocate eliminating the manual drafting of a network altogether must be condemned. This manual process is after all the key operation for developing the project network and the distillation of the various ideas and inputs of the team. In other words, it is the thinking part of network analysis. The number crunching can then be left to the computer.

Once the network has been numbered and the times or durations added, it must be analyzed. This means that the earliest starting and completion dates must be ascertained and the floats or 'spare times' calculated. There are three main types of analysis:

1. Arithmetical
2. Graphical
3. Computer

Since these three different methods (although obviously giving the same answers) require very different approaches, a separate chapter has been devoted to each technique (see Chapters 21, 22, and 24).

Constraints

By far the most common logical constraint of a network is as given in the examples on the previous pages, i.e., 'Finish to Start' (F-S) where activity B can only start when activity A is complete. However, it is possible to configure other restraints. These are: Start to Start (S-S), Finish to Finish (F-F), and Start to Finish (S-F). Figure 19.36 shows these less usual constraints, which are sometimes used when a lag occurs between the activities. Analysing a

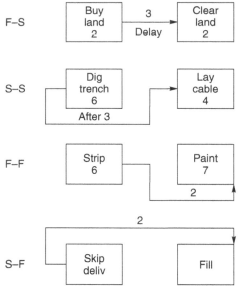

Figure 19.36
Dependencies

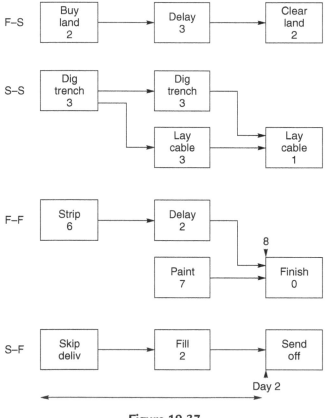

Figure 19.37
Alternative configurations

network manually with such restraints can be very confusing, and should there be a lag or delay between any two activities, it is better to show this delay as just another activity. In fact all these three less usual constraints can be redrawn in the more conventional Finish to Start (F-S) mode as shown in Figure 19.37.

When an activity can start before the previous one has been completed, i.e., when there is an overlap, it is known as *lead*. When an activity cannot start until part of the previous activity has been completed, it is called a *lag*.

Bar (Gantt) Charts

The bar chart was originally first invented by a Polish engineer, Karol Adamiecki, but it was an American production engineer, Henry Laurence Gantt, who developed it at the beginning of the 20th century, and the chart is consequently known now as the Gantt chart. The Gantt chart was used in a number of planning applications during the first world war and was the main planning tool for production and construction engineers until the invention of the critical path network. Each activity on the Gantt chart is represented by a straight horizontal line. The length of the line is proportional to its duration and the starting and finishing times specified by the planner are plotted against the calendar scale provided at the top (or bottom) of the paper. The original set of lines are called the base lines, and progress can be monitored against them. By performing this graphical representation for each activity, an overall picture of the required work on a project can be seen at a glance. Progress of an activity can be recorded either by drawing a second line beneath the base line or colouring up the base line to show the progress on a daily, weekly, or monthly basis.

If the schedule is first drafted manually on a network, it is very quick to produce a Gantt chart as all the thinking has been done at the network stage. However, when using a computer for scheduling, although the Gantt chart is generated automatically as the data is being inputted, the operation takes longer because the sequences and dependencies have to be considered and firmed up as the activities are being typed in. It is for this reason that the first draft of a schedule should be on a network, as the dependencies and basic logic as well as the durations can be very easily and quickly modified to produce the optimum (earliest) completion date.

A Gantt chart is most useful as a method for showing and allocating resources. When the quantity of a common resource, such as labourers, is recorded against each period of an activity, the vertical addition of the resources for any time period gives the total resources for this period. This can then be shown graphically as a vertical bar or column. By carrying out this operation for all the time periods, a series of vertical bars show what resources are required for the project, which is known as a resource histogram. Figure 19.38 shows a simple bar chart where progress is represented by the hatched lines below the activity lines. The resources (men) have been added for each activity and after vertical summation, converted into a histogram. This is discussed further in Chapter 30.

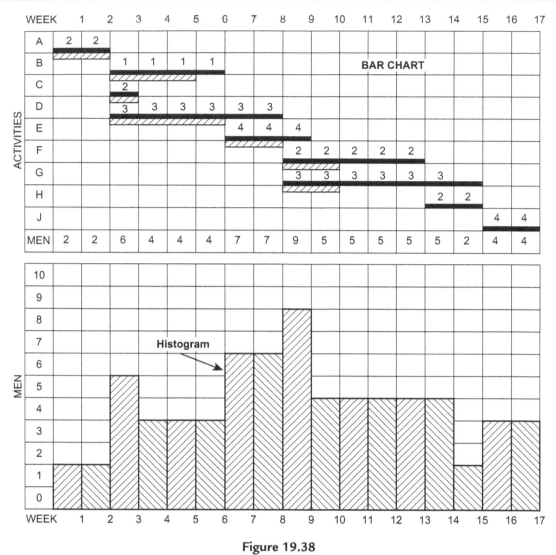

Figure 19.38
Gantt (bar) chart and histogram. Note: The horizantal scale on all the graphs must be in day numbers or week numbers. On no account must they be calender dates. Numbers must be written on the vertical lines, not in the spaces between the lines.

Time Scale Networks and Linked Bar Charts

When preparing presentation or tender documents, or when the likelihood of the programme being changed is small, the main features of a network and bar chart can be combined in the form of a time scale network, or a linked bar chart. A time scale network has the length of the arrows drawn to a suitable scale in proportion to the duration of the activities. The whole network can, in fact, be drawn on a gridded background where each square of the grid represents a period of time such as a day, week, or month. Free float is easily ascertainable by inspection, but total float must be calculated in the conventional manner.

By drawing the activities to scale and starting each activity at the earliest date, a type of bar chart is produced that differs from the conventional bar chart in that some of the activity bars are on the same horizontal line. The disadvantage of such a presentation is that part of the network has to be redrawn 'downstream' from any activity that changes its duration. It can be seen that if one of the early activities changes in either duration or starting point, the whole network has to be modified.

However, a time scale network (especially if restricted to a few major activities) is a clear and concise communication document for reporting up. It loses its value in communicating down because changes increase with detail and constant revision would be too time consuming.

A further development of the Gantt chart, called a linked bar chart, is very similar to a normal bar chart, i.e., each activity is on a separate line and the activities are listed vertically at the edge of the paper. However, by drawing vertical lines connecting the end of one activity with the start of another, one can show the dependencies as a Gantt chart format. Unfortunately, these links, like the original bar charts, only show the implied relationship, as opposed to a critical path network, which shows the logically absolute relationship.

Chapter 24 describes the graphical analysis of networks, and it can be seen that if the ends of the activities were connected by the dummies a linked bar chart would result. This would, however, be based on the logic of the original AoA or AoN network.

Figure 19.39 shows a small time scale network and Figure 19.40 shows the same programme drawn as a linked bar chart.

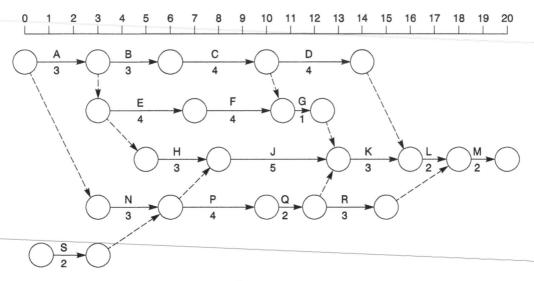

Figure 19.39
Time scale network

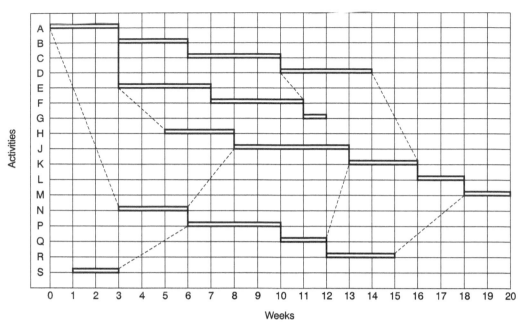

Figure 19.40
Linked bar chart

Planning Blocks and Subdivision of Blocks

Before any meaningful programme can be produced, it is essential that careful thought is given to the number and size of networks required. Not only is it desirable to limit the size of the network, but each 'block' of networks should be considered in relation to the following aspects:

1. The geographical location of the various portions or blocks of the project;
2. The size and complexity of each block;
3. The systems in each block;
4. The process or work being carried out in the block when the plant is complete;
5. The engineering disciplines required during the design and construction stage;
6. The erection procedures;
7. The stages at which individual blocks or systems have to be completed, i.e., the construction programme;
8. The site organization envisaged; and
9. Any design or procurement priorities.

For convenience, a block can be defined as a *geographical process area within a project*, which can be easily identified, usually because it serves a specific function. The importance of choosing the correct blocks, i.e., drawing the demarcation lines in the most advantageous way, cannot be overemphasized. This decision has an effect not only on the number and size of planning networks but also on the organization of the design teams and, in the case of large projects, on the organizational structure of the site management setup.

Because of its importance, a guide is given below which indicates the type of block distribution that may be sensibly selected for various projects. The list is obviously limited, but it should not be too difficult to abstract some firm guidelines to suit the project under consideration.

Pharmaceutical Factory

Block A   Administration block (offices and laboratories)
Block B   Incoming goods area, raw material store
Block C   Manufacturing area 1 (pills)
Block D   Manufacturing area 2 (capsules)
Block E   Manufacturing area 3 (creams)
Block F   Boiler house and water treatment
Block G   Air-conditioning plant room and electrical distribution control room
Block H   Finished goods store and dispatch

For planning purposes, general site services such as roads, sewers, fencing, and guard houses can be incorporated into Block A or, if extensive, can form a block of their own.

New Housing Estate

Block A   Low-rise housing area – North
Block B   Low-rise housing area – East
Block C   Low-rise housing area – South
Block D   Low-rise housing area – West
Block E   High-rise – Block 1
Block F   High-rise – Block 2
Block G   Shopping precinct
Block H   Electricity substation

Obviously, the number of housing areas or high-rise blocks can vary with the size of the development. Roads and sewers and statutory services are part of their respective housing blocks unless they are constructed earlier as a separate contract, in which case they would form their own block or blocks.

Portland Cement Factory

Block A   Quarry crushing plant and conveyor
Block B   Clay pit and transport of clay
Block C   Raw meal mill and silos
Block D   Nodulizer plant and precipitators
Block E   Preheater and rotary kiln
Block F   Cooler and dust extraction
Block G   Fuel storage and pulverization
Block H   Clinker storage and grinding
Block I   Cement storage and bagging
Block J   Administration, offices, maintenance workshops, and lorry park

Here again, the road and sewage system could form a block on its own incorporating the lorry park.

Oil Terminal

Block A   Crude reception and storage
Block B   Stabilization and desalting
Block C   Stabilized crude storage
Block D   NGL separation plant
Block E   NGL storage
Block F   Boiler and water treatment
Block G   Effluent and ballast treatment
Block H   Jetty loading
Block J   Administration block and laboratory
Block K   Jetty 1
Block L   Jetty 2
Block M   Control room 1
Block N   Control room 2
Block P   Control room 3

Here roads, sewers, and underground services are divided into the various operational blocks.

Multi-Storey Block of Offices

Block A   Basement and piling work
Block B   Ground floor
Block C   Plant room and boilers
Block D   Office floors 1–4
Block E   Office floors 5–8

Block F   Lift well and service shafts
Block G   Roof and penthouse
Block H   Substation
Block J   Computer room
Block K   External painting, access road, and underground services

Clearly, in the construction of a multi-storey building, whether for offices or flats, the method of construction has a great bearing on the programme. There is obviously quite a different sequence required for a block with a central core – especially if sliding formwork is used – than with a more conventional design using reinforced concrete or structural steel columns and beams. The degree of precasting will also have a great influence on the split of the network.

Colliery Surface Reconstruction

Block A   Headgear and airlocks
Block B   Winding house and winder
Block C   Mine car layout and heapstead building
Block D   Fan house and duct
Block E   Picking belt and screen building
Block F   Wagon loading and bunkering
Block G   Electricity substation, switch room, and lamp room
Block H   Administration area and amenities
Block J   Baths and canteen (welfare block)

Roads, sewers, and underground services could be part of Block J or be a separate block.

Bitumen Refinery

Block A   Crude line and tankage
Block B   Process unit
Block C   Effluent treatment and oil/water separator
Block D   Finished product tankage
Block E   Road loading facility, transport garage, and lorry park
Block F   Rail loading facility and sidings
Block G   Boiler house and water treatment
Block H   Fired heater area
Block J   Administration building, laboratory, and workshop
Block K   Substation
Block L   Control room

Depending on size, the process unit may have to be subdivided into more blocks, but it may be possible to combine K and L. Again, roads and sewers may be separate or part of each block.

Typical Manufacturing Unit

Block A   Incoming goods ramps and store
Block B   Batching unit
Block C   Production area 1
Block D   Production area 2
Block E   Production area 3
Block F   Finishing area
Block G   Packing area
Block H   Finished goods store and dispatch
Block J   Boiler room and water treatment
Block K   Electrical switch room
Block L   Administration block and canteen

Additional blocks will, of course, be added where complexity or geographical location dictates this.

It must be emphasized that these typical block breakdowns can, at best, be a rough guide, but they do indicate the splits that are possible. When establishing the boundaries of a block, the main points given on p. 46 must be considered.

The interrelationship and interdependence between blocks during the construction stage is, in most cases, remarkably small. The physical connections are usually only a number of pipes, conveyors, cables, underground services, and roads. None of these offers any serious interface problems and should not, therefore, be permitted to unduly influence the choice of blocks. Construction restraints must, of course, be taken into account, but they too must not be allowed to affect the basic block breakdown.

This very important point is only too frequently misunderstood. On a refinery site, for example, a delay in the process unit has hardly any effect on the effluent treatment plant except, of course, right at the end of the job.

In a similar way, the interrelationship at the design stage is often overemphasized. Design networks are usually confined to work in the various engineering departments and need not include such activities as planning and financial approvals or acceptance of codes and standards. These should preferably be obtained in advance by project management. Once the main flow sheets, plot plans, and piping and instrument diagrams have been drafted (i.e., they need not even have been completed), design work can proceed in each block with a considerable degree of independence. For example, the tank farm may be designed quite independently of the process unit or the NGL plant, etc., and the boiler house has little effect on the administration building or the jetties and loading station.

In the case of a single building being divided into blocks, the roof can be designed and detailed independently of the other floors or the basement, provided, of course, that the interface operations such as columns, walls, stairwell, lift shaft, and service ducts have been located and more or less finalized. In short, therefore, the choice of blocks is made as early as possible, taking into account all or most of the factors mentioned before, with particular attention being given to design and construction requirements.

This split into blocks or work areas is, of course, taking place in practice in any design office or site, whether the programme is geared to it or not. One is, in effect, only formalizing an already well-proven and established procedure. Depending on size, most work areas in the design office are serviced by squads or teams, even if they only consist of one person in each discipline who looks after that particular area. The fact that on a small project the person may look after more than one area does not change the principle; it merely means that the team is half an operator instead of one.

On-site, the natural breakdown into work areas is even more obvious. Most disciplines on a site are broken down into gangs, with a ganger or foreman in charge, and, depending again on size and complexity, one or more gangs are allocated to a particular area or block. On very large sites, a number of blocks are usually combined into a complete administrative centre with its own team of supervisors, inspectors, planners, subcontract administrators, and site engineers, headed by an area manager.

No difficulty should, therefore, be experienced in obtaining the cooperation of an experienced site manager when the type, size, and number of blocks are proposed. Indeed, this early discussion serves as an excellent opportunity to involve the site team in the whole planning process, the details of which are added later. By that time, the site team is at least aware of the principles and a potential communication gap, so frequently a problem with construction people has been bridged.

Subdivision of Blocks

One major point that requires stressing covers the composition of a string of activities. It has already been mentioned that the site should be divided into blocks that are compatible with the design networks. However, each block could in itself be a very large area and a complex operational unit. It is necessary, therefore, to subdivide each block into logical units. There are various ways of doing this. The subdivision could be by:

1. Similar items of equipment
2. Trades and disciplines
3. Geographical proximity
4. Operational systems
5. Stages of completion

Each subdivision has its own merits and justifies further examination.

Similar Items of Equipment

Here the network shows a series of strings that collect together similar items of equipment, such as pumps, tanks, vessels, boilers, and roads. This is shown in Figure 20.1.

Advantages:

(a) Equipment items are quickly found;
(b) Interface with design network is easily established.

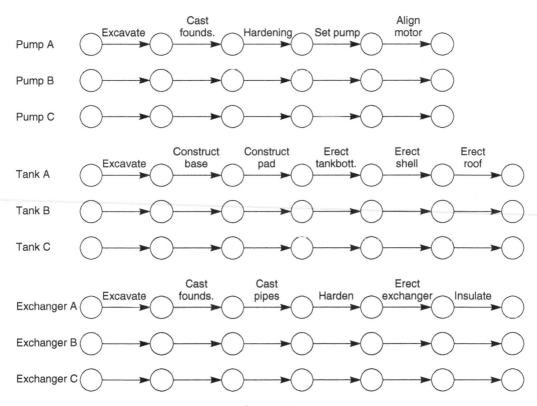

Figure 20.1
Similar items of equipment

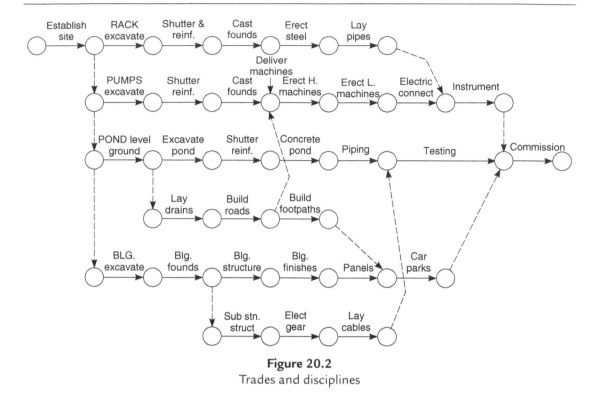

Figure 20.2
Trades and disciplines

Trades and Disciplines

This network groups the work according to type. It is shown in Figure 20.2.

Advantages:

(a) Suitable when it is desirable to clear a trade off the site as soon as completed;
(b) Eases resource loading of individual trades.

Geographical Proximity

It may be considered useful to group together activities that are geographically close to each other without further segregation into types or trades. This is shown in Figure 20.3.

Advantages:

(a) Makes a specific area self-contained and eases control;
(b) Coincides frequently with natural subdivision on site for construction management.

Operational Systems

Here the network consists of all the activities associated with a particular system such as the boiler plant, the crude oil loading, and the quarry crushing and screening. A typical system network is shown in Figure 20.4.

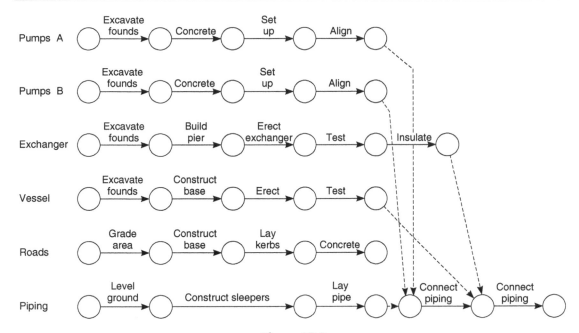

Figure 20.3
Geographical proximity

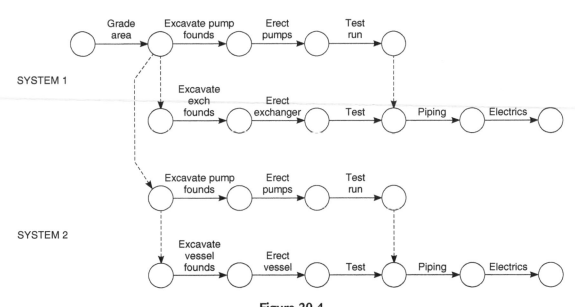

Figure 20.4
Operational system

Advantages:

(a) Easy to establish and monitor the essential interrelationships of a particular system;
(b) Particularly useful when commissioning is carried out by system since a complete 'package' can be programmed very easily;
(c) Ideal where stage completion is required.

Stages of Completion

If particular parts of the site have to be completed earlier than others (i.e., if the work has to be handed over to the client in well-defined stages), it is essential that each stage is programmed separately. There will, of course, be interfaces and links with preceding and succeeding stages, but within these boundaries the network should be self-contained.

Advantages:

(a) Attention is drawn to activities requiring early completion;
(b) Predictions for completion of each stage are made more quickly;
(c) Resources can be deployed more efficiently;
(d) Temporary shut-off and blanking-off operations can be highlighted.

In most cases a site network is in fact a combination of a number of the above subdivisions. For example, if the boiler plant and water treatment plant are required first to service an existing operational unit, it would be prudent to draw a network based on operational systems but incorporating also stages of completion. In practice, geographical proximity would almost certainly be equally relevant since the water treatment plant and boiler plant would be adjacent.

It must be emphasized that the networks shown in Figures 20.1 to 20.4 are representative only and do not show the necessary inter relationships or degree of detail normally shown on a practical construction network. The oversimplication on these diagrams may in fact contradict some of the essential requirements discussed in other sections of this book, but it is hoped that the main point, i.e., the differences between the various types of construction network formats, has been highlighted.

Banding

If we study Figure 20.1 we note that it is very easy to find a particular activity on the network. For example, if we wanted to know how long it would take to excavate the foundations of exchanger B, we would look down the column EXCAVATE until we found the line EXCHANGER B, and the intersection of this column and line shows the required excavation activity. This simple identification process was made possible because Figure 20.1 was drawn using very crude subdivisions or bands to separate the various operations.

For certain types of work this splitting of the network into sections can be of immense assistance in finding required activities. By listing the various types of equipment or materials vertically on the drawing paper and writing the operations to be performed horizontally, one produces a grid of activities that almost defines the activity. In some instances the line of operations may be replaced by a line of departments involved. For example, the electrical department involvement in the design of a piece of equipment can be found by reading across the equipment line until one comes to the electrical department column.

The principle is shown clearly in Figure 20.5, and it can be seen that the idea can be applied to numerous types of networks. A few examples of banding networks are given below, but these are for guidance only since the actual selection of bands depends on the type of work to be performed and the degree of similarity of operation between the different equipment items.

Vertical listing	*Horizontal listing*
(*Horizontal line*)	(*Vertical column*)
Equipment	Operations
Equipment	Departments
Material	Operations
Design stages	Departments
Construction stages	Subcontracts
Decision stages	Departments
Approvals	Authorities (clients)
Operations	Department responsibilities
Operations	Broad time periods

It may, of course, be advantageous to reverse the vertical and horizontal bands; when considering, for example, the fifth item on the list, the subcontracts could be listed vertically and the construction stages horizontally. This would most likely be the case when the subcontractors perform similar operations since the actual work stages would then follow logically across the page in the form of normally timed activities. It may indeed be beneficial to draw a small trial network of a few (say, 20–30) activities to establish the best banding configuration.

It can be seen that banding can be combined with the coordinate method of numbering by simply allocating a group of letters of the horizontal coordinates to a particular band.

Banding is particularly beneficial on master networks which cover, by definition, a number of distinct operations or areas, such as design, manufacture, construction, and commissioning. Figure 20.5 is an example of such a network.

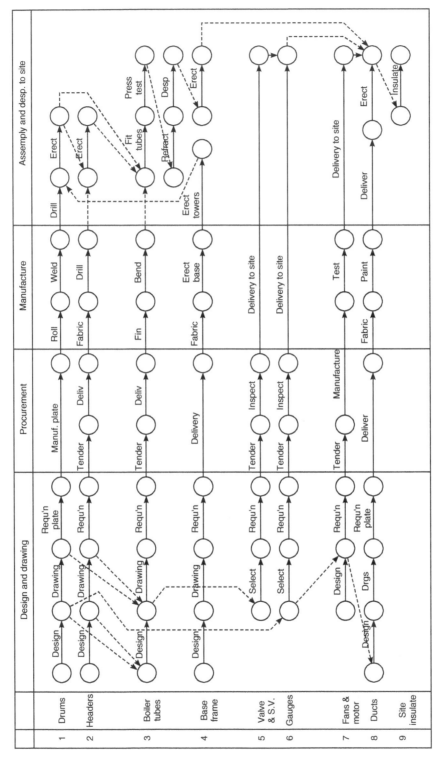

Figure 20.5
Simplified boiler network

CHAPTER 21

Arithmetical Analysis and Floats

Chapter Outline

Arithmetical Analysis

This method is the classical technique and can be performed in a number of ways. One of the easiest methods is to add up the various activity durations on the network itself, writing the sum of each stage in a square box at the end of that activity, i.e., next to the end event (Figure 21.1). It is essential that each route is examined separately and where the routes meet, the *largest* sum total must be inserted in the box. When the complete network has been summed in this way, the *earliest* starting will have been written against each event.

Now the reverse process must be carried out. The last event sum is now used as a base from which the activities leading into it are subtracted. The results of these subtractions are entered

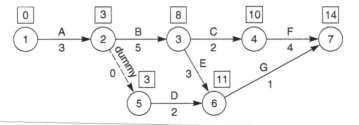

Figure 21.1
Forward pass

141

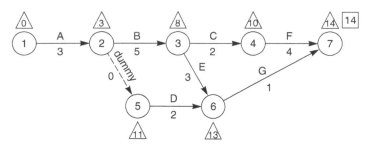

Figure 21.2
Backward pass

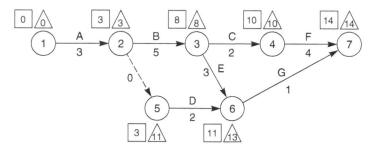

Figure 21.3

in triangular boxes against each event (Figure 21.2). As with the addition process for calculating the earliest starting times, a problem arises when a node is reached where two routes or activities meet. Since the *latest* starting times of an activity are required, the *smallest* result is written against the event.

The two diagrams are combined in Figure 21.3. The difference between the earliest and latest times gives the 'float', and if this difference is zero (i.e., if the numbers in the squares and triangles are the same) the event is on the critical path.

A table can now be prepared setting out the results in a concise manner (Table 21.1).

Slack

The difference between the latest and earliest times of any event is called 'slack'. Since each activity has two events, a beginning event and an end event, it follows that there are two slacks for each activity. Thus the slack of the beginning event can be expressed as $TL_B - TE_B$ and called beginning slack, and the slack of the end event, appropriately called end slack, is $TL_E - TE_E$. The concept of slack is useful when discussing the various types of float, since it simplifies the definitions.

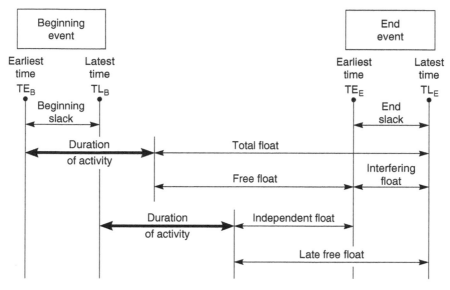

Figure 21.4
Floats

Table 21.1

a	b	c	d	e	f	g	h
Title	Activity	Duration, D	Latest time end event	Earliest time end event	Earliest time beginning event	Total float (d-f-c)	Free float (e-f-c)
A	1–2	3	3	3	0	0	0
B	2–3	5	8	8	3	0	0
DUMMY	2–5	0	11	3	3	8	0
C	3–4	2	10	10	8	0	0
E	3–6	3	13	11	8	2	0
F	4–7	4	14	14	10	0	0
D	5–6	2	13	11	3	8	6
G	6–7	1	14	14	11	2	2

Column a: activities by the activity titles. Column b: activities by the event numbers. Column c: activity durations, D. Column d: latest time of the activities' end event, TL_E. Column e: earliest time of the activities' end event, TE_E. Column f: earliest time of the activities' beginning event, TE_B. Column g: total float of the activity. Column h: free float of the activity.

Float

This is the name given to the spare time of an activity, and it is one of the more important by-products of network analysis. The four types of float possible will now be explained.

Total Float

It can be seen that activity 3–6 in Figure 21.3 *must* be completed after 13 time units, but can be started after 8 time units. Clearly, therefore, since the activity itself takes 3 time units, the activity *could* be completed in $8 + 3 = 11$ time units. Therefore there is a leeway of $13-11 = 2$ time units on the activity. This leeway is called total float and is defined as latest time of end event minus earliest time of beginning event minus duration, or $TL_E - TE_B - D$.

Figure 21.3 shows that total float is, in fact, the same as beginning slack. Also, free float is the same as total float minus end slack. The proof is given at the end of this chapter.

Total float has an important role to play in network analysis. By definition, it is the time between the anticipated start (or finish) of an activity and the latest permissible start (or finish).

The float can be either positive or negative. A positive float means that the operation or activity will be completed earlier than necessary, and a negative float indicates that the activity will be late. A prediction of the status of any particular activity is, therefore, a very useful and important piece of information for a manager. However, this information is of little use if not transmitted to management as soon as it becomes available, and every day of delay reduces the manager's ability to rectify the slippage or replan the mode of operation.

The reason for calling this type of float 'total float' is because it is the total of all the 'free floats' in a string of activities when working back from where this string meets the critical path to the activity in question.

For example, in Figure 21.10, the activities in the lowest string J to P have the following free floats: $J = 0$, $K = 10 - 9 = 1$, $L = 0$, $M = 15 - 14 = 1$, $N = 21 - 19 = 2$, $P = 0$. Total float for K is therefore $2 + 1 + 1 + 1 = 4$. This is the same as the 4 shown in the lower middle space of the node.

It is very easy to calculate the total floats and free floats in a precedence or Lester diagram. For any activity, the total float is the difference between the *latest finish* and *earliest finish* (or *latest start* and *earliest start)*. The free float is the difference between the *earliest finish* of the activity in question and the *earliest start* of the following activity. Figure 21.15 makes this clear.

Calculation of Float

By far the quickest way to calculate the float of a particular activity is to do it manually. In practice, one does not need to know the float of *all* activities at the same time. A list of floats is, therefore, unnecessary. The important point is that the float of a particular activity that is of immediate interest is obtainable quickly and accurately.

Consider the string of activities in a simple construction process. This is shown in Figure 21.5 in activity on arrow (AoA) format and in Figure 21.6 in the simplified activity on node (AoN) format.

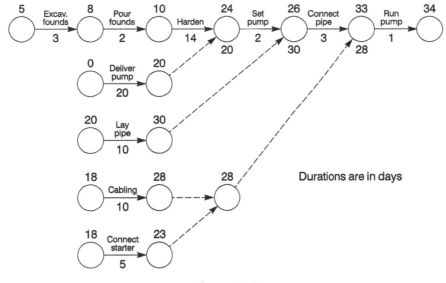

Figure 21.5

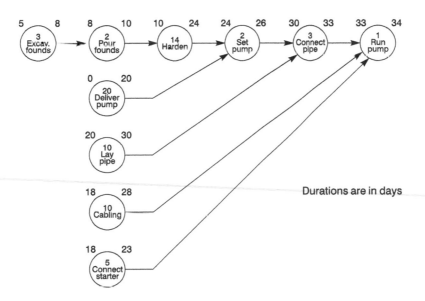

Figure 21.6

It can be seen that the total duration of the sequence is 34 days. By drafting the network in the method shown, and by using the day numbers at the end of *each* activity, including dummies, an accurate prediction is obtained immediately and the float of any particular activity can be seen almost by inspection. It will be noted that each activity has two dates or day numbers – one at the beginning and one at the end (Figure 21.7). Therefore, where two (or more) activities meet at a node, all the end day numbers are inserted (Figure 21.8). The highest number is now used

Figure 21.7

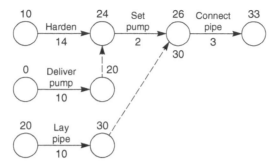

Figure 21.8

to calculate the overall project duration, i.e., 30 + 3 = 33, and the difference between the highest and the other number immediately gives the float of the other activity and *all* the activities in that string up to the previous node at which more than one activity meet. In other words, 'set pumps' (Figure 21.5) has a float of 30 − 26 = 4 days, as have all the activities preceding it except 'deliver pump', which has an additional 24 − 20 = 4 days.

If, for example, the electrical engineer needs to know for how long he can delay the cabling because of an emergency situation on another part of the site, without delaying the project, he can find the answer right away. The float is 33 − 28 = 5 days. If the labour he needs for the emergency can be drawn from the gang erecting the starters, he can gain another 28 − 23 = 5 days. This gives him a total of 10 days' grace to start the starter installation without affecting the total project time.

A few practice runs with small networks will soon emphasize the simplicity and speed of this method. We have in fact only dealt in this exposition with small – indeed, tiny – networks. How about large ones? It would appear that this is where the computer is essential, but, in fact, a well-drawn network can be analysed manually just as easily whether it is large or small. Provided the very simple base rules are adhered to, a very fast forward pass can be inserted. The float of any *string* can then be seen by inspection, i.e., by simply subtracting the lower node number from the higher node number, which forms the termination point of the string in question. This point can best be illustrated by the example given in Figure 21.9. For simplicity, the activities have been given letters instead of names, since the importance lies in understanding the principle, and the use of letters helps to identify the string of activities. In this example there are 50 activities. Normally, a practical network should have between 200 and 300 activities maximum (i.e., four to six times the number of activities shown), but this

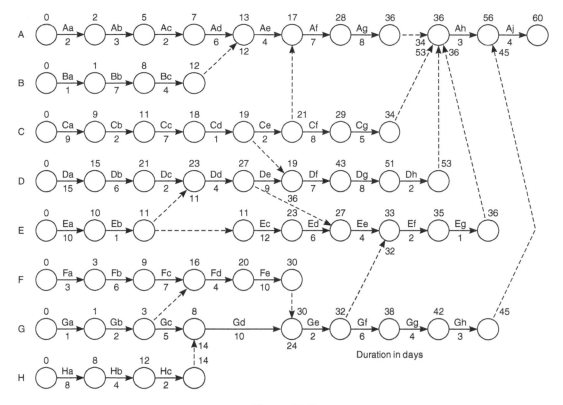

Figure 21.9

does not pose any greater problem. All the times (day numbers) were inserted, and the floats of activities in strings A, B, C, E, F, G, and H were calculated in 5 minutes. A 300-activity network would, therefore, take 30 minutes.

It can in fact be stated that any practical network can be 'timed', i.e., the forward pass can be inserted and the important float reported in 45 minutes. It is, furthermore, very easy to find the critical path. Clearly, it runs along the strings of activities with the highest node times. This is most easily calculated by working back from the end. Therefore the path runs through Aj, Ah, dummy, Dh, Dg, Df, De, Dd, Dc, Db, and Da.

An interesting little problem arises when calculating the float of activity Ce, since there are two strings emanating from the end node of that activity. By conventional backward pass methods – and indeed this is how a computer carries out the calculation – one would insert the backward pass in the nodes starting from the end (see Figure 21.10). When arriving at Ce, one finds that the latest possible time is 40 when calculating back along string Cg–Cf, while it is 38 when calculating back along string Ag–Af. Clearly, the actual float is the difference between the earliest date and the earliest of the two latest dates, i.e., day 38 instead of day 40. The float of Ce is therefore 38 – 21 = 17 days.

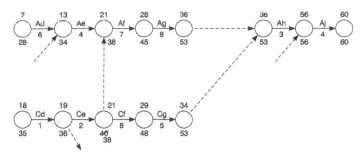

Figure 21.10

As described above, the calculation is tedious and time consuming. A far quicker method is available by using the technique shown in Figure 21.9, i.e., one simply inserts the various forward passes on each string and then looks at the end node of the activity in question – in our case, activity Ce. It can be seen that by following the two strings emanating from Ce that string Af–Ag joins Ah at day 36. String Cf–Cg, on the other hand, joins Ah at day 34. The float is, therefore, the *smallest* difference between the *highest* day number and one of the two day numbers just mentioned. Clearly, therefore, the float of activity Ce is 53 – 36 = 17 days. Cf and Cg, of course, have a float of 53 – 34 = 19 days.

The time to inspect and calculate the float by the second method is literally only a few minutes. All one has to do is to run through the paths emanating from the end node of the selected activity and note the *highest* day number where the strings *meet the critical path*. The difference between the day number of the critical string and the highest number on the tributary strings (emanating from the activity in question) is the float.

Supposing we now wish to find the float of activity Gb:

> Follow string Fd–Fe.
> Follow string Gc–Gd–Ge.
> Follow string Gf–Gg–Gh.
> Follow string Ef–Eg–Ah.
> Fe and Gd meet at Ge, therefore they can be ignored.

> String Gf–Gh meets Aj at day   45.
> String Ef–Eg meets Ah at day   36.
> Therefore the float is either   56 – 45 = 11
> or   53 – 36 = 17.

Clearly, the correct float is 11 since it is the smaller of the two. The time taken to inspect and calculate the float was exactly 21 seconds!

All the floats calculated above have been total floats. Free float can only occur on activities entering a node when more than one enters that node. It can be calculated very easily by

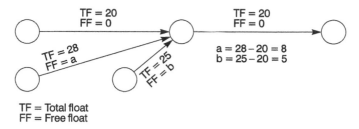

Figure 21.11

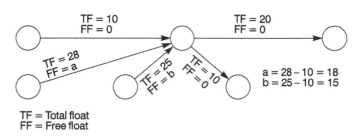

Figure 21.12

subtracting the total float of the incoming activity from the total float of the outgoing activity, as shown in Figure 21.11. It should be noted that one of the activities entering the node *must* have zero *free* float.

When more than one activity leaves a node, the value of the free float to be subtracted is the *lowest* of the outgoing activity floats, as shown in Figure 21.12.

Free Float

Some activities, e.g., 5–6, as well as having total float have an additional leeway. It will be noted that activities 3–6 and 5–6 both affect activity 6–7. However, one of these two activities will delay 6–7 by the same time unit by which it itself may be delayed. The remaining activity, on the other hand, may be delayed for a period without affecting 6–7. This leeway is called free float, and can only occur in one or more activities where several meet at one event, i.e., if x activities meet at a node, it is possible that $x - 1$ of these have free float. This free float may be defined as earliest time of end event minus earliest time of beginning event minus duration, or $TE_E - TE_B - D$.

The Concept of Free Float

Students often find it difficult to understand the concept of free float. The mathematical definitions are unhelpful, and the graphical representation of Figure 21.4 can be confusing. The easiest way to understand the difference between total float and free float is to inspect the

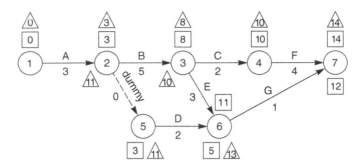

Figure 21.13

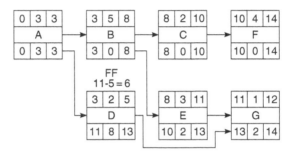

Figure 21.14

end node of the activity in question. As stated earlier, free float can only occur where two or more activities enter a node. If the *earliest* end times (i.e., the forward pass) for each individual activity are placed against the node, the free float is simply the difference between the highest number of the earliest time on the node and the number of the earliest time of the activity in question.

In the example given in Figure 21.13 the earliest times are placed in squares, so following the same convention it can be seen from the figure (which is a redrawing of Figure 21.1 with *all* the earliest and latest node times added) that:

Figure 21.14 shows the equivalent precedence (AoN) diagram from which the free float can be easily calculated by subtracting the early *finish* time of the *preceding* node from the early *start* time of the *succeeding* node.

$$\text{Free float of activity } D = 11 - 5 = 6.$$

$$\text{Free float of activity } G = 14 - 12 = 2.$$

Activity E, because it is not on the critical path, has total float of $13 - 11 = 2$ but has no free float.

The check of the free float by the formal definition is as follows:

$$\text{Free Float} = TE_E - TE_B - D$$
$$\text{For activity } D = 11 - 3 - 2 = 6$$
$$\text{For activity } G = 14 - 11 - 1 = 2$$

The check of the total float by the formal definitions is as follows:

$$\text{Total float} = TL_E - TE_B - D$$
$$\text{For activity } E = 13 - 8 - 3 = 2$$
$$D = 13 - 3 - 2 = 8$$
$$G = 14 - 11 - 2 = 2$$

It was stated earlier that total float is the same as beginning slack. This can be shown by rewriting the definition of total float $= TL_E - TE_B - D$ as total float $= TL_E - D - TE_B$ but $TL_E - D = TL_B$. Therefore:

$$\text{Total float} = TL_B - TE_B$$
$$= \text{Beginning slack}$$

To show that free float = total float − end slack, consider the following definitions:

$$\text{Free float} = TL_E - TE_B - D \qquad\qquad (21.A)$$

$$\text{Total float} = TL_E - TE_B - D \qquad\qquad (21.B)$$

$$\text{End slack} = TL_E - TE_E \qquad\qquad (21.C)$$

Subtracting equation (21.C) from equation (21.B):

$$= TL_E - TE_B - D - TL_E - TE_E$$
$$= TL_E - TE_B - D - TL_E + TE_E$$
$$= TE_E - TE_B - D \qquad\qquad (21.A)$$
$$= \text{Free float}$$

Therefore:

equation $(21.A) = $ equation $(21.B) - $ equation $(21.C)$ or free float $=$ total float $-$ end slack.

If a computer is not available, free float on an arrow diagram can be ascertained by inspection, since it can only occur where more than one activity meets a node. This was described in detail earlier with Figures 21.5 and 21.6. If the network is in the precedence format, the calculation of free float is even easier. All one has to do is to subtract the early finish time in the preceding node from the early start time of the succeeding

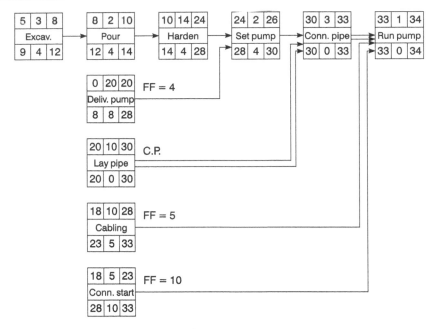

Figure 21.15
Bar chart of house

node. This is clearly shown in Figure 21.15, which is the precedence equivalent to Figure 21.5.

One of the phenomena of a computer printout is the comparatively large number of activities with free float. Closer examination shows that the majority of these are in fact dummy activities. The reason for this is, of course, obvious, since, by definition, free float can only exist when more than one activity enters a node. As dummies nearly always enter a node with another (real) activity, they all tend to have free float. Unfortunately, no computer program exists that automatically transfers this free float to the preceding real activity, so the benefit of the free float is not immediately apparent and is consequently not taken advantage of.

Interfering Float

The difference between the total float and the free float is known as interfering float. Using the previous notation, this can be expressed as

$$(TL_E - TE_B - D) - (TE_E - TE_B - D) = TL_E - TE_B - D - TE_E + TE_B + D$$
$$= TL_E - TE_E$$

i.e., as the latest time of the end event minus the earliest time of the end event. It is, therefore, the same as the end slack.

Independent Float

The difference between the free float and the beginning slack is known as independent float:

$$\begin{aligned}
\text{Since free float} &= TE_E - TE_B - D \\
\text{and beginning slack} &= TL_B - TE_B \\
\text{independent float} &= TE_E - TE_B - D - (TL_B - TE_B) \\
&= TE_E - TE_B - D
\end{aligned}$$

This problem does not exist with precedence diagrams as there are no dummy activities in this network format.

Thus independent float is given by the earliest time of the end event minus the latest time of the beginning event minus the duration.

In practice neither the interfering float nor the independent float find much application, and for this reason they will not be referred to in later chapters. The use of computers for network analysis enables these values to be produced without difficulty or extra cost, but they only tend to confuse the user and are therefore best ignored.

Summarizing all the above definitions, Figure 21.4 and the following expressions may be of assistance.

Notation

D_B = duration of activity
TE_B = earliest time of beginning event
TE_E = earliest time of end event
TL_B = latest time of beginning event
TL_E = latest time of end event

Definitions

$$\begin{aligned}
\text{beginning slack} &= TL_B - TE_B \\
\text{end slack} &= TL_E - TE_E \\
\text{total float} &= TL_E - TE_B - D
\end{aligned}$$

$$\begin{aligned}
\text{free float} &= TE_E - TE_B - D \\
\text{interfering float} &= TL_E - TL_B \ (= \text{end slack}) \\
\text{independent float} &= TE_E - TL_B - D \\
\text{last free float} &= TL_E - TL_B - D
\end{aligned}$$

Because float is such an important part of network analysis and because it is frequently quoted — or misquoted — by computer protagonists as another reason why computers *must* be used, a special discussion of the subject may be helpful to those readers not too familiar with its use in practice.

Of the three types of float shown on a printout, i.e., the total float, free float, and independent float, only the first – the total float – is in general use. Where resource smoothing is required, a knowledge of free float can be useful, since the activities with free float can be moved backwards or forwards in time without affecting any other activities. Independent float, on the other hand, is really quite a useless piece of information and should be suppressed (when possible) from any computer printout. Of the many managers, site engineers, and planners interviewed, none has been able to find a practical application of independent float.

Critical Path

Some activities have zero total float, i.e., no leeway is permissible for their execution, hence any delays incurred on the activities will be reflected in the overall project duration. These activities are therefore called critical activities, and every network has a chain of such critical activities running from the beginning event of the first activity to the end event of the last activity, without a break. This chain is called the critical path.

Frequently a project network has more than one critical path, i.e., two or more chains of activities all have to be carried out within the stipulated duration to avoid a delay to the completion date. In addition, a number of activity chains may have only one or two units of float, so that, for all intents and purposes, they are also critical. It can be seen, therefore, that it is important to keep an eye on all activity chains that are either critical or near-critical, since a small change in duration of one chain could quickly alter the priorities of another.

One disadvantage of the arithmetical method of analysis using the table or matrix shown in Table 21.1 is that all the floats must be calculated before the critical path can be ascertained. This method (which has only been included for historical interest) is very laborious and is therefore not recommended.

In any case, this drawback is eliminated when the method of analysis shown in Figure 21.9 is employed.

Critical Chain Project Management (CCPM)

This planning technique differs from conventional CPA in that resource restraints and dependencies are considered as well as operation logic. This requires buffers to be introduced to allow for possible resource issues. Progress monitoring can then carried out by measuring the rate at which the buffers are used up instead of just recording the individual task performance.

In practice this method is probably most appropriate where resources are not easily assessable at the planning stage. Contingencies in the form of buffers must therefore be incorporated, with the result that the overall duration will probably be longer, and as a consequence the planned completion date will more likely be met.

In the construction industry, where the program completion date is often set by the client, i.e., the opening of a venue or start of production, the project program will be based on the construction logic of the operational activities, which must then be provided with the necessary resources (plant, material, and labour) to carry out the work. It is inconceivable that at the start of a project a contractor in the construction industry would tell his or her client that the project cannot be completed by the specified date because the necessary resources cannot be obtained.

The Case for Manual Analysis

Although network analysis is applicable to almost every type of organization, as shown by the examples in Chapter 29, most of the planning functions described in this book have been confined to those related to engineering construction projects. The activities described cover the full spectrum of operations from the initial design stage, through detailing of drawings and manufacture, up to and including construction. In other words, from conception to handover.

In this age of specialization there is a trend to create specialist groups to do the work previously carried out by the members of more conventional disciplines. One example is teaching where teaching methods, previously devised and perfected by practising teachers, are now developed by a new group of people called educationalists.

Another example of specialization is planning. In the days of bar charts, planning was carried out by engineers or production staff using well-known techniques to record their ideas on paper and transmit them to other members of the team. Nowadays, however, the specialist planner or scheduler has come to the fore, leaving the engineer time to get on with his engineering.

The Planner

Planning in its own right does not exist. It is always associated with another activity or operation, i.e., design planning, construction planning, production planning, etc. It is logical, therefore, that a design planner should be or should have been a designer, a construction planner should be familiar with construction methods and techniques, and a production planner should be knowledgeable in the process and manufacturing operations of production – whether it be steelwork, motors cars, or magazines.

In construction, as long as the specialist planner has graduated from one of the accepted engineering disciplines and is familiar with the problems of a particular project, a realistic

Project Management, Planning, and Control. http://dx.doi.org/10.1016/B978-0-08-098324-0.00022-6

network will probably be produced. By calling in specialists to advise him in the fields with which he is not completely conversant, he can ensure that the network will be received with confidence by all the interested parties.

The real problem arises when the planner does not have the right background, i.e., when he has not spent a period in design or has not experienced the holdups and frustrations of a construction site. Strangely enough, the less familiar a planner is with the job he is planning, the less he is inclined to seek help. This may well be due to his inability to ask the right questions, or he may be reluctant to discuss technical matters for fear of revealing his own lack of knowledge. One thing is certain: A network that is not based on sound technical knowledge is not realistic, and an unrealistic network is dangerous and costly, since decisions may well be made for the wrong reasons.

All that has been said so far is a truism that can be applied not only to planning but to any human activity where experts are necessary in order to achieve acceptable results. However, in most disciplines it does not take long for the effects of an inexperienced assistant to be discovered, mainly because the results of his work can be monitored and assessed within a relatively short time period. In planning, however, the effects of a programme decision may not be felt for months, so it may be very difficult to ascertain the cause of the subsequent problem or failure.

The Role of the Computer

Unfortunately, the use of computers has enabled inexperienced planners to produce impressive outputs that are frequently utterly useless. There is a great danger in shifting the emphasis from the creation of the network to the analysis and report production of the machine, so that many people believe that to carry out an analysis of a network one must have a computer. In fact, of course, the computer is only a sophisticated number cruncher. It does not see the whole picture, including access problems, political or cultural restraints, labour issues, and staff idiosyncrasies. The kernel of network analysis is the drafting, checking, refining, and redrafting of the network itself, an operation that must be carried out by a team of experienced participants of the job being planned. To understand this statement, it is necessary to go through the stages of network preparation and subsequent updating.

Preparation of the Network

The first function of the planner in conjunction with the project manager is to divide the project into manageable blocks. The name is appropriate since, like building blocks, they can be handled by themselves and shaped to suit the job but are still only a part of the whole structure to be built.

The number and size of each block is extremely important since, if correctly chosen, a block can be regarded as an entity that suits both the design and the construction phases of a project. Ideally, the complexity of each block should be about the same, but this is rarely possible in practice since other criteria such as systems and geographical location have to be considered. If a block is very complex, it can be broken down further, but a more convenient solution may be to produce more than one network for such a block. The aim should be to keep the number of activities down to 200–300 so that they can be analysed manually if necessary.

As the planner sketches his logic roughly, and in pencil on the back of an old drawing, the construction specialists are asked to comment on the type and sequence of the activities. In practice, these sessions – if properly run – generate an enthusiasm that is a delight to experience. Often consecutive activities can be combined to simplify the network, thus easing the subsequent analysis. Gradually the job is 'built', difficulties are encountered and overcome, and even specialists who have never been involved in network planning before are carried away by this visual unfolding of the programme.

The next stage is to ask each specialist to suggest the duration of the activities in his discipline. These are entered onto the network without question. Now comes the moment of truth. Can the job be built on time? With all the participants present, the planner adds up the durations and produces his forward pass. Almost invariably the total time is longer than the deadlines permit. This is when the real value of network analysis emerges. Logics are re-examined, durations are reduced, and new construction methods are evolved to reduce the overall time. When the final network – rough though it may be – is complete, a sense of achievement can be felt pervading the atmosphere.

This procedure, which is vital to the production of a realistic programme, can, of course, only be carried out if the 'blocks' are not too large. If the network has more than 300 activities it may well pay the planner or project manager to re-examine that section of the programme with a view to dividing it into two smaller networks. If necessary, it is always possible to draw a master network, usually quite small, to link the blocks together.

One of the differences between the original PERT program and the normal CPM programs was the facility to enter three time estimates for every activity. The purpose of the three estimates is to enable the computer to calculate and subsequently use the most probable time, on the assumption that the planner is unwilling or unable to commit himself to one time estimate. The actual duration used is calculated from the expression known as the β distribution:

$$t_e = \frac{a + 4m + b}{6}$$

where t_e is the expected time, a the optimistic time, b the pessimistic time, and m the most likely time.

However, this degree of sophistication is not really necessary, since the planner himself can insert what he considers to be the most probable time. For example, a foreman, upon being pressed, estimated the times for a particular operation to be:

$$Optimistic = 5\ days$$
$$Pessimistic = 10\ days$$
$$Probable = 7\ days$$

The planner will probably insert 7 days or 8 days. The computer, using the above distribution, would calculate:

$$t_c = \frac{5 + (4 \times 7) + 10}{6}$$
$$= 7.16\ days$$

With the much larger variables found in real-life projects, such finesse is a waste of time.

A single time estimate by an experienced planner is all that is required.

Typical Site Problems

Once construction starts, problems begin to arise. Drawings or other data arrive late on-site, materials are delayed, equipment is held up, labour becomes scarce or goes on strike, underground obstructions are found, the weather deteriorates, etc.

Each new problem must be examined in the light of the overall project programme. It will be necessary to repeat the initial planning meeting to revise the network, to reflect on these problems, and to possibly help reduce their effects. It is at these meetings that ingenious innovations and solutions are suggested and tested.

For example, Figure 22.1 shows the sequence of a section of a pipe rack. Supposing the delivery of pipe will be delayed by four weeks, completion now looks like week 14. However, someone suggests that the pump bases can be cast early with starter bars bent down to bond the plinths at a later date. The new sequence appears in Figure 22.2. Completion time is now only week 11, a saving of three weeks.

This type of approach is the very heart of successful networking and keeps the whole programme alive. It is also very rapid. The very act of discussing problems in the company of interested and knowledgable colleagues generates an enthusiasm that carries the project forward. With good management, this momentum is passed right down the line to the people who are actually doing the work at the sharp end.

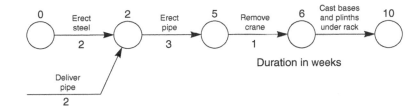

Figure 22.1

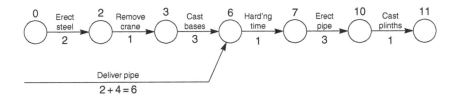

Figure 22.2

The NEDO (National Economic Development Office) Report

Perhaps the best evidence that networks are most effective when kept simple is given by the NEDO report, which is still applicable today even though it was produced way back in the early 1980s.

The relevant paragraphs are reproduced below by permission of HM Stationery Office.

1. Even if it is true that UK clients build more complex plants, it should still be possible to plan for and accommodate the extra time and resources this would entail. By and large the UK projects were more generously planned but, none the less, the important finding of the case studies is that, besides taking longer, the UK projects tended also to encounter more overrun against planned time. There was no correlation across the case studies between the sophistication with which programming was done and the end result in terms of successful completion on time. On the German power station the construction load represented by the size and height of the power station was considerable, but the estimated construction time was short and was achieved. This contrasts with the UK power stations, where a great deal of effort and sophistication went into programming, but schedules were overrun. On most of the case studies, the plans made at the beginning of the project were thought realistic at that stage, but they varied in their degree of sophistication and in the importance attached to them.

2. One of the British refineries provided the one UK example where the plan was recognized from the start by both client and contractor to be unrealistic. None the less, the contractor claimed that he believed planning to be very important, particularly in the circumstances

of the UK, and the project was accompanied by a wealth of data collection. This contrasts with the Dutch refinery project where planning was clearly effective but where there was no evidence of very sophisticated techniques. There is some evidence in the case studies to suggest that UK clients and contractors put more effort into planning, but there is no doubt that the discipline of the plan was more easily maintained on the foreign projects. Complicated networks are useful in developing an initial programme, but subsequently, though they may show how badly one has done, they do not indicate how to recover the situation. Networks need, therefore, to be developed to permit simple rapid updates, pointing where action must be taken. Meanwhile the evidence from the foreign case studies suggests that simple techniques, such as bar charts, can be successful.

3. The attitudes to planning on UK1 and the Dutch plant were very different, and this may have contributed to the delay of UK1 although it is impossible to quantify the effect. The Dutch contractor considered planning to be very important, and had two site planning engineers attached to the home office during the design stage. The programme for UK1 on the other hand was considered quite unrealistic by both the client and the contractor, not only after the event but while the project was under way, but neither of them considered this important in itself.

On UK 1 it was not until the original completion date arrived that construction was rescheduled to take 5 months longer. At this point construction was only 80% complete and in the event there was another eight month's work to do. Engineering had been 3 months behind schedule for some time. A wealth of progress information was being collected but no new schedule appears to have been made earlier.

Progress control and planning was clearly a great deal more effective on the Dutch project; the contractor did not believe in particularly sophisticated control techniques, however.

Clearly, modern computer programs are more sophisticated and user-friendly and have far greater functionality, but it is precisely because these programs are so attractive that there is the risk of underestimating, or even ignoring, the fundamental and relatively simple planning process described earlier in this chapter.

CHAPTER 23

Lester Diagram

With the development of the *network grid*, the drafting of an arrow diagram enables the activities to be easily organized into disciplines or work areas and eliminates the need to enter reference numbers into the nodes. Instead the grid reference numbers (or letters) can be fed into the computer. The grid system also makes it possible to produce acceptable arrow diagrams on a computer that can be used 'in the field' without converting them into the conventional bar chart. An example of a computer-generated arrow diagram is shown in Figure 23.1. It will be noticed that the link lines never cross the nodes.

A grid system can, however, pose a problem when it becomes necessary to insert an activity between two existing ones. In practice, resourceful planners can overcome the problem by combining the new activity with one of the existing activities.

If, for example, two adjoining activities were 'Cast Column, 4 days' and 'Cast Beam, 2 days' and it were necessary to insert 'Strike Formwork, 2 days' between the two activities, the planner would simply restate the first activity as 'Cast Column and Strike Formwork, 6 days' (Figure 23.2).

While this overcomes the drafting problem, it may not be acceptable from a cost control point of view, especially if the network is geared to an EVA system (see Chapter 32). Furthermore the fact that the grid numbers were on the nodes meant that when it was necessary to move a string along one or more grid spaces, the relationship between the grid number and the activity changed. This could complicate the EVA analysis. To overcome this, the grid number was placed *between* the nodes (Figure 23.3).

It can be argued that a precedence network lends itself admirably to a grid system, because the grid number is always and permanently related to the activity and is therefore ideal for EVA. However, the problem of the congested link lines (especially the vertical ones) remains.

Now, however, the perfect solution has been found. It is in effect a combination of the arrow diagram and the precedence diagram, and like the marriage of Henry VII, which ended the Wars of the Roses, this marriage should end the war of the networks!

The new diagram, which could be called the 'Lester' diagram, is simply an arrow diagram where each activity is separated by a short link in the same way as in a precedence network (Figure 23.4).

Project Management, Planning, and Control. http://dx.doi.org/10.1016/B978-0-08-098324-0.00023-8

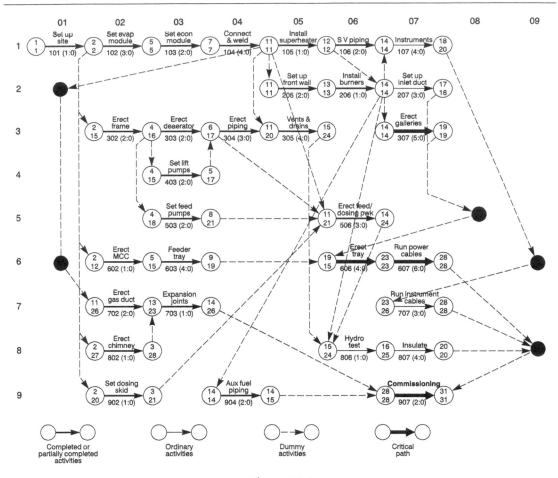

Figure 23.1
AoA network drawn on grid

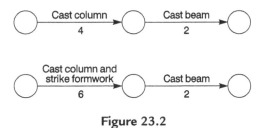

Figure 23.2

In this way it is possible to eliminate or at least reduce logic errors, show total float, and free float as easily as on a precedence network, but it has the advantages of an arrow diagram in speed of drafting, clarity of link presentation, and the ability to insert new activities in a grid system without altering the grid number/activity relationship. Figure 23.5 shows all these features.

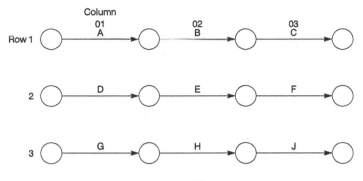

Figure 23.3

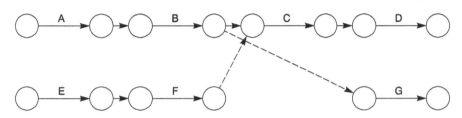

Figure 23.4
Lester diagram principle

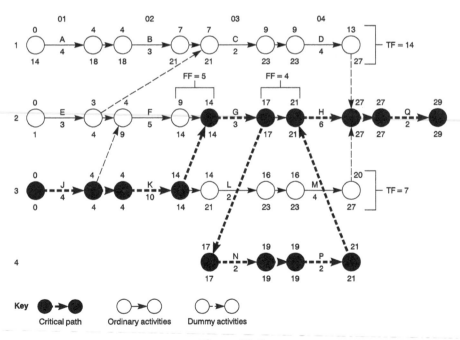

Figure 23.5
Lester diagram

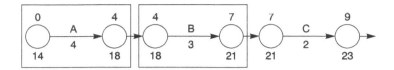

Figure 23.6

If a line or box is drawn around any activity, the similarity between the Lester diagram and the precedence diagram becomes immediately apparent. (See Figure 23.6.)

Although all the examples in subsequent chapters use arrow diagrams, precedence diagrams or 'Lester' diagrams could be substituted in most cases. The choice of technique is largely one of personal preference and familiarity. Provided the user is satisfied with one system and is able to extract the maximum benefit, there is little point in changing to another.

Summary

The advantages of a Lester diagram are:

1. Faster to draw than precedence diagram – about the same speed as an arrow diagram;
2. As in a precedence diagram:
   • total float is vertical difference
   • free float is horizontal difference;
3. Room under arrow for duration and total float value;
4. Logic lines can cross the activity arrows;
5. Requires less space on paper when drafting the network;
6. Good for examinations due to speedy drafting and elimination of node boxes;
7. Can be updated for progress by 'redding' up activity arrows as arrow diagram;
8. Uses same procedures for computer inputting as precedence networks;
9. Output from computer similar to precedence network;
10. Can be used on a grid;
11. Less chance of error when calculating backward pass due to all lines emanating from one node point instead of one of four sides of a rectangular node;
12. Shows activity as flow lines rather than points in time;
13. Looks like an arrow diagram, but is in fact more like a precedence diagram;
14. No risk of individual link lines being merged into a thick black line when printed out; and
15. No possibility of creating the type of logic error often associated with ladders.

Graphical and Computer Analysis

Graphical Analysis

It is often desirable to present the programme of a project in the form of a bar chart, and when the critical path and floats have been found by either the arithmetical or computer methods, the bar chart has to be drawn as an additional task. (Most computer programs can actually print a bar chart, but these often run to several sheets.)

As explained in Chapter 30, bar charts, while they are not as effective as networks for the actual planning function, are still one of the best methods for allocating and smoothing resources. If resource listing and subsequent smoothing is an essential requirement, graphical analysis can give the best of both worlds. Naturally, any network, however analysed, can be converted very easily into a bar chart, but if the network is analysed graphically the bar chart can be 'had for free', as it were.

Modern computer programs will of course produce bar charts (or Gantt charts) from the inputs almost automatically. Indeed the input screen itself often generates the bar chart as the data are entered. However, when a computer is not available or the planner is not conversant with the particular computer program, the graphical method becomes a useful alternative.

The following list gives some of the advantages over other methods, but before the system is used on large jobs planners are strongly advised to test it for themselves on smaller contracts so that they can appreciate the shortcut methods and thus save even more planning time.

1. The analysis is extremely rapid, much quicker than the arithmetical method. This is especially the case when, after some practice, the critical path can be found by inspection.

2. As the network is analysed, the bar chart is generated automatically and no further labour need be expended to do this at a later stage.

3. The critical path is produced *before* the floats are known. (This is in contrast to the other methods, where the floats have to be calculated first before the critical path can be seen.) The advantage of this is that users can see at once whether the project time is within the specified limits, permitting them to make adjustments to the critical activities without bothering about the non-critical ones.

4. Since the results are shown in bar chart form, they are more readily understood by persons familiar with this form of programme. The bar chart will show more vividly than a printout the periods of heavy resource loading, and highlights periods of comparative inactivity. Smoothing is therefore much more easily accomplished.

5. By marking the various trades or operational types in different colours, a rapid approximate resource requirement schedule can be built up. The resources in any one time period can be ascertained by simply adding up vertically, and any smoothing can be done by utilizing the float periods shown on the chart.

6. The method can be employed for single or multi-start projects. For multi-project work, the two or more bar charts can (provided they are drawn to the same time and calendar scale) be superimposed on transparent paper and the amount of resource overlap can be seen very quickly.

Limitations

The limitations of the graphical method are basically the size of the bar chart paper and therefore the number of activities. Most programmes are drawn on either A1 or A0 size paper and the number of different activities must be compressed into the 840 mm width of this sheet. (It may, of course, be possible to divide the network into two, but then the interlinking activities must be carefully transferred.) Normally, the divisions between bars is about 6 mm, which means that a maximum of 120 activities can be analysed. However, bearing in mind that in a normal network 30% of the activities are dummies, a network of 180 to 200 activities could be analysed graphically on one sheet.

Briefly, the mode of operation is as follows:

1. Draw the network in arrow diagram or precedence format and write in the activity titles (Figure 24.1 or 24.2). Although a forward pass has been carried out on both these diagrams, this is not necessary when using the graphical method of analysis.
2. Insert the durations.
3. List the activities on the left-hand vertical edge of a sheet of graph paper (Figure 21.11) showing:
   (a) Activity title
   (b) Duration (in days, weeks, etc.)
   (c) Node no. (only required when using these for bar chart generation)

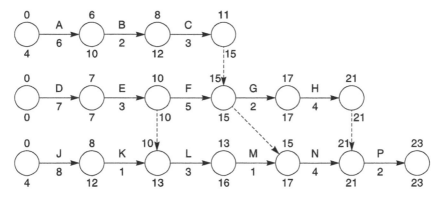

Figure 24.1
AoA Network

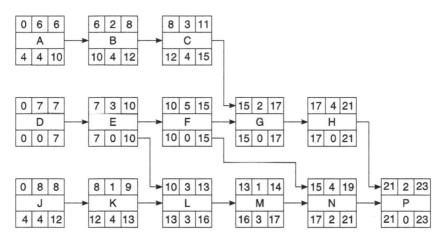

Figure 24.2
AoN Network

4. Draw a time scale along the bottom horizontal edge of the graph paper.
5. Draw a horizontal line from day 0 of the first activity, which is proportional to the duration (using the time scale selected), e.g., 6 days would mean a line six divisions long (Figure 24.3). To ease identification an activity letter or no. can be written above the bar.
6. Repeat this operation with the next activity on the table starting on day 0.
7. When using arrow (AoA) networks, mark dummy activities by writing the end time of the dummy next to the start time of the dummy, e.g., 4 → 7 would be shown as 4,7 (Figure 21.13).
8. All subsequent activities must be drawn with their start time (start day no.) directly below the end time (end day no.) of the previous activity having the same time value (day no.).
9. If more than one activity has the same end time (day no.), draw the new activity line from the activity end time (day no.) furthest to the *right*.

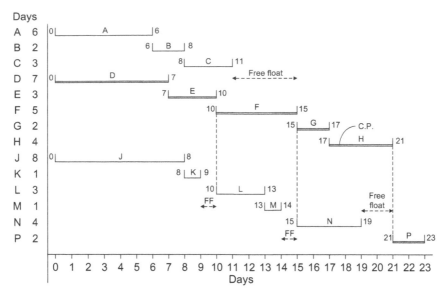

Figure 24.3
Bar chart

10. Proceed in this manner until the end of the network.
11. The critical path can now be traced back by following the line (or lines), which runs back to the start without a horizontal break.
12. The break between consecutive activities on the bar chart is the *free float* of the preceding activity.
13. The summation of the free floats in one string, before that string meets the critical path, is the *total float* of the activity from which the summation starts, e.g., in Figure 21.11 the total float of activity K is $1 + 1 + 2 = 4$ days, the total float of activity M is $1 + 2 + 3$ days, and the total float of activity N is 2 days.

The advantage of using the start and end times (day nos.) of the activities to generate the bar chart is that there is no need to carry out a forward pass. The correct relationship is given automatically by the disposition of the bars. This method is therefore equally suitable for arrow and precedence diagrams.

An alternative method can however be used by substituting the day numbers by the node numbers. Clearly this method, which is sometimes quicker to draw, can only be used with arrow diagrams, because precedence diagrams do not have node numbers. When using this method, the node numbers are listed next to the activity titles (Figure 24.5) and the bars are drawn from the starting node of the first activity with a length equal to the duration. The next bar starts vertically below the end node with the same node number as the starting node of the activity being drawn.

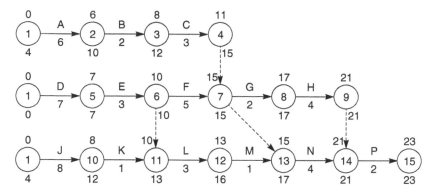

Figure 24.4

Arrow diagram

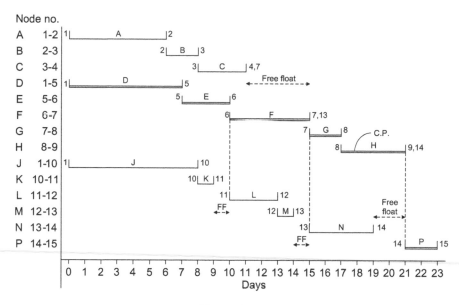

Figure 24.5

Bar chart

As with the day No. method, if more than one activity has the same end node number, the one furthest to the right must be used as a starting time. Figure 24.4 shows the same network with the node numbers inserted, and Figure 24.5 shows the bar chart generated using the node numbers.

Figure 24.6 shows a typical arrow diagram, and Figure 24.7 shows a bar chart generated using the starting and finishing node numbers. Note that these node numbers have been listed on the left-hand edge together with the durations to ease plotting.

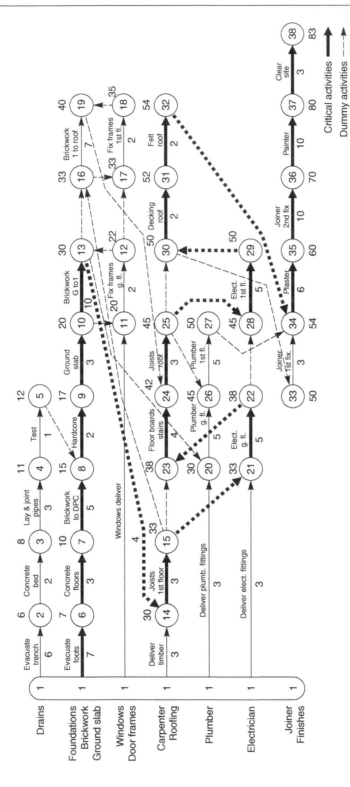

Figure 24.6
Arrow diagram of house

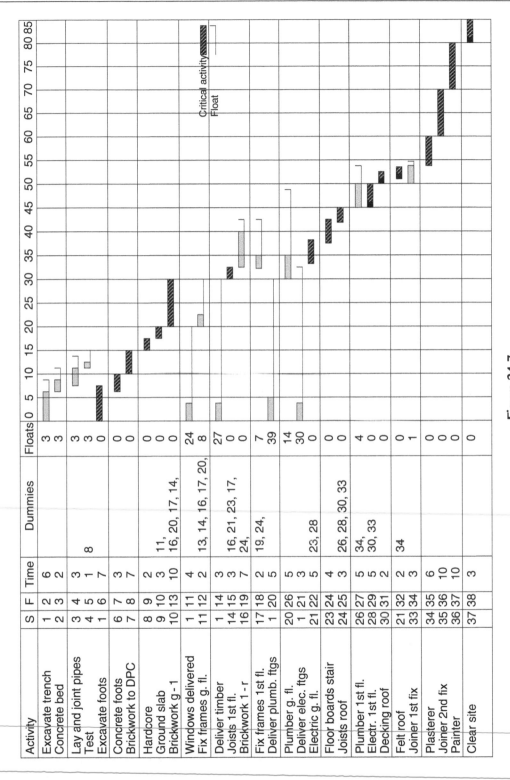

Activity	S	F	Time	Dummies	Floats	0	5	10	15	20	25	30	35	40	45	50	55	60	65	70	75	80	85
Excavate trench	1	2	6		3																		
Concrete bed	2	3	2		3																		
Lay and joint pipes	3	4	3		3																		
Test	4	5	1	8	3																		
Excavate foots	1	6	7		0																		
Concrete foots	6	7	3		0																		
Brickwork to DPC	7	8	7		0																		
Hardcore	8	9	2	11,	0																		
Ground slab	9	10	3	20, 17, 14,	0																		
Brickwork g - 1	10	13	10	16, 20, 17, 14,	0																		
Windows delivered	1	11	4	13, 14, 16, 17, 20,	24																		
Fix frames g. fl.	11	12	2	13, 14, 16, 17, 20,	8																		
Deliver timber	1	14	3	16, 21, 23, 17,	27																		
Joists 1st fl.	14	15	3	24,	0																		
Brickwork 1 - r	16	19	7	24,	0																		
Fix frames 1st fl.	17	18	2	19, 24,	7																		
Deliver plumb. ftgs	1	20	5	19, 24,	39																		
Plumber g. fl.	20	26	5	23, 28	14																		
Deliver elec. ftgs	1	21	3	23, 28	30																		
Electric g. fl.	21	22	5	23, 28	0																		
Floor boards stair	23	24	4	26, 28, 30, 33	0																		
Joists roof	24	25	3	26, 28, 30, 33	0																		
Plumber 1st fl.	26	27	5	34,	4																		
Electr. 1st fl.	28	29	5	30, 33	0																		
Decking roof	30	31	2	30, 33	0																		
Felt roof	21	32	2	34	0																		
Joiner 1st fix	33	34	3	34	1																		
Plasterer	34	35	6		0																		
Joiner 2nd fix	35	36	10		0																		
Painter	36	37	10		0																		
Clear site	37	38	3		0																		

Figure 24.7
Bar chart of house

Time for Analysis

Probably the most time-consuming operation in bar chart preparation is the listing of the activity titles, and for this there is no shortcut. The same time, in fact, must be expended typing the titles straight into the computer. However, in order to arrive at a quick answer it is only necessary at the initial stage to insert the node numbers, and once this listing has been done (together with the activity times) the analysis is very rapid. It is possible to determine the critical path for a 200-activity network (after the listing has been carried out) in less than an hour. The backward pass for ascertaining floats takes about the same time.

Computer Analysis

Most manufacturers of computer hardware, and many suppliers of computer software, have written programs for analysing critical path networks using computers. While the various commercially available programs differ in detail, they all follow a basic pattern, and give, by and large, a similar range of outputs. In certain circumstances a contractor may be obliged by his contractual commitments to provide a computerized output report for his client. Indeed, when a client organization has standardized on a particular project management system for controlling the overall project, the contractor may well be required to use the same proprietary system so that the contractor's reports can be integrated into the overall project control system on a regular basis.

History

The development of network analysis techniques more or less coincided with that of the digital computer. The early network analysis programs were, therefore, limited by the storage and processing capacity of the computer as well as the input and output facilities.

The techniques employed mainly involved producing punched cards (one card for each activity) and feeding them into the machine via a card reader. These procedures were time consuming and tedious, and, because the punching of the cards was carried out by an operator who usually understood little of the program or its purpose, mistakes occurred, which only became apparent after the printout was produced.

Even then, the error was not immediately apparent – only the effect. It then often took hours to scan through the reams of printout sheets before the actual mistake could be located and rectified. To add to the frustration of the planner, the new printout may still have given ridiculous answers because a second error was made on another card. In this way it often required several runs before a satisfactory output could be issued.

In an endeavour to eliminate punching errors, attempts were made to use two separate operators, who punched their own set of input cards. The cards were then automatically compared and, if not identical, were thrown out, indicating an error. Needless to say, such a practice cost twice as much in manpower.

Because these early computers were large and very expensive, usually requiring their own air-conditioning equipment and a team of operators and maintenance staff, few commercial companies could afford them. Computer bureaux were therefore set up by the computer manufacturers or special processing companies, to whom the input sheets were delivered for punching, processing, and printing.

The cost of processing was usually a lump sum fee plus x pence per activity. Since the computer could not differentiate between a real activity and a dummy one, planners tended to go to considerable pains to reduce the number of dummies to save cost. The result was often a logic sequence, which may have been cheap in computing cost but was very expensive in application, since frequently important restraints were overlooked or eliminated. In other words, the tail wagged the dog – a painful phenomenon in every sense. It was not surprising, therefore, that many organizations abandoned computerized network analysis or, even worse, discarded the use of network analysis altogether as being unworkable or unreliable.

There is no doubt that manual network analysis is a perfectly feasible alternative to using computers. Indeed, one of the largest petrochemical complexes in Europe was planned entirely using a series of networks, all of which were analysed manually.

The PC

The advent of the personal computer (PC) significantly changed the whole field of computer processing. In place of the punched card or tape we now have the computer keyboard and video screen, which enable the planner to input the data directly into the computer without filling in input sheets and relying on a punch operator. The information is taken straight from the network and displayed on the video screen as it is 'typed' in. In this way, the data can be checked or modified almost instantaneously.

Provided sufficient information has been entered, trial runs and checks can be carried out at any stage to test the effects and changes envisaged. Modern planning programs (or project management systems, as they are often called) enable the data to be inputted in a random manner to suit the operator, provided, of course, that the relationship between the node numbers (or activity numbers) and duration remains the same.

There are some programs that enable the network to be produced graphically on the screen as the information – especially the logic sequence – is entered. This, it is claimed, eliminates the need to draw the network manually. Whether this practice is as beneficial as suggested is very doubtful.

For a start, the number of activities that can be viewed simultaneously on a standard video screen is very limited, and the scroll facility that enables larger networks to be accommodated does not enable an overall view to be obtained at a glance. The greatest drawback of this practice, however, is the removal from the network planning process of the team spirit, which is engendered when a number of specialists sit down with the planner around a conference table to 'hammer out' the basic shape of the network. Most problems have more than one solution, and the discussions and suggestions, both in terms of network logic and durations, are invaluable when drafting the first programs. These meetings are, in effect, a brainstorming session at which the ideas of the various participants are discussed, tested, and committed to paper. Once this draft network has been produced, the planner can very quickly input it into the computer and call up a few test runs to see whether the overall completion date can, in fact, be achieved. If the result is unsatisfactory, logic and/or duration changes can be discussed with the project team before the new data are processed again by the machine. The speed of the new hardware makes it possible for the computer to be part of the planning conference, so that (provided the planner/operator is quick enough) the 'what if' scenarios can be tested while the meeting is in progress. A number of interim test runs can be carried out to establish the optimum network configuration before proceeding to the next stage. Even more important, errors and omissions can be corrected and durations of any or all activities can be altered to achieve a desired interim or final completion date.

The relatively low cost of the modern PCs has enabled organizations to install planning offices at the head office and sites as well as at satellite offices, associate companies, and offices of vital suppliers, contractors, and sub contractors. All these PCs can be linked to give simultaneous printouts as well as supply up-to-date information to the head office where the master network is being produced. In other words, the information technology (IT) revolution has made an important impact on the whole planning procedure, irrespective of the type or size of organization.

The advantages of PCs are:

1. The great reduction in the cost of the hardware, making it possible for small companies, or even individuals, to purchase their own computer system.
2. The proliferation of inexpensive, proven software of differing sophistication and complexity, enabling relatively untrained planners to operate the system.
3. The ability to allow the planner to input his or her own program or information via a keyboard and VDU.
4. The possibility to interrogate and verify the information at any stage on the video screen.
5. The speed with which information is processed and printed out either in numerical (tabular) or graphical form.

Programs

During the last few years a large number of proprietary programs have been produced and marketed. All these programs have the ability to analyse networks and produce the standard output of early and late starts and the three main types of float, i.e., total, free, and independent. Most programs can deal with either arrow diagrams or precedence diagrams, although the actual analysis is only carried out via one type of format.

The main differences between the various programs available at the time of writing are the additional facilities available and the degree of sophistication of the output. Many of the programs can be linked with 'add-on' programs to give a complete project management system covering not only planning but also cost control, material control, site organization, procurement, stock control, EVA, etc. It is impossible to describe the many intricacies of all the available systems within the confines of this chapter, nor is it the intention to compare one system with another. Such comparison can be made in terms of cost, user-friendliness, computing power, output sophistication, functionality, or range of add-ons. Should such surveys be required, it is best to consult the Internet or some of the specialist computer magazines or periodicals, which carry out such comparisons from time to time.

Most of the programs more commonly available can be found on the Internet, but to give a better insight into the versatility of a modern program, one of the more sophisticated systems is described in some detail in Chapter 51. The particular system was chosen because of its ability to be linked with the EVA system described in Chapter 32 of this book. The chosen system, Primavera P6, is capable of fully integrating critical path analysis with earned value analysis and presenting the results on one sheet of A4 paper.

At the time of publication, about 140 project management programs were listed and compared for functionality in Wikipedia on the Internet. Many of these will probably not exist anymore by the time this book is being read, while no doubt many more will have been created to take their place. It is futile therefore to even attempt to list them. The cost of these systems can vary between \$150 and \$6000, and the reader is therefore advised to investigate each 'offer' in some depth to ensure value for money. A simple inexpensive system may be adequate for a small organization running small projects or wishing to become familiar with computerized network analysis. Larger companies, whose clients may demand more sophisticated outputs and reports, may require the more expensive systems. Indeed, the choice of a particular system may well be dictated by the lead company of a consortium or the client, as described earlier.

It is recommended that the decision to produce any but the most basic printouts, as well as any printouts of reports or summaries, be delayed until the usefulness and degree of detail of a report has been studied and discussed with department managers. There is always a danger with computer outputs that recipients request more reports than they can digest, merely because they know they are available at the press of a button. Too much information or paper becomes self-defeating, since the very bulk frightens the reader to the extent of it not being read at all.

With the proliferation of the PC and the expansion of IT, especially the Internet, many of the project management techniques can now be carried out online. The use of e-mail and intranets allows information to be distributed to the many stakeholders of a project almost instantaneously. Where time is important – and it nearly always is – such a fast distribution of data or instructions can be of enormous benefit to the project manager. It does, however, require all information to be carefully checked before dissemination, precisely because so many people receive it at the same time. (See Chapter 52 on BIM.) It is an unfortunate fact that computer errors are more serious for just this reason as well as the naive belief that computers are infallible. Unfortunately, as with all computer systems, RIRO (Rubbish In, Rubbish Out) applies.

CHAPTER 25

Milestones and Line of Balance

Chapter Outline

Milestones

Important deadlines in a project programme are highlighted by specific points in time called *milestones*. These are timeless activities usually at the beginning or end of a phase or stage and are used for monitoring purposes throughout the life of the project. Needless to say, they should be SMART, which is an acronym for Specific, Measurable, Achievable, Realistic, Timebound. Often milestones are used to act as trigger points for progress payments or deadlines for receipt of vital information, permits, or equipment deliveries.

Milestone reports are a succinct way of advising top management of the status of the project and should act as a spur to the project team to meet these important deadlines. This is especially important if they relate to large tranches of progress payments.

Milestones are marked on bar charts or networks by a triangle or diamond and can be turned into a monitoring system in their own right when used in milestone *slip charts*, sometimes also known as *trend charts*.

Figure 25.1 shows such a slip chart, which was produced at reporting period 5 of a project. The top scale represents the project calendar, and the vertical scale represents the main reporting periods in terms of time. If both calendars are drawn to the same scale, a line drawn from the top left-hand corner to the bottom right-hand corner will be at 45° to the two axes.

The pre-planned milestones at the start of the project are marked on the top line with a black triangle (▼).

As the project progresses, the predicted or anticipated dates of achievement of the milestones are inserted so that the slippage (if any) can be seen graphically. This should then prompt management action to ensure that the subsequent milestones do not slip! At each reporting stage, the anticipated slippages of milestones as given by the programme are re-marked with an × while those that have not been re-programmed are marked with an

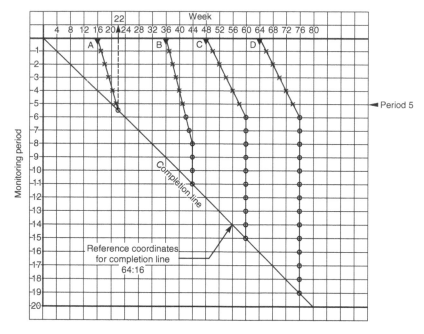

Figure 25.1
Milestone slip chart

O. Milestones that *have* been met will be on the diagonal and will be marked with a triangle (∇).

As the programmed slippage of each milestone is marked on the diagram, a pattern emerges, which acts not only as a historical record of the slippages, but can also be used to give a crude prediction of future milestone movements.

A slip chart showing the status at reporting period 11 is shown in Figure 25.2. It can be seen that milestone A was reached in week 22 instead of the original prediction of week 16. Milestones B, C, and D have all slipped, with the latest prediction for B being week 50, for C being week 62, and for D being week 76. It will be noticed that before reporting period 11, the programmed predictions are marked X, and the future predictions, after week 11, are marked O.

If a milestone is not on the critical path, it may well slip on the slip chart without affecting the next milestone. However, if two adjacent milestones on the slip chart *are* on the critical path, any delay on the first one must cause a corresponding slippage on the second. If this is then marked on the slip chart, it will in effect become a prediction, which will then alert the project manager to take action.

Once the milestone symbol meets the diagonal line, the required deadline has been achieved.

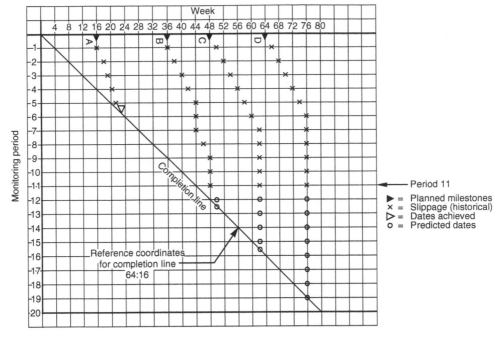

Figure 25.2
Milestone slip chart

Line of Balance

Network analysis is essentially a technique for planning one-off projects, whether this is a construction site, a manufacturing operation, a computer software development, or a move to a new premises. When the overall project consists of a number of identical or batch operations, each of which may be a subproject in its own right, it may be of advantage to use a technique called *line of balance* (LoB).

The quickest way to explain how this planning method works is to follow a simple example involving the construction of four identical, small, single-storey houses of the type shown in Chapter 47, Figure 47.1. For the sake of clarity, only the first five activities will be considered, and it will be seen from Figure 47.2 that the last of the five activities, *E –'floor joists'*, will be complete in week 9.

Assuming one has sufficient resources and space between the actual building plots, it is possible to start work on every house at the same time and therefore finish laying all the floor joists by week 9. However, in real life this is not possible, so the gang laying the foundations to house No. 1 will move to house No. 2 when foundation No. 1 is finished. When foundation No. 2 is finished, the gang will start No. 3, and so on. The same procedure will be carried out by all the following trades, until all the houses are finished.

Table 25.1

Activity letter	Activity description	Adjusted duration (weeks)	Dependency	Total float (weeks)	Buffer (weeks)
A	Clear ground	2.0	Start	0	0.0
B	Lay foundations	2.8	A	0	0.2
C	Build dwarf walls	1.9	B	0	0.1
D	Oversite concrete	0.9	B	1	0.1
E	Floor joists	1.8	C and D	0	0.2

Another practical device is to allow a time buffer between the trades so as to give a measure of flexibility and introduce a margin of error. Frequently such a buffer will occur naturally for such reasons as hardening time of concrete, setting time of adhesive, or drying time of plaster or paint.

Table 47.1 can now be partially redrawn showing the buffer time, which was originally included in the activity duration. The new table is now shown as Table 25.1.

Figure 25.3 shows the relationship between the trades involved. Each trade (or activity) is represented by two lines. The distance between these lines is the duration of the activity. The distance between the activities is the buffer period. As can be seen, all the work of activities A to E is carried out at the same rate, which means that for every house, enough resources are available for every trade to start as soon as its preceding trade is finished. This is shown to be the case in Figure 25.3.

However, if only one gang is available on the site for each trade, e.g., if only one gang of concreters laying the foundations (activity B) is available, concreting on house 2 cannot start until ground clearance (activity A) has been completed. The chart would then be as shown in Figure 25.4. If instead there were two gangs of concreters available on the site, the foundations for house 2 could be started as soon as the ground has been cleared.

Building the dwarf wall (activity C) requires only 1.9 weeks per house, which is a faster rate of work than laying foundations. To keep the bricklaying gang going smoothly from one house to the next, work can only start on house 1 in week 7.2, i.e., after the buffer of about 2.5 weeks following the completion of the foundations of house 1. In this way, by the time the dwarf walls are started on house 4, the foundations (activity B) of house 4 will just have been finished. (In practice of course there would be a further buffer to allow the concrete to harden sufficiently for the bricklaying to start.)

As the oversite concreting (activity D) only takes 0.9 weeks, the one gang of labourers doing this work will have every oversite completed well before the next house is ready for them. Their start date could be delayed if necessary by as much as 3.5 weeks, since apart from the buffer, this activity (D) has also 1 week float.

It can be seen therefore from Figure 25.4 that by plotting these operations with the time as the horizontal axis and the number of houses as the vertical axis, the following becomes apparent.

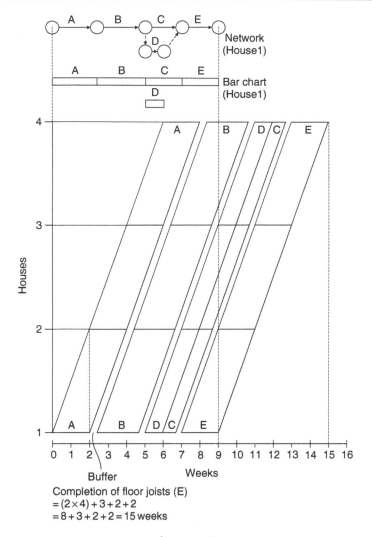

Figure 25.3
Line of balance

If the slope of an operation is less (i.e., flatter) than the slope of the preceding operation, the chosen buffer is shown at the *start* of the operation. If, on the other hand, the slope of the succeeding operation is *steeper*, the buffer must be inserted at the *end* of the previous operation, since otherwise there is a possibility of the trades clashing when they get to the last house.

What becomes very clear from these diagrams is the ability to delay the start of an operation (and use the resources somewhere else) and still meet the overall project programme.

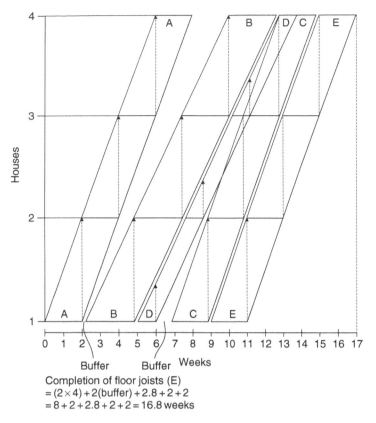

Figure 25.4
Line of balance

When the work is carried out by trade gangs, the movement of the gangs can be shown on the LoB chart by vertical arrows as indicated in Figure 25.4.

Simple Examples

Chapter Outline

To illustrate the principles set out in Chapter 20, let us now examine two simple examples.

Example 1

For the first example let us consider the rather mundane operation of getting up in the morning, and let us look at the constituent activities between the alarm going off and boarding our train to the office.

The list of activities – not necessarily in their correct sequence – is roughly as follows:

		Time (min)
A	switch off alarm clock	0.05
B	lie back and collect your thoughts	2.0
C	get out of bed	0.05
D	go to the bathroom	0.10
E	wash or shower	6.0
F	brush teeth	3.0
G	brush hair	3.0
H	shave (if you are a man)	4.0
J	boil water for tea	2.0
K	pour tea	0.10
L	make toast	3.0
M	fry eggs	4.0
N	serve breakfast	1.0
P	eat breakfast	8.0
Q	clean shoes	2.0
R	kiss wife goodbye	0.10
S	don coat	0.05

Project Management, Planning, and Control. http://dx.doi.org/10.1016/B978-0-08-098324-0.00026-3

		Time (min)
T	walk to station	8.0
U	queue and buy ticket	3.0
V	board train	1.0
		50.45

The operations listed above can be represented diagrammatically in a network. This would look something like that shown in Figure 26.1.

It will be seen that the activities are all joined in one long string, starting with A (switch off alarm) and ending with V (board train). If we give each activity a time duration, we can easily calculate the total time taken to perform the complete operation by simply adding up the individual durations. In the example given, this total time – or project duration – is 50.45 minutes. In theory, therefore, if any operation takes a fraction of a minute longer, we will miss our train. Consequently, each activity becomes critical and the whole sequence can be seen to be on the critical path.

In practice, however, we will obviously try to make up the time lost on an activity by speeding up a subsequent one. Thus, if we burn the toast and have to make a new piece, we can make up the time by running to the station instead of walking. We know that we can do this because we have a built-in margin or float in the journey to the station. This float is, of course, the difference between the time taken to walk and run to the station. In other words, the path is not as critical as it might appear, i.e., we have not in our original sequence – or network – pared each activity down to its minimum duration. We had something up our sleeve.

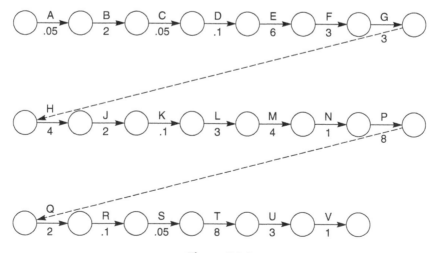

Figure 26.1

However, let us suppose that we cannot run to the station because we have a bad knee; how then can we make up lost time? This is where network analysis comes in. Let us look at the activities succeeding the making of toast (L) and see how we can make up the lost time of, say, two minutes. The remaining activities are:

		Times (min)
M	fry eggs	4.0
N	serve breakfast	1.0
P	eat breakfast	8.0
Q	clean shoes	2.0
R	kiss wife goodbye	0.10
S	don coat	0.05
T	walk to station	8.0
U	queue and buy ticket	3.0
V	board train	1.0
		27.15

The total time taken to perform these activities is 27.15 minutes.

The first question therefore is, have we any activity which is unnecessary? Yes. We need not kiss the wife goodbye. But this only saves us 0.1 minute, and the saving is of little benefit.

Besides, it could have serious repercussions. The second question must therefore be, are there any activities which we can perform simultaneously? Yes. We can clean our shoes while the eggs fry. The network shown in Figure 26.2 can thus be redrawn as demonstrated in Figure 26.3. The total now from M to V adds up to 25.15 minutes. We have, therefore, made up our lost two minutes without apparent extra effort. All we have to do is to move the shoe-cleaning box to a position in the kitchen where we can keep a sharp eye on the eggs while they fry.

Figure 26.2

Figure 26.3

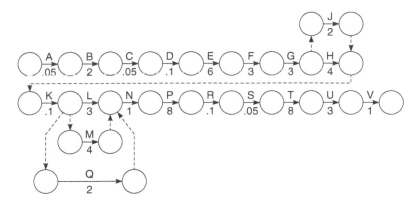

Figure 26.4

Encouraged by this success, let us now re-examine the whole operation to see how else we can save a few minutes, since a few moments extra in bed are well worth saving. Let us therefore see what other activities can be performed simultaneously:

1. We could brush our teeth under the shower;
2. We could put the kettle on before we shaved so that it boils while we shave;
3. We could make the toast while the kettle boils or while we fry the eggs;
4. We could forget about the ticket and pay the ticket collector at the other end; or
5. We could clean our shoes while the eggs fry as previously discussed.

Having considered the above list, we eliminate (1) since it is not nice to spit into the bathtub, and (4) is not possible because we have an officious guard on our barrier. So we are left with (2), (3), and (5). Let us see what our network looks like now (Figure 26.4). The total duration of the operation or programme is now 43.45 minutes, a saving of seven minutes or over 13% for no additional effort. All we did was to resequence the activities. If we moved the washbasin near the shower and adopted the 'brush your teeth while you shower' routine, we could save another three minutes, and if we bought a season ticket we would cut another three minutes off our time. It can be seen, therefore, that by a little careful planning we could well spend an extra 13 minutes in bed – all at no extra cost or effort.

If a saving of over 25% can be made on such a simple operation as getting up, it is easy to see what tremendous savings can be made when planning complex manufacturing or construction operations.

Let us now look at our latest network again. From A to G the activities are in the same sequence as on our original network. H and J (shave and boil water) are in parallel. H takes four minutes and J takes two. We therefore have two minutes float on activity J in relation to H. To get the total project duration we must, therefore, use the four minutes of H in our adding-up process, i.e., the *longest* duration of the parallel activities.

Similarly, activities L, M, and Q are being carried out in parallel and we must, therefore, use M (fry eggs) with its duration of four minutes in our calculation. Activity L will, therefore, have one minute float while activity Q has two minutes float. It can be seen, therefore, that activities H, L, and Q could all be delayed by their respective floats without affecting the overall programme. In practice, such a float is absorbed by extending the duration to match the parallel critical duration or left as a contingency for disasters. In our example it may well be prudent to increase the toast-making operation from three minutes to four by reducing the flame on the grill in order to minimize the risk of burning the bread!

Example 2

Let us now look at another example. Suppose we decide to build a new room into the loft space of our house. We decide to coordinate the work ourselves because the actual building work will be carried out by a small jobbing builder, who has little idea of planning, while the drawings will be prepared by a freelance architect who is not concerned with the meaning of time. If the start of the programme is the brief to the architect and the end is the fitting of carpets, let us draw up a list of activities we wish to monitor to ensure a speedy completion of the project. The list would be as follows:

		Days
A	brief architect	1
B	architect produces plans for planning permission	7
C	obtain planning permission	60
D	finalize drawings	10
E	obtain tenders	30
F	adjudicate bids	2
G	builder delivers materials	15
H	strip roof	2
J	construct dormer	2
K	lay floor	2
L	tile dormer walls	3
M	felt dormer roof	1
N	fit window	1
P	move CW tank	1
Q	fit doors	1
R	fit shelves and cupboards	4
S	fit internal lining and insulation	4
T	lay electric cables	2
U	cut hole in existing ceiling	1
V	fit stairs	2

		Days
W	plaster walls	2
X	paint	2
Y	fit carpets	1
		156

Rather than draw out all these activities in a single long string, let us make a preliminary analysis on which activities can be carried out in parallel. The following immediately spring to mind:

1. Final drawings can be prepared while planning permission is obtained.
2. It may even be possible to obtain tenders during the planning permission period, which is often extended.
3. The floor can be laid while the dormer is being tiled.

The preliminary network would, therefore, be as shown in Figure 26.5.

If all the activities were carried out in series, the project would take 156 days. As drawn in Figure 26.5 the duration of the project is 114 days. This shows already a considerable saving by utilizing the planning permission period for finalizing drawings and obtaining tenders.

However, we wish to reduce the overall time even further, so we call the builder in before we start work and go through the job with him. The first question we ask is how many men he will employ. He says between two and four. We then make the following suggestions:

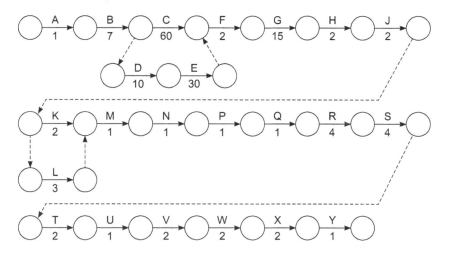

Figure 26.5

1. Let the electrician lay the cables while the joiners fit the stairs.
2. Let the plumber move the tank while the roof of the dormer is being constructed.
3. Let the glazier fit the windows while the joiner fits the shelves.
4. Let the roofer felt the dormer while the walls are being tiled.
5. Fit the doors while the cupboards are being built.

The builder may object that this requires too many men, but you can tell him that his overall time will be reduced and he will probably gain in the end. The revised network is, therefore, shown in Figure 26.6. The total project duration is now reduced to 108 days. The same network in precedence format (AoN) is shown in Figure 26.7.

If we now wish to reduce the period even further, we may have to pay the builder a little extra. However, let us assume that time is of the essence since our rich old uncle will be coming to stay and an uncomfortable night on the sofa in the sitting room might prejudice our chances in his will. It is financially viable, therefore, to ensure that the room will be complete.

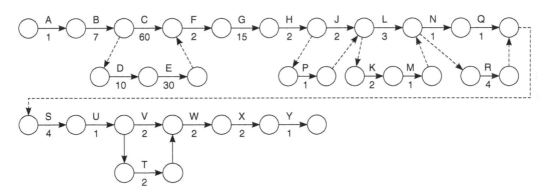

Figure 26.6

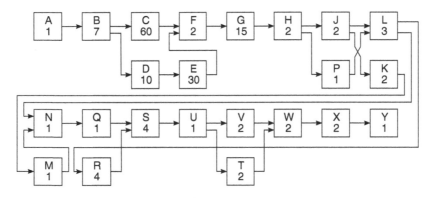

Figure 26.7
Precedence network

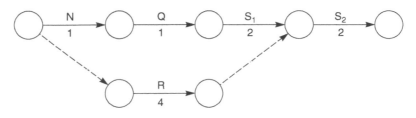

Figure 26.8

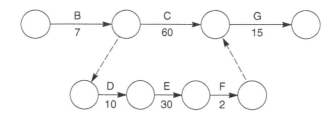

Figure 26.9

Suppose we have to cut the whole job to take no longer than 96 days. Somehow we have to save another 12 days. First, let us look at those activities that have float. N and Q together take two days while R takes four. N and Q have, therefore, two days float. We can utilize this by splitting the operation S (fit internal lining) and doing two days' work while the shelves and cupboards are being built. The network of this section would, therefore, appear as in Figure 26.8. We have saved two days provided that labour can be made available to start insulating the rafters.

If we adjudicate the bids (F) before waiting for planning permission, we can save another two days. This section of the network will, therefore, appear as in Figure 26.9.

Total saving to this stage is 2 + 2 = 4 days. We have to find another eight days, so let us look at the activities that take longest: C (obtaining planning permission) cannot be reduced since it is outside our control. It is very difficult to hurry a local authority. G (builder delivers materials) is difficult to reduce since the builder will require a reasonable mobilization period to buy materials and allocate resources. However, if we select the builder before planning permission has been received, and we do, after all, have 18 days float in loop D-E-F, we may be able to get him to place preliminary orders for the materials required first, and thus enable work to be started a little earlier. We may have to guarantee to pay the cost for this material if planning permission is not granted, but as time is of the essence we are prepared to take the risk. The saving could well be anything from one to 15 days.

Let us assume we can realistically save five days. We have now reduced the programme by 2 + 2 + 5 = 9 days. The remaining days can now only be saved by reducing the actual durations of some of the activities. This means more resources and hence more money. However, the rich uncle cannot be put off, so we offer to increase the contract sum if the builder can manage to reduce V, T, W, and X by one day each, thus saving three days altogether. It should be noted that we only save three days although we have reduced the time of four activities by one day each. This is, of course, because V and T are carried out in parallel, but our overall period – for very little extra cost – is now 96 days, a saving of 60 days, or 38%.

Example 3

This example from the IT industry uses the AoN (precedence) method of network drafting. This is now the standard method for this industry, probably because of the influence of MS Project and because networks in IT are relatively small compared to the very large networks in construction, which can have between two hundred and several thousand activities. The principles are of course identical.

A supermarket requires a new stock control system linked to a new check-out facility. This involves removing the existing check-out, designing and manufacturing new hardware, and writing new software for the existing computer, which will be retained.

The main activities and durations (all in days) for this project are as follows:

		Days
A	obtain brief from client (the supermarket owner)	1
B	discuss the brief	2
C	conceptual design	7
D	feasibility study	3
E	evaluation	2
F	authorization	1
G	system design	12
H	software development	20
J	hardware design	40
K	hardware manufacture	90
L	hardware delivery (transport)	2
M	removal of existing check-out	7
N	installation of new equipment	6
P	testing on site	4
Q	handover	1
R	trial operation	7
S	close out	1

The network for this project is shown in Figure 26.10, from which it can be seen that there are virtually no parallel activities, so only two activities, M (Removal of existing check-out)

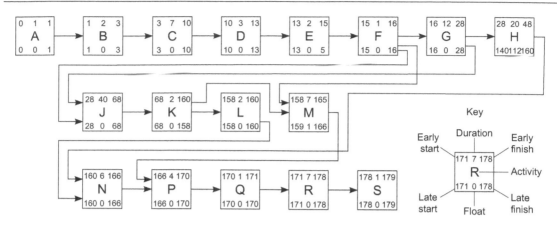

Figure 26.10
(Duration in days)

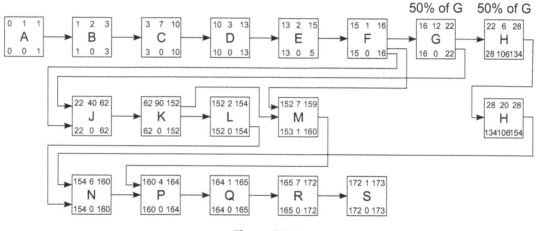

Figure 26.11
(Duration in days)

and H (Software development), have any float. However, the float of M is only one day, so for all intents and purposes it is also critical. It may be possible, however, to start J (Hardware design) earlier, after G (System design) is 50% complete. This change is shown on the network in Figure 26.11. As a result of this change, the overall project period has been reduced from 179 days to 173 days. It could be argued that the existing check-out (M) could be removed earlier, but the client quite rightly wants to make sure that the new equipment is ready for dispatch before removing the old one. As the software developed under H is only required in time for the start of the installation (N), there is still plenty of float (106 days), even after the earlier start of hardware design (J) to make sure everything is ready for the installation of the new equipment (N).

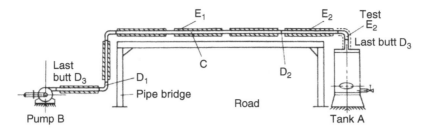

Figure 26.12
Pipe bridge

In practice, this means that the start of software development (H) could be delayed if the resources allocated to H are more urgently required by another project.

Summary of Operation

The three examples given are, of course, very small, simple programmes, but they do show the steps that have to be taken to get the best out of network analysis. These are:

1. Draw up a list of activities and anticipated durations;
2. Make as many activities as possible run in parallel;
3. Examine new sequences after the initial network has been drawn;
4. Start a string of activities as early as possible and terminate as late as possible;
5. Split activities into two or more steps if necessary;
6. If time is vital, reduce durations by paying more for extra resources; and
7. Always look for new techniques in the construction or operation being programmed.

It is really amazing what savings can be found after a few minutes' examination, especially after a good night's sleep.

Example 4 (Using Manual Techniques)

An example of how the duration of a small project can be reduced quite significantly using manual techniques is shown by following the stages of Figure 26.13.

The project involves the installation of a pump, a tank, and the interconnecting piping, which has to be insulated. Figure 26.12 shows the diagrammatic representation of the scheme, which does not include the erection of the pipe bridge over which the line has to run. All the networks in Figure 26.13 are presented in activity on arrow (AoA), activity on node (AoN), and bar chart format, which clearly show the effect of overlapping activities. Figure 26.13(a) illustrates all the five operations in sequence. This is quite a realistic procedure, but it takes the maximum amount of time – 16 days. By erecting the tank and pump at the same time (Figure 26.13(b)), the overall duration has been reduced to 14 days. Figure 26.13(c) shows a

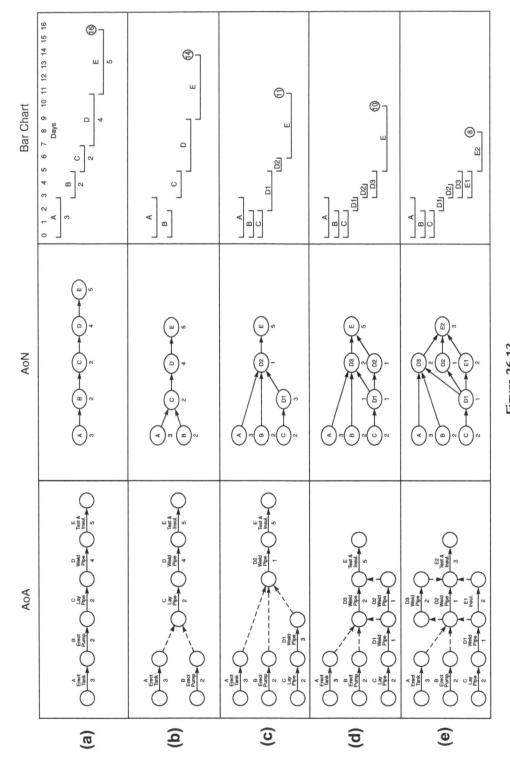

Figure 26.13
Small pipeline project

further saving of three days by erecting the pipe over the bridge while also erecting the pump and tank, giving an overall time of 11 days. When the pipe laying is divided into three sections (D_1; D_2; and D_3) it is possible to weld the last two sections at the same time, thus reducing the overall time to 10 days (Figure 26.13(d)). Further investigation shows that while the last two sections of pipe are being welded it is possible to insulate the already completed section. This reduces the overall duration to eight days (Figure 26.13(e)).

It can be argued, of course, that an experienced planner can foresee all the possibilities right from the start and produce the network and bar chart shown in Figure 26.13(e) without going through all the previous stages. However, most mortals tend to find the optimum solution to a problem by stages, using the logical thought processes as outlined above. A sketch pad and pocket calculator are all that is required to run through these steps. A computer at this stage would certainly not be necessary.

It must be pointed out that although the example shown is only a very small project, such problems occur almost daily, and valuable time can be saved by just running through a number of options before the work actually starts. In many cases the five activities will be represented by only one activity, e.g., 'Install lift pump system' on a larger construction network, and while this master network may be computerized, the small 'problem networks' are far more easily analysed manually.

CHAPTER 27

Progress Reporting

Chapter Outline
Feedback 200

Having drawn the network programme, it is now necessary to develop a simple but effective system of recording and reporting progress. The conventional method of recording progress on a bar (Gantt) chart is to thicken up or hatch in the bars, which are purposely drawn 'hollow' to allow this to be done. When drafting the network, activities are normally represented by single solid lines (Figure 27.1), but the principle of thickening up can still be applied. The simplest way is to thicken up the activity line and black in the actual node point (Figure 27.2). If an activity is only partially complete (say, 50%) this can be easily represented by only blacking in 50% of the activity (Figure 27.2). It can be seen, therefore, that in the case of the string of activities shown in Figure 27.2 the first activity is complete while the second one is half complete. By rights, therefore, the week number at that stage should be 4 + 50% of 6 = 7. However, this presupposes that the first activity has not been delayed and finished on week 4 as programmed.

How, then, can one represent the case of the first activity finishing, say, two weeks late (week 6)? The simple answer is to cross out the original week number (4) and write the revised week number next to it, as shown in Figure 27.3. If the duration of the second activity cannot be reduced, i.e., if it still requires six weeks as programmed, it will be necessary to amend all the subsequent week numbers as well (Figure 27.4).

This operation will, of course, have to be carried out every time there is a slippage, and it is prudent, therefore, to leave sufficient space over the node point to enable this to be done. Alternatively, it may be more desirable to erase the previous week numbers and insert the new ones, provided, of course, the numbers are written in pencil and not ink. At first sight, the job of erasing some 200 node numbers on a network may appear to be a tedious and time-consuming exercise. However, in practice, such an updating function poses no problems. A reasonably

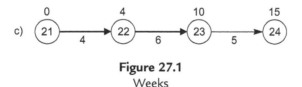

Figure 27.1
Weeks

Project Management, Planning, and Control. http://dx.doi.org/10.1016/B978-0-08-098324-0.00027-5

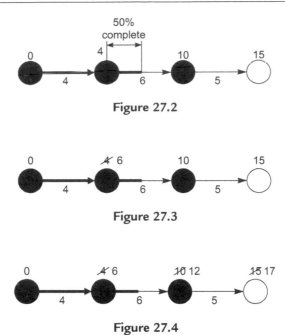

experienced planner can update a complete network consisting of about 200 activities in less than one hour. When one remembers that in most situations only a small proportion of the activities on a network require updating, the speed of the operation can be appreciated.

Naturally, only the earliest dates are calculated, since this answers the most important questions, i.e.,

1.   When can a particular activity start?
2.   When will the whole project be completed?

There is no need at this stage to calculate floats since these can be ascertained rapidly as and when required, as explained in Chapter 21.

Precedence (AoN) networks can be updated as shown in Chapter 19, Figures 19.2 and 19.3.

Feedback

Apart from reporting progress, it is also necessary to update the network to reflect logic changes and delays. This updating, which has to be on a regular basis, must reflect two main types of information:

1.   What progress, if any, has been achieved since the last update or reporting stage?
2.   What logic changes have to be incorporated to meet the technical or programme requirements?

To enable planners to incorporate this information on a revised or updated network they must be supplied with data in an organized and regular manner. Many schemes – some very complex and some very simple – have been devised to enable this to happen. Naturally, the simpler the scheme, the better, and the less paper, the more the information on the paper will be used.

The ideal situation is, therefore, one where no additional forms whatsoever are used, and this ideal can indeed be reached by using the latest IT systems.

However, unless the operatives in the field have the facilities to transmit the latest updated position direct electronically to the planning engineer's computer, a paper copy of the updated network will still have to be produced on site and sent back to the planning engineer. Provided that:

1. The networks have been drawn on small sheets, i.e., A3 or A4, or have been photographically reduced to these sizes.
2. A photocopier is available, updating the network is merely a question of thickening the completed or partially completed activities, amending any durations where necessary, and taking a photostat copy. This copy is then returned to the planner, preferably electronically. When a logic change is necessary, the amendment is made on a copy of the last network and this too is returned to the planner. If all the disciplines or departments do this, and return their feedback regularly to the planner, a master network incorporating all these changes can be produced and the effects on other disciplines calculated and studied.

There may be instances where a department manager may want to change a sequence of activities or add new items to his or her particular part of the network. Such logic changes are most easily transmitted to the planner electronically, or, in the absence of such facilities, by placing an overlay over that portion of the network that has to be changed and sketching in the new logic freehand.

Where logic changes have been proposed – for this is all a department can do in isolation at this stage – the effect on other departments only becomes apparent when a new draft network has been produced by the planner. Before accepting the situation, the planner must either inform the project manager or call a meeting of all the interested departments to discuss the implications of the proposed logic changes. In other words, the network becomes what it should always be – a focal point for discussion, a means by which the job can be seen graphically, so that it can be amended to suit any new restraints or requirements.

In many instances it will be possible for the planner to visit the various departments and update the programme by asking a few pertinent questions. This reduces the amount of paper even more and has, of course, the advantage that logic changes can be discussed and provisionally agreed right away. On a site, where the contract has been divided into a number of operational areas, this method is particularly useful since area managers are notorious for shunning paperwork – especially reports. Even very large projects can be controlled in this manner, and the personal contact again helps to generate the close relationship and involvement so necessary for good morale.

Where an efficient cost reporting system is in operation, and provided that this is geared to the network, the feedback for the programme can be combined with the weekly cost report information issued in the field or shop.

A good example of this is given in Chapter 32, which describes the EVA Cost Control System. In this system, the cost control and cost reporting procedures are based on the network so that the percentage complete of an operation can be taken from the site returns and entered straight onto the network. The application of EVA is particularly interesting, since the network can be either electronically or manually analysed while the cost report is produced by a computer, both using the same database.

One of the greatest problems found by main contractors is the submission of updated programmes from sub-suppliers or subcontractors. Despite clauses in the purchase order or subcontract documents, requiring the vendor to return a programme within a certain number of weeks of order date and update it monthly, many suppliers just do not comply. Even if programmes are submitted as requested, they vary in size and format from a reduced computer printout to a crude bar chart, which shows activities possibly useful to the supplier but quite useless to the main contractor or client.

One reason for this production of unsatisfactory information is that the main contractor (or consultant) was not specific enough in the contract documents setting out exactly what information is required and when it is needed. To overcome this difficulty, the simplest way is to give the vendor a pre-printed bar chart form as part of the contract documents, together with a list of suggested activities which *must* appear on the programme.

A pre-printed table, as drawn in Figure 27.5, shows by the letter X which activities are important for monitoring purposes, for typical items of equipment or materials. The list can be modified by the supplier if necessary, and obviously each main contractor can draw up his own requirements depending on the type of industry he is engaged in, but the basic requirements from setting out drawings to final test certificates are included. The dates by which some of the key documents are required should, of course, be given in the purchase order or contract document, since they may be linked to stage payments and/or penalties such as liquidated damages.

The advantage of the main contractor requesting the programme (in the form of a bar chart) to be produced to his own format, a copy of which is shown in Figure 27.6, is that:

1. All the returned programmes are of the same size and type and can be more easily interpreted and filed by the main contractor's staff.
2. Where the supplier is unsophisticated, the main contractor's programme is of educational value to the supplier.
3. Since the format is ready-made, the supplier's work is reduced and the programme will be returned by him earlier.

	Pumps	Heat exchanger	Air fins	Compress and turbines	Vessels towers	Valves	Struct. steel	Instr. panels	Large motors	Switchgear MCC invertors	Transformers	Fans	Pipe work
Drawings A – Setting plans	X	X	X	X	X	X	X	X	X	X	X	X	X
Drawings B – As specified	X	X	X	X	X	X	X	X	X	X	X	X	X
Drawings C – (Final)	X	X	X	X	X	X	X	X	X	X	X	X	X
Foster Wheeler Eng. cut-off	X	X	X	X	X	X	X	X	X	X	X		X
Place sub-orders	X	X	X	X	X	X	X	X	X	X	X	X	X
Receive forgings		X											
Receive plate					X								
Receive seals	X			X		X	X		X			X	
Receive couplings	X		X	X							X	X	
Receive gauges/instrum.					X			X		X	X		X
Receive tubes/fittings		X	X	X	X	X						X	X
Receive bearings	X		X	X		X			X			X	
Receive motor/actuator	X		X	X		X			X			X	
Casting of casing	X			X								X	
Casting impeller	X			X					X			X	
Casting bedplate	X			X								X	
Machine casting	X			X					X			X	
Machine impeller	X			X					X			X	
Machine flanges	X	X	X	X	X	X						X	
Machine gears				X					X			X	
Machine shaft	X		X	X		X			X			X	
Assemble rotor	X		X	X					X			X	
Assemble equipment	X	X	X	X	X	X	X	X	X	X		X	
Weld frame/supports		X	X	X			X	X	X	X		X	
Roll and weld shell		X	X	X	X	X						X	
Drill tube plate		X											
Form dished ends		X			X								
Weld/roll tubes		X	X		X								X
Weld nozzles	X	X	X	X	X	X							
Fit internals					X						X		
Access platforms		X			X		X				X		
Light presswork/guards	X		X					X		X		X	
Heat treatment				X		X		X		X		X	
Wiring								X	X	X	X		
Windings								X	X	X	X		
Lube-oil system	X		X	X				X	X			X	
Control system								X	X	X			
Galvanizing/plating	X	X	X	X	X	X	X	X	X	X		X	X
Painting/priming	X	X	X	X	X	X		X	X	X		X	X
Testing pressure/mech.	X	X	X	X	X	X		X	X	X		X	X
Testing witness/perform.	X	X	X	X	X	X		X	X	X		X	X
Prepare despatch	X	X	X	X	X	X	X	X	X	X		X	X
Data books/oper. instructions	X	X	X	X	X	X		X	X	X		X	
Weld procedures		X	X		X								
Spares schedules	X	X	X	X	X	X			X	X		X	X
Test certs	X	X	X	X	X	X	X	X	X	X		X	X

Figure 27.5

Suggested activities for a manufacturer's bar chart

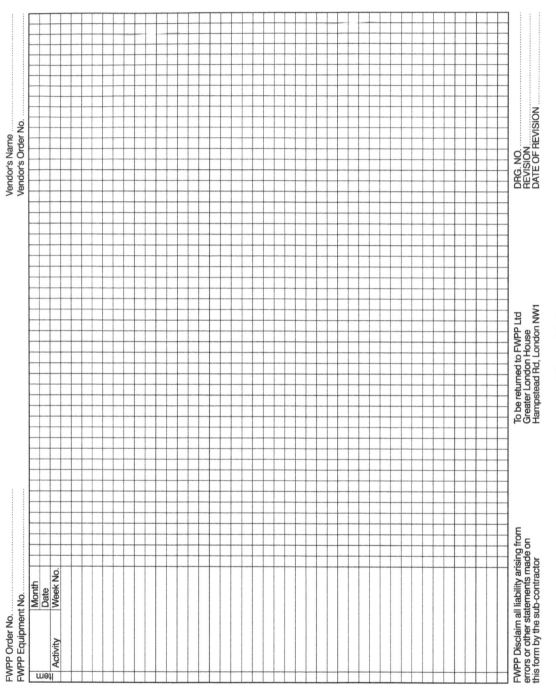

Figure 27.6
Manufacturer's bar chart

4. Since all the programmes are on A4 size paper, they can be reproduced and distributed more easily and speedily.

To ensure that the supplier has understood the principles and uses the correct method for populating the completed bar chart, an instruction sheet as shown in Figure 27.7 should be attached to the blank bar chart.

Foster Wheeler Power Products Ltd

Instructions to vendors for completing FWPP's standard programme format

1 Vendors are required to complete a Manufacturing Programme using the FWPP Standard Bar Chart form enclosed herewith.

2 The block on the top at the page given the FWPP Order Number, FWPP Equipment Number, Vendor's Name and Vendor's Order Number will be filled in by FWPP Purchasing Department at time of order issue.

3 Where a starting date is not known, Vendors must give the programme in week numbers with Week 1 as the date of the order. Subsequently, after order has been placed, the correct FWPP Week Number must be substituted together with the corresponding calendar date.

4 The left-hand column headed 'Activity' must be filled in by the Vendor showing the various stages of the manufacturing process. This should start with production of the necessary drawings requested in the Purchase Order document and continue through various stages of materials arriving at the Vendor's works, manufacuturing stages, assembly stages, testing stages and ending with actual delivery date.

5 For the benefit of vendors the attached Table shows some typical stages which FWPP Expeditors will be monitoring but it must be emphasized that these are for guidance only and must be amended or augmented by the Vendor to suit his method of production.

The Table consists of eleven (11) common items of equipment normally associated with Petrochemical Plants and where an item of equipment does not fall into one of these categories, vendors are required to build up their own detailed lists.

6 Activities with a duration of one (1) week or more should be represented by a thick line

thus: ████████████

while shorter activities or specific events such as cut-off dates or despatch dates should be shown by a triangle

despatch
thus: ▽

7 This programme must be returned to FWPP within three (3) weeks of receiving the Purchase Order.

Figure 27.7

Project Management and Network Planning

Responsibilities of the Project Managers

It is not easy to define the responsibilities of a project manager, mainly because the scope covered by such a position varies not only from industry to industry but also from one company to another. Three areas of responsibility, however, are nearly always part of the project manager's brief:

1. He must build the job to specification and to satisfy the operational (performance, quality, and safety) requirements.
2. He must complete the project on time.
3. He must build the job within previously established budgetary constraints.

The last two are, of course, connected; generally, it can be stated that if the job is on schedule, the cost has either not exceeded the budget, or additional resources have been supplied by the contractor to rectify his own mistakes, or good grounds exist for claiming any extra costs from the client. It is far more difficult to obtain extra cash if the programme has been exceeded and the client has also suffered loss due to the delay.

Time, therefore, is vitally important, and the control of time, whether at the design stage or the construction stage, should be a matter of top priority with the project manager. It is surprising, therefore, that so few project managers are fully conversant with the mechanics of

Project Management, Planning, and Control. http://dx.doi.org/10.1016/B978-0-08-098324-0.00028-7

network analysis and its advantages over other systems. Even if it had no other function but to act as a polarizing communication document, it would justify its use in preference to other methods.

Information from Network

A correctly drawn network, regularly updated, can be used to give vital information and has the following beneficial effects on the project.

1. It enables the interaction of the various activities to be shown graphically and clearly.
2. It enables spare time or float to be found where it exists so that advantage can be taken to reduce resources if necessary.
3. It can pinpoint potential bottlenecks and trouble spots.
4. It enables conflicting priorities to be resolved in the most economical manner.
5. It gives an up-to-date picture of progress.
6. It acts as a communication document between all disciplines and stakeholders.
7. It shows all interested parties the intent of the method of construction.
8. It acts as a focus for discussion at project meetings.
9. It can be expanded into subnets showing greater detail or contracted to show the chief overall milestones.
10. If updated in coloured pencil, it can act as a spur between rival gangs of workers.
11. It is very rapid and cheap to operate and is a base for EVA.
12. It is quickly modified if circumstances warrant it.
13. It can be used when formulating claims, as evidence of disruption due to late decisions or delayed drawings and equipment.
14. Networks of past jobs can be used to draft proposal networks for future jobs.
15. Networks stimulate discussion provided everyone concerned is familiar with them.
16. It can assist in formulating a cash-flow chart to minimize additional funding.

To get the maximum benefit from networks, a project manager should be able to read them as a musician reads music. He should feel the slow movements and the crescendos of activities and combine these into a harmonious flow until the grand finale is reached.

To facilitate the use of networks at discussions, the sheets should be reduced photographically to A3 (approximately 42 cm × 30 cm). In this way, a network can be folded once and kept in a standard A4 file, which tends to increase its usage. Small networks can, of course, be drawn on A3 or A4 size sheets in the first place, thus saving the cost of subsequent reduction in size.

It is often stated that networks are not easily understood by the man in the field, the area manager, or the site foreman. This argument is usually supported by statements that the field men were brought up on bar charts and can, therefore, understand them fully, or that they are confused by all the computer printouts, which take too long to digest. Both statements are true. A bar chart is easy to understand and can easily be updated by hatching or colouring in the bars. It is also true that computer output sheets can be overwhelming by their complexity. Even if the output is restricted to a discipline report, only applicable to the person in question, confusion is often caused by the mass of data. As is so often the case, network analysis and computerization are regarded as being synonymous, and the drawbacks of the latter are then invoked (often quite unwittingly) to discredit the former.

The author's experience, however, contradicts the argument that site people cannot or will not use networks. On the contrary, once the foreman or chargehand understands and appreciates what a network can do, he will prefer it to a bar chart. This is illustrated by the following example, which describes an actual situation on a contract.

Site-Preparation Contract

The job described was a civil engineering contract comprising the construction of oversite base slabs, roads, footpaths, and foul and stormwater sewers for a large municipal housing scheme consisting of approximately 250 units. The main contractor, who confined his site activities to the actual house building, was anxious to start work as soon as possible to get as much done before the winter months. It was necessary, therefore, to provide him with good roads and a fully drained site.

Contract award was June and the main contractor was programmed to start building operations at the end of November the same year. To enable this quite short civil-engineering stage to be completed on time, it was decided to split the site into four main areas that could be started at about the same time. The size and location of these areas was dictated by such considerations as access points, site clearance (including a considerable area of woodland), natural drainage, and house-building sequence.

Once this principle was established by management, the general site foreman was called in to assist in the preparation of the network, although it was known that he had never even heard of, let alone worked to, a critical path programme.

After explaining the basic principles of network techniques, the foreman was asked where he would start work, what machines he would use, which methods of excavation and construction he intended to follow, etc. As he explained his methods, the steps were recorded on the back of an old drawing print by the familiar method of lines and node points (arrow diagram). Gradually a network was evolved which grew before his eyes and his previous fears and scepticism began to melt away.

When the network of one area was complete, the foreman was asked for the anticipated duration of each activity. Each answer was religiously entered on the network without query, but when the forward pass was made, the overall period exceeded the contract period by several weeks. The foreman looked worried, but he was now involved. He asked to be allowed to review some of his durations and reassess some of the construction methods. Without being pressurized, the man, who had never used network analysis before, began the process that makes network analysis so valuable, i.e., he reviewed and refined the plan until it complied with the contractual requirements. The exercise was repeated with the three other areas, and the following day the whole operation was explained to the four chargehands who were to be responsible for those areas.

Four separate networks were then drawn, together with four corresponding bar charts. These were pinned on the wall of the site hut with the instruction that one of the programmes, either network or bar chart, be updated daily. Great stress was laid on the need to update regularly, since it is the monitoring of the programme that is so often neglected once the plan has been drawn. The decision on which of the programmes was used for recording progress was left to the foreman, and it is interesting to note that the network proved to be the format he preferred.

Since each chargehand could compare the progress in his area with that of the others, a competitive spirit developed quite spontaneously to the delight of management. The result was that the job was completed four weeks ahead of schedule without additional cost. These extra weeks in October were naturally extremely valuable to the main contractor, who could get more units weatherproof before the cold period of January to March. The network was also used to predict cash flow, which proved to be remarkably accurate. (The principles of this are explained in Chapter 31.)

It can be seen, therefore, that in this instance a manual network enabled the project manager to control both the programme (time) and the cost of the job with minimum paperwork. This was primarily because the men who actually carried out the work in the field were involved and were convinced of the usefulness of the network programme.

Confidence in Plan

It is vitally important that no one, but no one, associated with a project must lose faith in the programme or the overall plan. It is one of the prime duties of a project manager to ensure that this faith exists. Where small cracks do appear in this vital bridge of understanding between the planning department and the operational departments, the project manager must do everything in his power to close them before they become chasms of suspicion and despondency. It may be necessary to re-examine the plan, or change the planner, or hold a meeting explaining the situation to all parties, but a plan in which the participants have no faith is not worth the paper it is drawn on.

Having convinced all parties that the network is a useful control tool, the project manager must now ensure that it is kept up to date and the new information transmitted to all the interested parties as quickly as possible. This requires exerting a constant pressure on the planning department, or planning engineer, to keep to the 'issue deadlines', and equally learning on the operational departments to return the feedback documents regularly. To do this, the project manager must use a combination of education, indoctrination, charm, and rank pulling, but the feedback *must* be returned as regularly as the issue of the company's pay cheque.

The returned document might only say 'no change', but if this vital link is neglected, the network ceases to be a live document. The problem of feedback for the network is automatically solved when using the EVA cost control system (explained in Chapter 32), since the man-hour returns are directly related to activities, thus giving a very accurate percentage completion of each activity.

It would be an interesting and revealing experience to carry out a survey among project managers of large projects to obtain their unbiased opinion on the effectiveness of networks. Most of the managers with whom this problem was discussed felt that there was some merit in network techniques, but, equally, most of them complained that too much paper was being generated by the planning department.

Network and Method Statements

More and more clients and consultants require contractors to produce method statements as part of their construction documentation. Indeed, a method statement for certain complex operations may be a requirement of ISO 9000 Part I. A method statement is basically an explanation of the sequence of operations augmented by a description of the resources (i.e., cranes and other tackle) required for the job. It must be immediately apparent that a network can be of great benefit, not only in explaining the sequence of operations to the client but also for concentrating the writer's mind when the sequence is committed to paper. In the same way as the designer produces a freehand sketch of his ideas, so a construction engineer will be able to draw a freehand network to crystallize his thoughts.

The degree of detail will vary with the complexity of the operation and the requirements of the client or consultant, but it will always be a clear graphical representation of the sequences, which can replace pages of narrative. Any number of activities can be 'extracted' from the network for further explanation or in-depth discussion in the accompanying written statement.

The network, which can be produced manually or by computer, will mainly follow conventional lines and can, of course, be in arrow diagram or precedence format. For certain operations, however, such as structural steelwork erection, it may be advantageous to draw the network in the form of a table, where the operations (erect column, erect beam, plumb, level, etc.) are in horizontal rows. In this way, a highly organized, easy-to-read network can be produced. Examples of such a procedure are shown in Figures 28.1 and 28.2. There are doubtless other situations where this system can be adopted, but the prime objective must always be clarity and ease of understanding. Complex networks only confuse clients and reflect a lack of appreciation of the advantages of method statements.

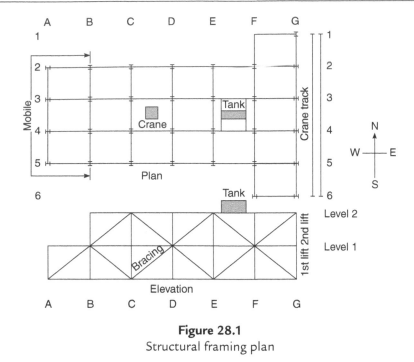

Figure 28.1
Structural framing plan

Integrated Systems

The trend is to produce and operate integrated project management systems. By using the various regular inputs generated by the different operating departments, these systems can, on demand, give the project manager an up-to-date status report of the job in terms of time, cost, and resources. This facility is particularly valuable once the project has reached the construction stage. The high cost of mainframe machines and the unreliability of regular feedback – even with the use of terminals – has held back the full utilization of computing facilities in the field, especially in remote sites. The PCs, with their low cost, mobility, and ease of operation, have changed all this so that effective project control information can be generated on the spot.

The following list shows the types of management functions that can be successfully carried out either in the office, in the workshop, or on a site by a single computer installation:

- Cost accounting
- Material control
- Plant movement
- Machine loading
- Man-hour and time sheet analysis
- Progress monitoring
- Network analysis and scheduling
- Risk analysis
- Technical design calculations, etc.

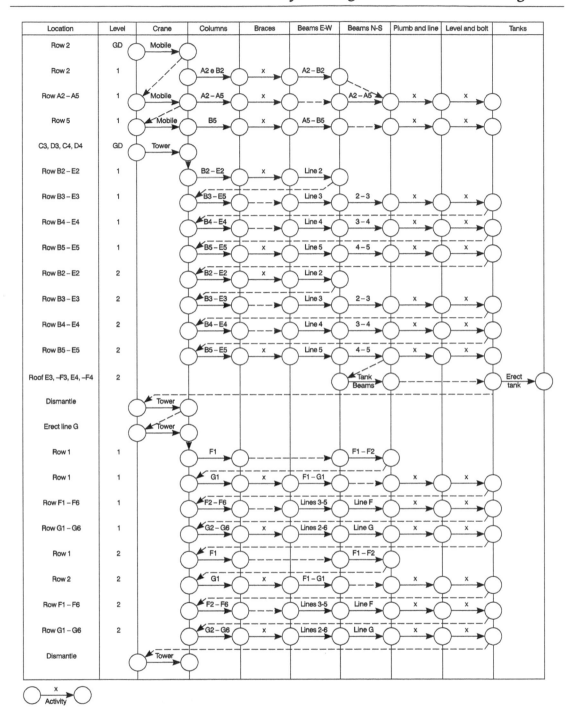

Figure 28.2
Network of method statement

Additional equipment is available to provide presentation in graphic form such as bar charts, histograms, S-curves, and other plots. If required, these can be in a number of colours to aid in identification.

The basis of all these systems is, however, still a good planning method based on well-defined and realistic networks and budgets. If this base is deficient, all comparisons and controls will be fallacious. The procedures described in Chapters 20 to 21 therefore still apply. In fact, the more sophisticated the analysis and data processing, the more accurate and meaningful the base information has to be. This is because the errors tend to be multiplied by further manipulation, and the wider dissemination of the output will, if incorrect, give more people the wrong data on which to base management decisions.

Networks and Claims

From the contractor's point of view, one of the most useful (and lucrative) applications of network presentation arises when it is necessary to formulate claims for extension of time, disruption to anticipated sequences, or delays of equipment deliveries. There is no more convincing system than a network to show a professional consultant how his late supply of design information has adversely affected progress on-site, or how a late delivery has disrupted the previously anticipated and clearly stated method of construction.

It is, of course, self-evident that to make the fullest use of the network for claim purposes, the method of construction *must* have been previously stated, preferably also in network form. The wise contractor will include a network showing the anticipated sequences with his tender, and indicate clearly the deadlines by which drawings, details, and equipment are required.

In most cases the network will be accepted as a fair representation of the construction programme, but it is possible that the client or consultant will try to indemnify himself by such statements as that he (the consultant) does not necessarily accept the network as the only logical sequence of operations, etc. Therefore it is up to the contractor to use his skills and experience to construct the works in light of circumstances prevailing at the time.

Such vague attempts to forestall genuine claims for disruption carry little weight in a serious discussion among reasonable people, and count even less should the claim be taken to arbitration or adjudication. The contractor is entitled to receive his access, drawings, and free issue equipment in accordance with his stated method of construction, as set out in his tender, and all the excuses or disclaimers by the client or consultant cannot alter this right. Those contractors who have appreciated this facility have undoubtedly profited handsomely by making full use of network techniques, but these must, of course, be prepared accurately.

To obtain the maximum benefit from the network, the contractor must show that:

1. The programme was reasonable and technically feasible;
2. It represented the most economical construction method;
3. Any delays in the client's drawings or materials will either lengthen the overall programme or increase costs, or both;
4. Any acceleration carried out by him to reduce the delay caused by others resulted in increased costs;
5. Any absorption of float caused by the delay increased the risk of completion on time and had to be countered by acceleration in other areas or by additional costs.

The last point is an important one, since 'float' belongs to the contractor. It is the contractor who builds it into his programme. It is the contractor who assesses the risks and decides which activities require priority action. The mere fact that a delayed component only reduced the float of an activity, without affecting the overall programme, is not a reason for withholding compensation if the contractor can show increased costs were incurred.

Examples of Claims for Delays

The following examples show how a contractor could incur (and probably reclaim) costs by late delivery of drawings or materials by the employer.

Example 1

To excavate a foundation the network in Figure 28.3 was prepared by the contractor. The critical path obviously runs through the excavation, giving the path through the reinforcing steel supply and fabrication a float of four days. If the drawings are delayed by four days, both paths become critical and, in theory, no delays occur. However, in practice, the contractor may now find that the delay in the order for reinforcing steel has lost him his place in the queue of the steel supplier, since he had previously advised the supplier that information would be available by day 10. Now that the information was only given to the supplier on

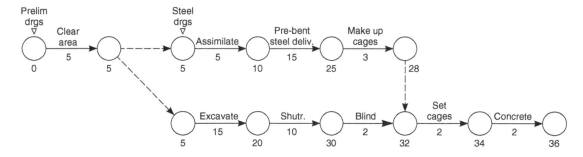

Figure 28.3

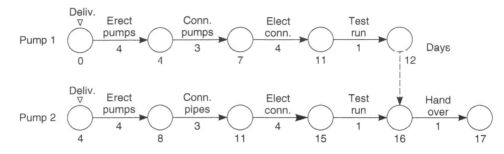

Figure 28.4

day 14, labour for the cages was diverted to another contract and, to meet the new delivery of day 29, overtime will have to be worked. These overtime costs are claimable.

In any case, the four-day float, which the contractor built in as an insurance period, has now disappeared, so even if the steel had arrived by day 29 and the cage fabrication took longer than three days, a claim would have been justified.

Example 2

The network in Figure 28.4 shows a sequence for erecting and connecting a set of pumps. The first pump was promised to be delivered by the client on a 'free issue' basis in week 0. The second pump was scheduled for delivery in week 4. In the event, both pumps were delivered together in week 4. The client argued that since there was a float of 4 days on pump 1, there was no delay to the programme since handover could still be effected by week 16.

What the programme does *not* show, and what it *need not show*, are the resource restraints imposed by the contractor to give him economical working. A network submitted as a contractual document need only show the logic from an *operational* point of view. Resource restraints are *not* logic restraints since they can be overcome by merely supplying additional resources.

The contractor rightly pointed out that he always intended to utilize the float on the first set of pumps to transfer the pipe fitters and electricians to the second pump as soon as the first pump was piped up and electrically connected. The *implied* network, utilizing the float economically, was therefore as shown in Figure 28.5.

Now, to meet the programme, the contractor has to employ two teams of pipefitters and electricians, which may have to be obtained at additional cost from another site and certainly requiring additional supervision if the two pumps are geographically far apart. Needless to say, if the contractor shows the *resource* restraints in his contract network, his case for a reimbursement of costs will be that much easier to prove.

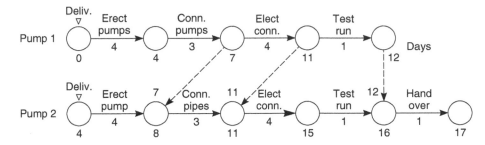

Figure 28.5

Force Majeure Claims

The causes giving rise to force majeure claims are usually specified in the contract, and there is generally no difficulty in claiming extension of time for the period of a strike or (where permitted) the duration of extraordinary bad weather. What is more difficult to prove is the loss of time caused by the *effect* of a force majeure situation. It is here where a network can help the contractor to state his case.

Example 3

A boiler manufacturer has received two orders from different clients and has programmed the two contracts through his shops in such a way that as one boiler leaves the assembly area, the parts of the second boiler can be placed into position ready for assembly. The simplified network is shown in Figure 28.6. Because the factory had only one assembly bay, Boiler No. 2 assembly had to await completion of Boiler No. 1, and the delivery promises of Boiler No. 2 reflected this.

Unfortunately the plate for the drum of Boiler No. 1, which was ordered from abroad, was delayed by a national dock strike that lasted 15 days. The result was that both boilers were delayed by this period, although the plate for Boiler No. 2 arrived as programmed.

The client of Boiler No. 2 could not understand why his boiler should be delayed because of the late delivery of a plate for another boiler, but when shown the network, he appreciated the position and granted an extension. Had the assembly of Boiler No. 2 started first, Boiler No. 1 would have been delayed 70 days instead of only 15, while Boiler No. 2 would have incurred storage costs for 60 days. Clearly, such a situation was seen to be unrealistic by all parties. The revised network is shown in Figure 28.7.

Example 4

The contract for large storage tanks covered supply erection and painting. Bad weather was a permissible force majeure claim. During the erection stage, high winds slowed down the work because the cranes would not handle the large plates safely. The winds delayed the erection

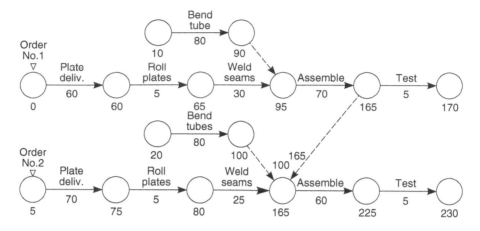

Figure 28.6

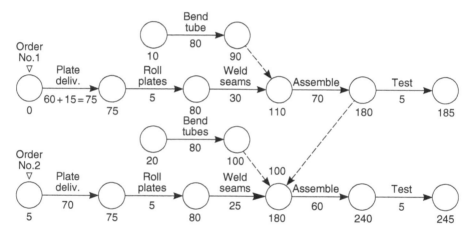

Figure 28.7

by four weeks, but by the time the painting stage started, the November mists set in and the inspector could not allow painting to start on the damp plate. The contractor submitted a network with the contract to show that the painting would be finished *before* November. Because of the high winds, the final coat of paint was, in fact, delayed until March, when the weather permitted painting to proceed.

Figure 28.8 illustrates the network submitted which, fortunately, clearly showed the non-painting month, so that the client was aware of the position before contract award. The same point could obviously have been made on a bar chart, but the network showed that no acceleration was possible after the winds delayed the erection of the side plates. To assist in relating the week numbers to actual dates, a week number/calendar date table should be provided on the network.

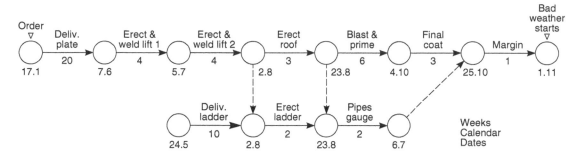

Figure 28.8

The above examples may appear to be rather negative, i.e., it looks as if network analysis is advocated purely as a device with which the contractor can extract the maximum compensation from the client or his advisers. No doubt, in a dispute both sides will attempt to field whatever weapons are at their disposal, but a more positive interpretation is surely that network techniques put *all* parties on their mettle. Everyone can see graphically the effects of delays on other members of the construction team and the cost or time implications that can develop. The result is, therefore, that all parties will make sure that they will not be responsible for the delay, so that in the end everyone – client, consultant, and contractor – will benefit: the client, because he gets his job on time; the consultant, because his reputation is enhanced; and the contractor, because he can make a fair profit.

Fortunately the trend is for claims to be reduced due to the introduction of partnering. In these types of contracts, which are usually a mixture of firm price and reimbursable costs, an open book policy by the contractor allows the employer to see how and where his money is being expended, so that there are no hidden surprises at the end of the contract ending up as a claim.

Frequently any cost savings are shared by a predetermined ratio so that all parties are encouraged to minimize delays and disruptions as much as possible. In such types of contracts network analysis can play an important part, in that, provided the network is kept up to date and reflects the true and latest position of the contract, all parties can jointly see graphically where the problem lies and can together hammer out the most economical solution.

CHAPTER 29

Network Applications Outside the Construction Industry

Most of the examples of network analysis in this book are taken from the construction industry, mainly because network techniques are particularly suitable for planning and progressing the type of operations found in either the design office or on a site. However, many operations outside the construction industry that comprise a series of sequential and/or parallel activities can benefit from network analysis – indeed, the Polaris project is an example of such an application.

The following examples are included, therefore, to show how other industries can make use of network analysis, but as can be seen from Chapter 26, even the humble task of getting up in the morning can be networked. When network analysis first became known, one men's magazine even published a critical path network of a seduction!

Bringing a New Product onto the Market

This operations involved in launching a new product and required careful planning and coordination. The example shows how network techniques were used to plan the development, manufacture, and marketing of a new type of water meter for use in countries where these are installed on all premises.

The list of operations are first grouped into five main functions:

A   Management
B   Design and development
C   Production
D   Purchasing and supply
E   Sales and marketing

Project Management, Planning, and Control. http://dx.doi.org/10.1016/B978-0-08-098324-0.00029-9
221

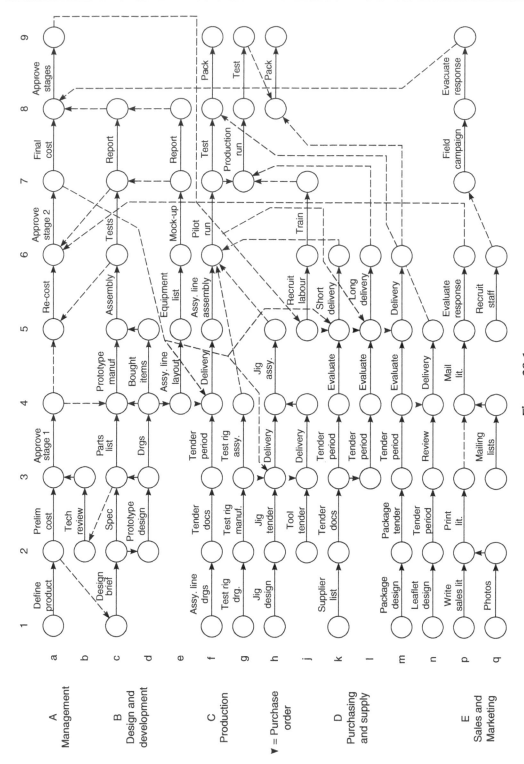

Figure 29.1
New product

Each main function is then divided into activities that have to be carried out in defined sequences and by specific times. The management function would therefore include the following of product:

A–1	Definition of product	– size, range, finish, production rate, etc.
2	Costing	– selling price, manufacturing costs.
3	Approvals for expenditure	– plant materials, tools and jigs, storage advertising, training, etc.
4	Periodic reviews	
5	Instruction to proceed with stages	

The design and development function would consist of:

B–1	Product design brief
2	Specification and parts list
3	Prototype drawings
4	Prototype manufacture
5	Testing and reports
6	Preliminary costing

Once the decision has been made to proceed with the meter, the production department will carry out the following activities:

C–1	Production planning
2	Jig tool manufacture
3	Plant and machinery requisition
4	Production schedules
5	Materials requisitions
6	Assembly-line installation
7	Automatic testing
8	Packing bay
9	Inspection procedures
10	Labour recruitment and training
11	Spares schedules

The purchasing and supply function involves the procurement of all the necessary raw materials and bought-out items and includes the following activities:

D–1	Material enquiries
2	Bought-out items enquiries
3	Tender documents

4   Evaluation of bids
5   Long delivery orders
6   Short delivery orders
7   Carton and packaging
8   Instruction leaflets, etc.
9   Outside inspection

The sales and marketing function will obviously interlink with the management function and consists of the following activities:

E–1   Sales advice and feedback

2   Sales literature   – photographs, copying, printing, films,
                                                              displays, packaging.

3   Recruitment of sales staff
4   Sales campaign and public relations
5   Technical literature   – scope and production.
6   Market research

Obviously, the above breakdowns are only indicative and the network shown in Figure 29.1 gives only the main items to be programmed. The actual programme for such a product would be far more detailed and would probably contain about 120 activities.

The final presentation for those who prefer it could then be in bar chart form covering a time span of approximately 18 months from conception to main production run.

Moving a Factory

One of the main considerations in moving the equipment and machinery of a manufacturing unit from one site to another is to carry out the operation with the minimum loss of production. Obviously, at some stage manufacturing must be halted unless certain key equipment is duplicated, but if the final move is carried out during the annual works' holiday period the loss of output is minimized.

Consideration must therefore be given to the following points:

1. Identifying equipment or machines which can be temporarily dispensed with;
2. Identifying essential equipment and machines;
3. Dismantling problems of each machine;
4. Re-erection on the new site;
5. Service connections;
6. Transport problems – weight, size, fragility, route restrictions;

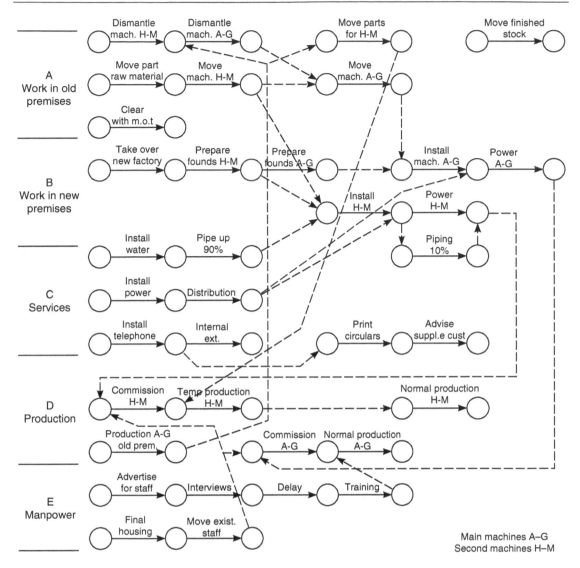

Figure 29.2
Moving a factory

7. Orders in pipeline;
8. Movement of stocks;
9. Holiday periods;
10. Readiness of new premises;
11. Manpower availability;
12. Overall cost;
13. Announcement of move to customers and suppliers;

14. Communication equipment (telephone, e-mail, fax);
15. Staff accommodation during and after the move;
16. Trial runs;
17. Recruitment and staff training.

By collecting these activities into main functions, a network can be produced that will facilitate the organization and integration of the main requirements. The main functions would therefore be:

A   Existing premises and transport;
B   New premises – commissioning;
C   Services and communications;
D   Production and sales;
E   Manpower, staffing.

The network for the complete operation is shown in Figure 29.2. It will be noticed that, as with the previous example, horizontal banding (as described in Chapter 19) is of considerable help in keeping the network disciplined.

By transferring the network onto a bar chart, it will be possible to arrange for certain activities to be carried out at weekends or holidays. This may require a rearrangement of the logic, which, though not giving the most economical answer in a physical sense, is still the best overall financial solution when production and marketing considerations are taken into account.

Centrifugal Pump Manufacture

The following network shows the stages required for manufacturing centrifugal pumps for the process industry. The company providing these pumps has no foundry, so the un-machined castings have to be bought in.

Assuming that the drawings for the pump are complete and the assembly line set up, a large order for a certain range of pumps requires the following main operations:

1. Order castings – bodies, impellers, etc.;
2. Order raw materials for shafts, seal plates, etc.;
3. Order seals, bearings, keys, bolts;
4. Machine castings, impellers, shafts;
5. Assemble;
6. Test;
7. Paint and stamp;
8. Crate and dispatch;
9. Issue installation and maintenance instructions and spares list.

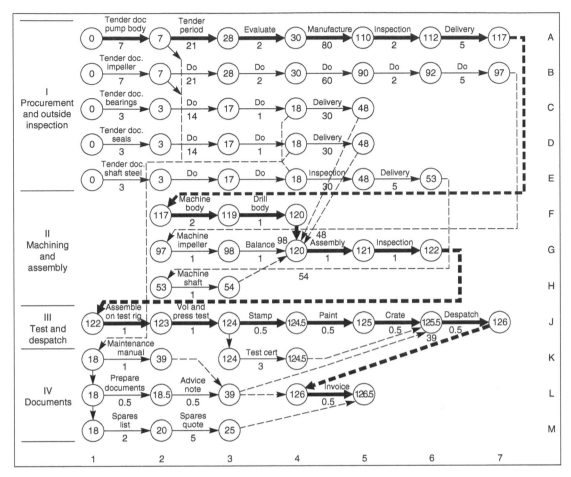

Figure 29.3
Pump manufacture (duration in days)

Figure 29.3 shows the network of the various operations complete with coordinate node numbers, durations, and earliest start times. The critical path is shown by a thickened line and total float can be seen by inspection. For example, the float of all the activities on line C is $120 - 48 = 72$ days. Similarly, the float of all activities on line D is $120 - 48 = 72$ days.

Figure 29.4 is the network redrawn in bar-chart form, on which the floats have been indicated by dotted lines. It is apparent that the preparation of documents such as maintenance manuals, spares lists, and quotes can be delayed without ill effect for a considerable time, thus releasing these technical resources for more urgent work such as tendering for new enquiries.

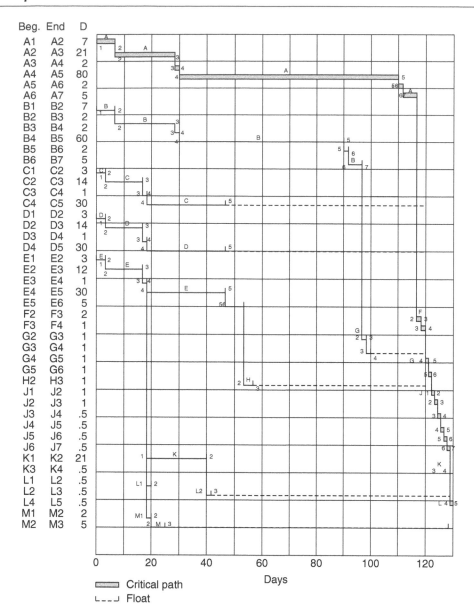

Figure 29.4
Pump manufacture – critical path analysis

Planning a Mail Order Campaign

When a mail order house decides to promote a specific product, a properly coordinated sequence of steps has to be followed to ensure that the campaign will have the maximum impact and success. The following example shows the activities required for promoting a new set of record albums and involves both the test campaign and the main sales drive.

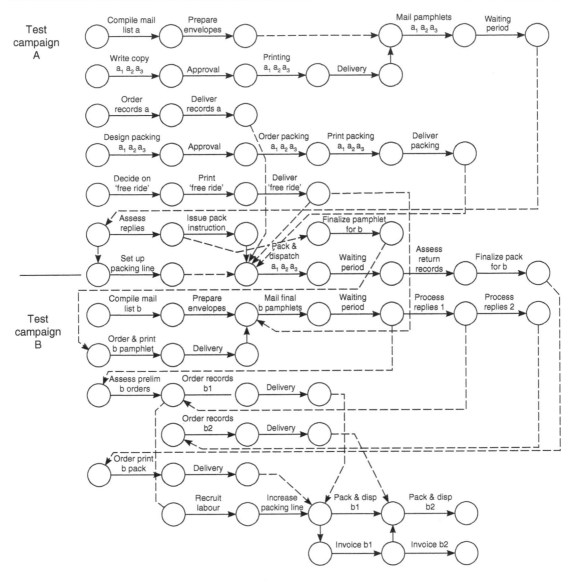

Figure 29.5
Mail order campaign

The two stages are shown separately on the network (Figure 29.5) since they obviously occur at different times, but in practice intermediate results could affect management decisions on packaging and text on the advertising leaflet. At the end of the test shot management will have to decide on the percentage of records to be ordered to meet the initial demand.

In practice, the test shot will consist of three or more types of advertising leaflet and record packaging, and the result of each type will have to be assessed before the final main campaign leaflets are printed.

Depending on the rate of return of orders, two or more record-ordering and dispatch stages will have to be allowed for. These are shown on the network as B1 and B2.

Manufacture of a Package Boiler

The programme in this example covers the fabrication and assembly of a large package boiler of about 75,000 kg of superheated steam per hour at 30 bar g and a temperature of 300 °C. The separate economizer is not included.

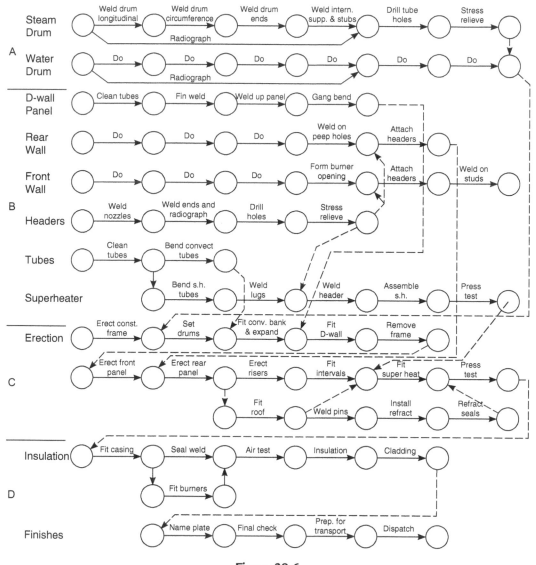

Figure 29.6
Boiler manufacture

The drum shells, drum ends, tubes, headers, doors, and nozzles are bought out, leaving the following manufacturing operations:

1. Weld drums (longitudinal and circumferential seams);
2. Weld on drum ends;
3. Weld on nozzles and internal supports;
4. Drill drum for tubes;
5. Stress relieve top and bottom drums;
6. Bend convection bank tubes;
7. Fit and expand tubes in drums – set up erection frame;
8. Weld fins to furnace tubes; pressure test;
9. Produce waterwall panels;
10. Gang bend panels;
11. Erect wall panels;
12. Weld and drill headers; stress relieve;
13. Weld panels to headers;
14. Weld on casing plates;
15. Attach peepholes, access doors, etc.;
16. Pressure test;
17. Seal-weld furnace walls;
18. Fit burners and seals;
19. Air test – inspection;
20. Insulate;
21. Prepare for transport;
22. Dispatch.

There are four main bands in the manufacturing programme:

A Drum manufacture;
B Panel and tube manufacture;
C Assembly;
D Insulation and preparation for dispatch.

The programme assumes that all materials have been ordered and will be available at the right time. Furthermore, in practice, sub-programmes would be necessary for panel fabrication, which includes blast cleaning the tubes and fin bar, automatic welding, interstage inspection, radiography, and stress relieving. Figure 29.6 shows the main production stages covering a period of approximately seven months.

Manufacture of a Cast Machined Part

The casting, machining, and finishing of a steel product can be represented in network form as shown in Figure 29.7. It can be seen that the total duration of the originally planned operation is 38 hours, but the aim was to reduce this manufacturing period to make this product more competitive. By incorporating the principle, that efficiency can be increased if some of the operations on a component can be performed while it is on the move between workstations, it is obviously possible to reduce the overall manufacturing time. The obvious activities that can be carried out while the component is actually being transported (usually on a conveyor system) is cooling off, painting, and paint drying. As can be seen from Figure 29.8 such a change in the manufacturing procedure saves three hours.

Any further time savings now require a reduction in duration of some of the individual activities. The first choice must obviously be those with the longest durations, i.e.,

1. Make pattern (8 hours);
2. Cool off (6 hours);
3. Dry paint (8 hours).

These operations require new engineering solutions. For example, in (1), the pattern may have to be split, with each component being made by a separate pattern maker. It may also be possible to subcontract the pattern to a firm with more resources. Activity (2) can be reduced in time by using forced-draught air to cool the casting before fettling. Care must, of course, be taken not to cool it at such a rate that it causes cracking or other metallurgical changes. Conversely to (2), the paint drying in (3) can be speeded up by blowing warm air over the finished component. If the geographical layout permits it, it may be possible to take the heated air from the cooling process, pass it through a filter, and use it to dry the paint!

Further time reductions are possible by reducing the machining time of the milling and drilling operations. This may mean investing in cutters or drills that can withstand higher cutting speeds. It may also be possible to increase the speed of the different conveyors, which, even on the revised network, make up one hour of the cycle time.

For those planners who are familiar with manufacturing flow charts it may be an advantage to draw the network in precedence (AoN) format (see Chapter 19). Such a representation of the initial and revised networks is shown in Figure 29.9 and 29.10, respectively.

It is important to remember that the network itself does not reduce the overall durations. Its first function is to show in a graphic way the logical interrelationship of the production processes and the conveying requirement between the manufacturing stages. It is

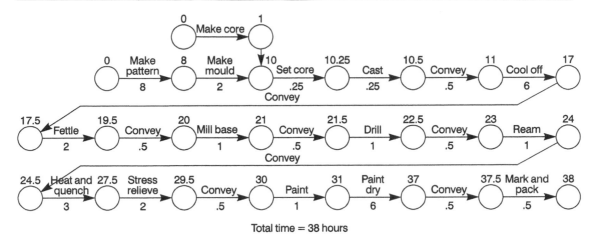

Figure 29.7
(Original)

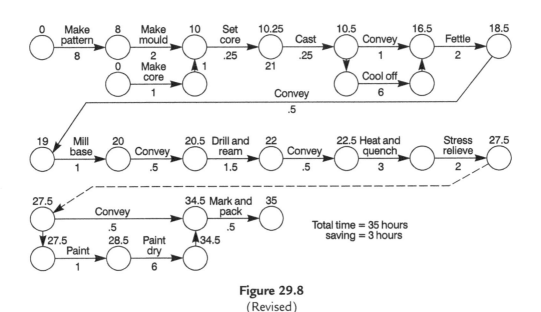

Figure 29.8
(Revised)

then up to the production engineer or controller to examine the network to see where savings can be made. This is, in fact, the second function of the network – to act as a catalyst for the thought processes of the user to give him the inspiration to test a whole series of alternatives until the most economical or fastest production sequence has been achieved.

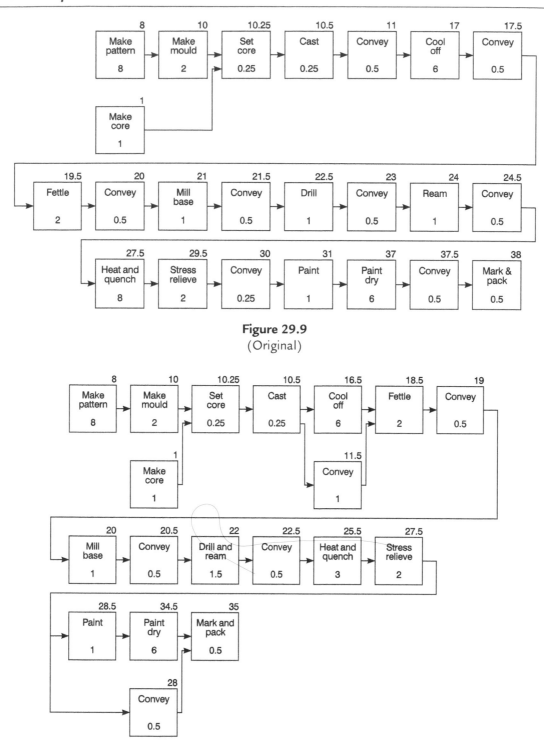

Figure 29.9
(Original)

Figure 29.10
(Revised)

The use of a PC at this stage will, of course, enable the various trial runs to be carried out quite rapidly, but, as can be seen, even a manual series of tests takes no longer than a few minutes. As explained in Chapter 21, the first operation is to calculate the shortest forward pass – a relatively simple operation – leaving the more complex calculations of float to the computer when the final selection has been made.

Resource Loading

Chapter Outline

Most modern computer programs incorporate facilities for resource loading, resource allocation, and resource smoothing. Indeed, the Primavera P6 program shown in Chapter 51 features such a capability.

In principle, the computer aggregates a particular resource in any time period and compares this with a previously entered availability level for that resource. If the availability is less than the required level, the program will either:

1. Show the excess requirement in tabular form, often in a different colour to highlight the problem; or
2. Increase the duration of the activity requiring the resource to spread the available resources over a longer period, thus eliminating the unattainable peak loading.

The more preferable action by the computer is option (1), i.e., the simple report showing the overrun of resources. It is then up to management to make the necessary adjustments by either extending the time period – if the contractual commitments permit – or mobilizing additional resources. In practice, of course, the problem is complicated by such issues as available access or working space as well as financial, contractual, or even political restraints. Often it may be possible to make technical changes that alter the resource mix. For example, a shortage of carpenters used for formwork erection may make it necessary to increase the use of pre-cast concrete components with a possible increase in cost but a decrease in time. Project management is more than just writing and monitoring programs. The so-called project management systems marketed by software companies are really only there to present to the project manager on a regular basis the position of the project to date and the possible consequences unless some form of remedial action is taken. The type of action and the timing of it rests fairly and squarely on the shoulders of management.

The options by management are usually quite wide, provided sufficient time is taken to think them out. It is in such situations that the 'what if' scenarios are a useful facility on a computer. However, the real implication can only be seen by 'plugging' the various alternatives into the

Project Management, Planning, and Control. http://dx.doi.org/10.1016/B978-0-08-098324-0.00030-5

network on paper and examining the down-stream effects in company with the various specialists, who, after all, have to do the actual work. There is no effective substitute for good teamwork!

The Alternative Approach

Resource smoothing can, of course, be done very effectively without a computer – especially if the program is not very large. Once a network has been prepared it is very easy to convert it into a bar chart, since all the 'thinking' has already been completed. Using the earliest starting and finishing times, the bars can be added to the gridded paper in minutes. Indeed, the longest operation in drawing a bar chart (once a network has been completed) is writing down the activity descriptions on the left-hand side of the paper. By leaving sufficient vertical space between the bars and dividing the grid into week (or day) columns, the resource levels for each activity can be added. Generally, there is no need to examine more than two types of resources per chart, since only the potentially restrictive or quantitatively limited ones are of concern. When all the activity bars have been marked with the resource value, each time period is added up vertically and the total entered in the appropriate space. The next step is to draw a histogram to show the graphical distribution of the resources. This will immediately highlight the peaks and troughs and trigger off the next step – resource smoothing.

Manual resource smoothing is probably the most practical method, since unprogrammable factors such as access, working space, hard-standing for cranes, personality traits of foremen, etc., can only be considered by a human when the smoothing is carried out. Nevertheless, the smoothing operation must still follow the logical pattern given below:

1. Advantage should be taken of float. In theory, activities with free float should be the first to be extended, so that a limited resource can be spread over a longer time period. In practice, however, such opportunities are comparatively rare, and for all normal operations, all activities with total float can be used for the purpose of smoothing. The floats can be indicated on the bars by dotted line extensions, again read straight off the network by subtracting the earliest from the latest times of the beginning node of the activity.
2. When the floats have been absorbed and the resources are distributed over the longer activity durations, another vertical addition is carried out from which a new histogram can be drawn. A typical network, bar chart, and histogram is shown in Figure 30.1.
3. If the peaks still exceed the available resources for any time period, logic changes will be required. These changes are usually carried out on the network, but it may be possible to make some of them by 'sliding' the bars on the bar chart. For example, a common problem when commissioning a process or steam-raising plant is a shortage of suitably qualified commissioning engineers. If the bars of the bar chart are cut out and pasted onto cardboard with the resources written against each time period on the activity bar, the various operations can be moved on the time-scaled bar chart until an acceptable resource level is obtained. The reason it is not always necessary to use the network is that in a commissioning operation

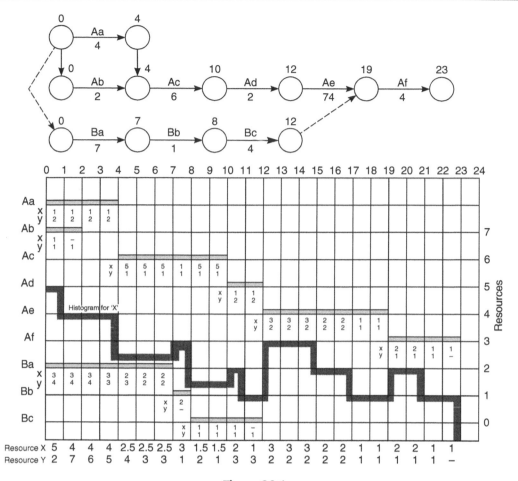

Figure 30.1
Bar chart and histogram

there is often considerable flexibility as to which machine is commissioned first. Whether pump A is commissioned before or after compressor B is often a matter of personal choice rather than logical necessity. When an acceptable solution has been found, the strips of bar can be held on to the backing sheet with an adhesive putty (Blu-Tack) and (provided the format is of the necessary size) photo-copied for distribution to interested parties.

4. If the weekly (or daily) aggregates are totalled cumulatively it is sometimes desirable to draw the cumulative curve (usually known as the S-curve, because it frequently takes the shape of an elongated letter S), which gives a picture of the build-up (and run-down) of the resources over the period of the project. This curve is also useful for showing the cumulative cash flow, which, after all, is only another resource. An example of such a cash flow curve is given in Chapter 45.

The following example shows the above steps in relation to a small construction project where there is a resource limitation. Figure 30.2 shows the AoA configuration and Figure 30.3 shows the same network in AoN configuration. Figure 30.4 shows their translation into a bar (or Gantt) chart where the bars are in fact a string of resource numbers. For simplicity, all the resources shown are of the same type (e.g., welders). By adding up the resources of each week a totals table can be drawn, from which it can be seen that in week 9 the resource requirement is 14. This amount exceeds the availability, which is only 11 welders, and an adjustment is therefore necessary. Closer examination of the bar chart reveals a low resource requirement of only 6 in week 12. A check on the network

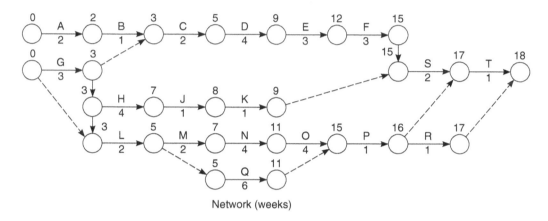

Network (weeks)

Figure 30.2

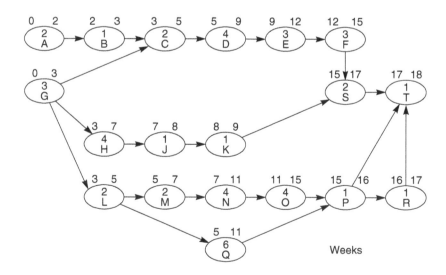

Weeks

Figure 30.3

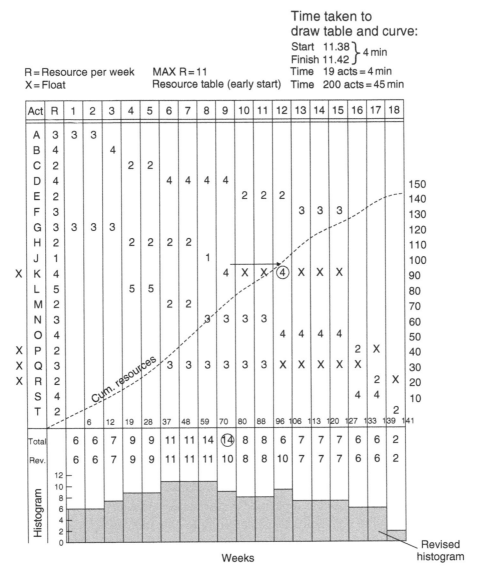

Time taken to
draw table and curve:
Start 11.38 ⎱ 4 min
Finish 11.42 ⎰
Time 19 acts = 4 min
Time 200 acts = 45 min

R = Resource per week   MAX R = 11
X = Float   Resource table (early start)

Figure 30.4
Histogram and 'S' curve

(Figure 30.2) shows that there is 15 − 9 = 6 weeks' float on activity K. This activity can therefore be used to smooth the resources. By delaying activity K by three weeks, the resource requirement is now:

Week 9, −10
Week 12, −10

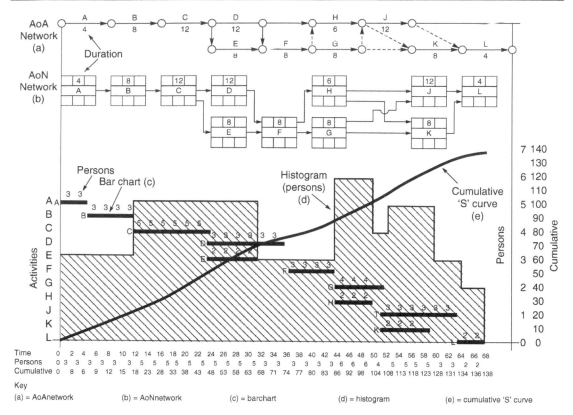

Figure 30.5
Network, bar chart, histogram and 'S' curve

From the revised totals table a histogram has been drawn as well as a cumulative resource curve. The latter can also be used as a planned performance curve since the resources (if men) are directly proportional to manhours. It is interesting to note that any 'dip' or 'peak' in the cumulative resource curve indicates a change of resource requirement, which should be investigated. A well-planned project should have a smooth resource curve following approximately the shape of a letter S.

The method described may appear to be lengthy and time consuming, but the example given by Figures 30.2 (or 30.3) and 30.4, including the resource smoothing and curve plotting, took exactly six minutes. Once the activities and resources have been listed on graph paper, the bar chart draughting and resource smoothing of a practical network of approximately 200 activities can usually be carried out in about one hour.

Most modern computers' project management programs have resource smoothing facilities, which enable the base to be re-positioned on the screen to give the required resource total for any time period.

However, it is advisable not to do this automatically as the machine cannot make allowances for congestion of work area, special skills of operators, clients' preferences, and other factors only apparent to the people on the job.

Figure 30.5 shows the relationship between the networks, bar chart, histogram, and cumulative 'S' curve.

It should be noted that the term used for redistributing the resources was 'resource smoothing'. Some authorities also use the term 'resource levelling', by which they mean flattening out the histogram to keep the resource usage within the resource availability for a particular time period. However, to do this without moving the position of some of the activities is just about impossible. Whether the resources are 'levelled' to reduce the unacceptable peaks due to resource restraints or 'smoothed' to produce a more even resource usage pattern, activities have to be readjusted along the time scale by utilizing the available float. To ascribe different meanings to the terms smoothing and levelling is therefore somewhat hair-splitting, since in both cases the operations to be carried out are identical. If the resource levels are so restricted that even the critical activities have to be extended, the project completion will inevitably be delayed.

Cash Flow Forecasting

Chapter Outline

It has been stated in Chapter 30 that it is very easy to convert a network into a bar chart, especially if the durations and week (or day) numbers have been inserted. Indeed, the graphical method of analysis actually generates the bar chart as it is developed.

If we now divide this bar chart into a number of time periods (say, weeks or months) it is possible to see, by adding up vertically, what work has to be carried out in any time period. For example, if the time period is in months, then in any particular month we can see that one section is being excavated, another is being concreted, and another is being scaffolded and shuttered, etc.

From the description we can identify the work and can then find the appropriate rate (or total cost) from the bills of quantities. If the total period of that work takes six weeks and we have used up four weeks in the time period under consideration, then approximately two-thirds of the value of that operation has been performed and could be certificated.

By this process it is possible to build up a fairly accurate picture of anticipated expenditure at the beginning of the job, which in itself might well affect the whole tendering policy. Provided the job is on programme, the cash flow can be calculated, but, naturally, due allowance must be made for the different methods and periods of retentions, billing, and reimbursement. The cost of the operation must therefore be broken down into six main constituents:

- Labour
- Plant
- Materials and equipment
- Subcontracts
- Site establishment
- Overheads and profit

By drawing up a table of the main operations as shown on the network, and splitting up the cost of these operations (or activities) into the six constituents, it is possible to calculate the average percentage that each constituent contains in relation to the value. It is very important, however, to deduct the values of the subcontracts from any operation and treat these subcontracts separately. The reason for this is, of course, that a subcontract is self-contained and is

Project Management, Planning, and Control. http://dx.doi.org/10.1016/B978-0-08-098324-0.00031-7

often of a specialized nature. To break up a subcontract into labour, plant, materials, etc., would not only be very difficult (since this is the prerogative of the subcontractor) but would also seriously distort the true distribution of the remainder of the project.

Example of Cash Flow Forecasting

The simplest way to explain the method is to work through the example described in Figures 31.1 to 31.6. This is a hypothetical construction project of three identical simple unheated warehouses with a steel framework on independent foundation blocks, profiled steel roof and side cladding, and a reinforced-concrete ground slab. It has been assumed that as an area of site has been cleared, excavation work can start, and the sequences of each warehouse are identical. The layout is shown in Figure 31.1 and the network for the three warehouses is shown in Figure 31.2.

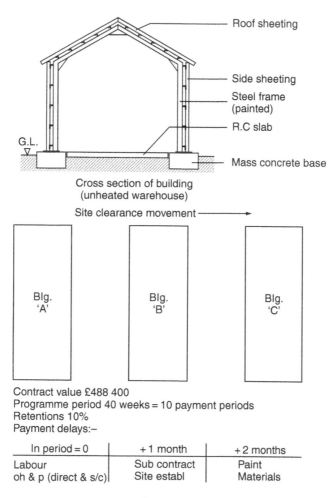

Contract value £488 400
Programme period 40 weeks = 10 payment periods
Retentions 10%
Payment delays:–

In period = 0	+ 1 month	+ 2 months
Labour	Sub contract	Paint
oh & p (direct & s/c)	Site establ	Materials

Figure 31.1
Warehouse building

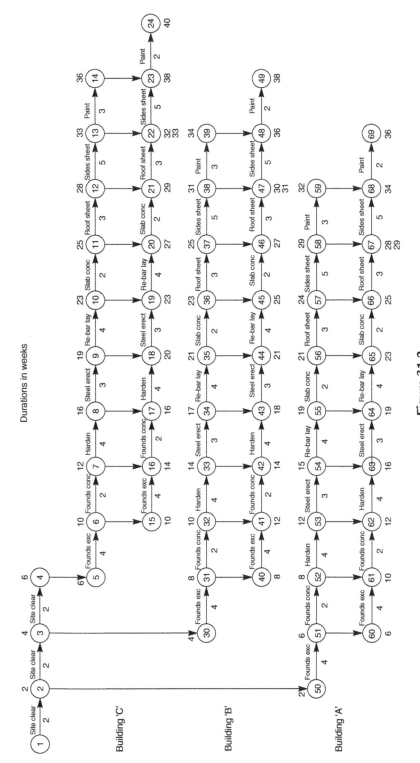

Figure 31.2
Construction network

Figure 31.3 shows the graphical analysis of the network separated for each building. The floats can be easily seen by inspection, e.g., there is a two-week float in the first paint activity (58–59) since there is a gap between the following dummy 59–68 and activity 68–69. The speed and ease of this method soon become apparent after a little practice.

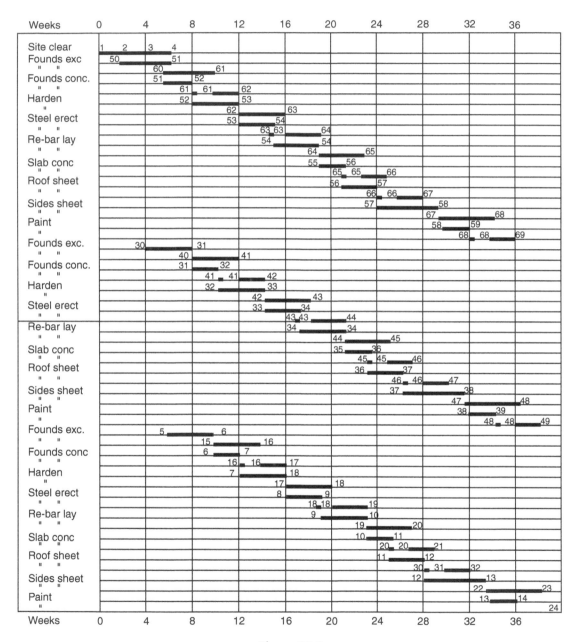

Figure 31.3
Graphical analysis

One building   Units in $ × 100

Activity	Duration weeks	Total value	Labour Value	%	Plant Value	%	Materials Value	%	Sub Contr Value	%	Site Estb. Value	%	OH & P Value	%
Clear site	2	62	30	48	20	32	–		3	5	4	7	5	8
Founds exc.	4	94	40	43	40	43	–		–		6	6	8	8
" =	4	94	40	43	40	43	–		–		6	6	8	8
Founds conc.	2	71	20	28	10	14	30	42	–		5	8	6	8
" =	2	71	20	28	10	14	30	42	–		5	8	6	8
Steel erect	3	220	–		–		–		200	91	–		20	9
" =	3	220	–		–		–		200	91	–		20	9
Re-bar lay	4	106	30	28	–		60	56	–		7	7	9	9
" =	4	106	30	28	–		60	56	–		7	7	9	9
Slab conc.	2	71	20	28	10	14	30	42	–		5	8	6	8
" =	2	71	20	28	10	14	30	42	–		5	8	6	8
Roof sheet	3	66	–		–		–		60	91	–		6	9
" =	3	66	–		–		–		60	91	–		6	9
Sides sheet	5	100	–		–		–		90	90	–		10	10
" =	5	100	–		–		–		90	90	–		10	10
Paint	3	66	–		–		–		60	91	–		6	9
" =	2	44	–		–		–		40	91	–		4	9
Total direct		743	250	34	140	19	240	32			50	7	63	8
Total sub-contr.		885							803	91			82	9
Grand total		1628												
For 3 blgs		4884												

Figure 31.4
Earned value table

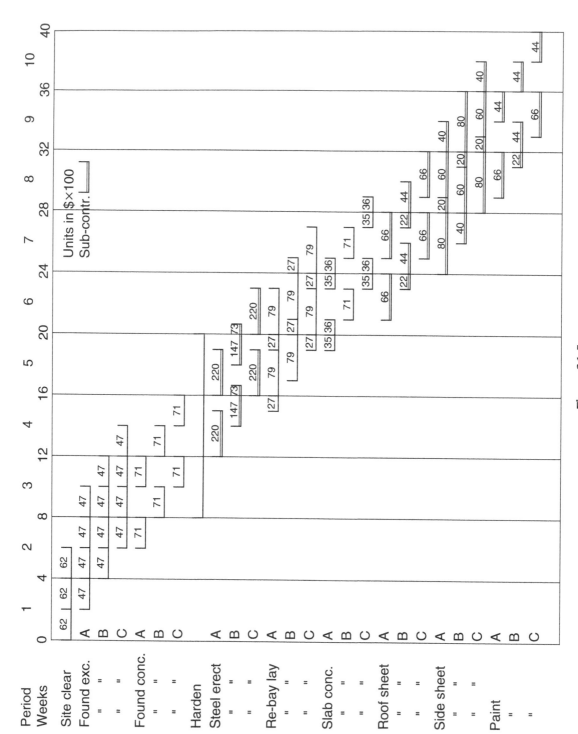

Figure 31.5
Bar charts and costs

All monetary values are $ ×100.

Item	% / Delay	0	1	2	3	4	5	6	7	8	9	10	11
Period		0	1	2	3	4	5	6	7	8	9	10	11
Week		0	4	8	12	16	20	24	28	32	36	40	44
Total S/C			—	—	—	367	660	381	318	438	354	128	
S/C	91					334	600	347	289	399	322	116	
OH & P	9					33	60	34	29	39	32	12	
Direct	%		171	368	448	216	247	468	284	36			
Labour	34		58	125	153	74	84	159	97	12			
Plant	19		33	70	85	41	47	89	54	7			
Material	32		55	118	143	69	79	150	91	11			
Site est.	7		12	26	31	15	17	33	20	3			
OH & P	8		13	29	36	17	20	37	22	3			
Total value			171	368	448	583	907	849	602	474	354	128	
Outflow	Delay												
Labour	0		58	125	153	74	84	159	97	12			
Plant	2				33	70	85	41	47	89	54	7	
Material	2				55	118	143	69	79	150	91	11	
S/C	1						334	600	347	289	399	322	116
Site est.	1			12	26	31	15	17	33	20	3		
OH & P	0		13	29	36	17	20	37	22	3			
S/C OH&P	0					33	60	34	29	39	32	12	
Out			71	166	303	343	741	957	654	602	579	352	116
In 90%	1			154	331	403	525	816	764	542	427	319	115
Net flow			(71)	(12)	28	60	(216)	(141)	110	(60)	(152)	(33)	(1)
Cumul. out	−		71	237	540	883	1624	2581	3235	3837	4416	4768	4884
Cumul. in	+			154	485	888	1413	2229	2993	3535	3962	4281	4396
Cumul. net			−71	−83	−55	+5	−211	−352	−242	−302	−454	−487	−488

Figure 31.6
Cash flow chart

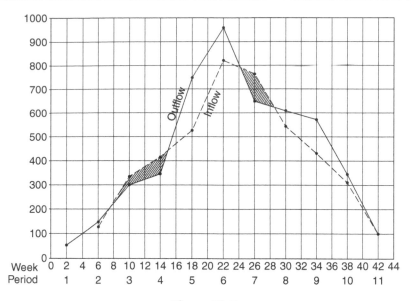

Figure 31.7
Cash flow graph

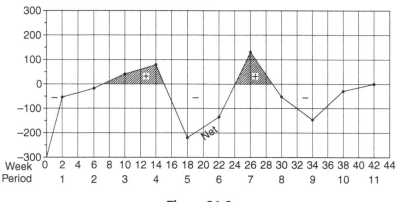

Figure 31.8
Cash flow graph

The bar chart in Figure 31.5 has been drawn straight from the network (Figure 31.2) and the costs in $100 units added from Figure 31.4. For example, in Figure 31.4 the value of foundation excavation for any one building is $9,400 per four-week activity. Since there are two four-week activities, the total is $18,800. To enable the activity to be costed in the corresponding measurement period, it is convenient to split this up into two-weekly periods of $4,700. Hence in Figure 31.5, foundation excavation for building A is shown as

$$47 \text{ in period 1}$$
$$47 + 47 = 94 \text{ in period 2}$$
$$47 \text{ in period 3}$$

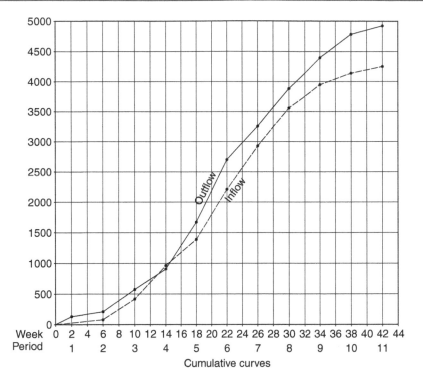

Figure 31.9
Cumulative curves

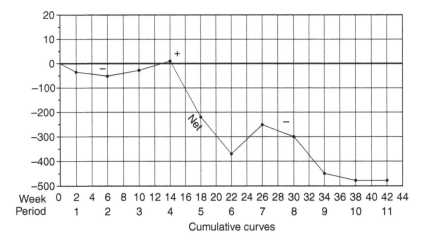

Figure 31.10

The summation of all the costs in any period is shown in Figure 31.6.

Figure 31.6 clearly shows the effect of the anticipated delays in payment of certificates and settlement of contractor's accounts. For example, material valued at 118 in period 2 is paid to the contractor after one month in period 3 (part of the 331, which is 90% of 368, the total value of period 2), and is paid to the supplier by the contractor in period 4 after the two-month delay period.

From Figure 31.6 it can be seen that it has been decided to extract overhead and profit monthly as the job proceeds, but this is a policy that is not followed by every company. Similarly, the payment delays may differ in practice, but the principle would be the same.

Figure 31.6 shows the total outflow and inflow for each time period and the net differences between the two. When these values are plotted on graphs as in Figures 31.7 and 31.8, it can be seen that there are only three periods of positive cash flow, i.e., periods 3, 4, and 7. However, while this shows the actual periods when additional moneys have to be made available to fund the project, it does not show, because the gap between the outflow and inflow is so large for most of the time, that for all intents and purposes the project has a negative cash flow throughout its life.

This becomes apparent when the cumulative outflows and inflows, which are tabulated in the last three lines of Figure 31.6, are plotted on a graph as in Figure 31.9 and 31.10. From these it can be seen that cumulatively, the Chapter_27 13.01.38 13.01.38 13.01.38 13.01.38.pages a positive cash flow (a mere $500) is in period 4.

This example shows that the project is not self-financing and will possibly only show a profit when the 10% retention moneys have been released. To restore the project to a positive cash flow, it would be necessary to negotiate a sufficiently large mobilization fee at the start of the project to ensure that the contract is self-financing.

Cost Control and EVA

Chapter Outline

Apart from ensuring that their project is completed on time, all managers, whether in the office, workshop, factory, or on-site, are concerned with cost. There is little consolation in finishing on time, when, from a cost point of view, one wished the job had never started!

Cost control has been a vital function of management since the days of the pyramids, but only too frequently is the term confused with mere cost reporting. The cost report is usually part of every manager's monthly report to his superiors, but an account of the past month's expenditure is only stating historical facts. What the manager needs is a regular and up-to-date monitoring system that enables him to identify the expenditure with specific operations or stages, determine whether the expenditure was cost-effective, plot or calculate the trend, and then take immediate action if the trend is unacceptable.

Network analysis forms an excellent base for any cost-control system, since the activities can each be identified and costed, so that the percentage completion of an activity can also give the proportion of expenditure, if that expenditure is time related. The system is ideal, therefore, for construction sites, drawing offices, or factories where the basic unit of control is the manhour.

SMAC – Manhour Control

Site manhours and cost (SMAC)[*] is a cost control system developed in 1978 specifically on a critical path network base for either manual or computerized cost and progress monitoring, which enables performance to be measured and trends to be evaluated, thus providing the project manager with an effective instrument for further action. The system, which is now known as earned value analysis (EVA), can be used for all operations where manhours or

---

[*]SMAC is the proprietary name given to the cost-control program developed by Foster Wheeler.

Project Management, Planning, and Control. http://dx.doi.org/10.1016/B978-0-08-098324-0.00032-9
255

costs have to be controlled, and since most functions in an industrial (and now more and more commercial) environment are based on manhours and can be planned with critical path networks, the utilization of the system is almost limitless.

The following operations or activities could benefit from the system:

1.   Construction sites
2.   Fabrication shops
3.   Manufacturing (batch production)
4.   Drawing offices
5.   Removal services
6.   Machinery commissioning
7.   Repetitive clerical functions
8.   Road maintenance

The criteria laid down when the system was first mooted were:

1.   *Minimum site (or workshop) input.* Site staff should spend their time managing the contract and not filling in unnecessary forms.
2.   *Speed.* The returns should be monitored and analysed quickly so that action can be taken.
3.   *Accuracy.* The manhour expenditure must be identifiable with specific activities that are naturally logged on time sheets.
4.   *Value for money.* The useful manhours on an activity must be comparable with the actual hours expended.
5.   *Economy.* The system must be inexpensive to operate.
6.   *Forward looking.* Trends must be seen quickly so that remedial action can be taken when necessary.

The final system satisfied all these criteria with the additional advantage that the percentage complete returns become a simple but effective feedback for updating the network programme.

One of the most significant differences between EVA and the conventional progress-reporting systems is the substitution of 'weightings' given to individual activities, by the concept of 'value hours'. If each activity is monitored against its budget hours (or the hours allocated at the beginning of the contract, to that activity) then the 'value hour' is simply the percentage complete of that activity multiplied by its budget hours. In other words, it is the useful hours against the actual hours recorded on the time sheets.

If all the value hours of a project are added up and the total divided by the total budget hours, the overall per cent complete of the project is immediately seen.

The advantage of this system over the weighting system is that activities can be added or eliminated without having to 're-weight' all the other activities. Furthermore, the value

Table 32.1: Weighting System

1 Activity No.	2 Activity	3 Budget × 100	4 Weighting	5 % Complete	6 % Weighted	7 Actual Hours × 100
1	A	1000	0.232	100	23.2	1400
2	B	800	0.186	50	9.3	600
3	C	600	0.140	60	8.4	300
4	D	1200	0.279	40	11.2	850
5	E	300	0.070	70	4.9	250
6	F	400	0.093	80	7.4	600
Total		4300	1.000		64.4	4000

Overall % complete = 64.4%.

Predicted final hours $\dfrac{4000}{0.644} = 6211 \times 100$ hours

Efficiency $= \dfrac{4300 \times 0.644}{4000} = 69.25\%$

Table 32.2: Value Hours (Earned Value) System

1 Activity No.	2 Activity	3 Budget × 100	4 % Complete	5 Value Hours × 100	6 Actual Hours × 100
1	A	1000	100	1000	1400
2	B	800	50	400	600
3	C	600	60	360	300
4	D	1200	40	480	850
5	E	300	70	210	250
6	F	400	80	320	600
Total		4300		2770	4000

Overall % complete $\dfrac{2770}{4300} = 64.4\%$.

Predicted final hours $\dfrac{4000}{0.644} = 6211 \times 100$ hours

Efficiency $= \dfrac{2770}{4000} = 69.25\%$

hours are a tangible parameter, which, if plotted on a graph against actual hours, budget hours, and predicted final hours, gives the manager a 'feel' of the progress of the job that is second to none. The examples in Table 32.1 and 32.2 show the difference between the two systems.

Summary of Advantages

Comparing the weighting and value hours systems, the following advantages of the value hours system are immediately apparent:

1. The basic value hours system requires only six columns against the weighting system's seven.
2. There is no need to carry out a preliminary time-consuming 'weighting' at the beginning of the job.
3. Activities can be added or removed or have the durations changed without the need to recalculate the weightings of each activity. This saves hundreds of manhours on a large project.
4. The value hours are easily calculated and can even, in many cases, be assessed by inspection.
5. Errors are easily seen, as the value can never be more than the budget.
6. Budget hours, actual hours, value hours, and forecast hours can all be plotted on one graph to show trends.
7. The method is ideal for assessing the value of work actually completed for progress payments to main and subcontractors. Since it is based on manhours, it truly represents construction progress independently of material or plant costs, which so often distort the assessment.

The efficiency (output/input) for each activity is obtained by dividing the value hours by the actual hours. This is also known as the cost performance index (CPI).

The analysis can be considerably enhanced by calculating the efficiency and forecast final hours for each activity and adding these to the table.

The forecast final hours are obtained by either:

1. dividing the budget hours by the efficiency; or
2. dividing the actual hours by the % complete.

Both these methods give the same answer as the following proof (using the same abbreviations) shows:

(a) Final hours $= \dfrac{\text{Budget}}{\text{Efficiency}} = \dfrac{B}{G}$

Efficiency (CPI) $= \dfrac{\text{Value}}{\text{Actual}} = \dfrac{\text{Earned Value}}{\text{Actual}} = \dfrac{E}{C}$ (value is always the numerator)

Hence, Final hours $= \dfrac{\text{Budget}}{\text{Value/Actual}} = \dfrac{\text{Budget}}{\text{Value}} \times \text{Actual} = \dfrac{B \times C}{E}$

But Value $=$ Budget $\times$ % complete $= B \times D$

(b) Hence, Final hours $= \dfrac{\text{Budget} \times \text{Actual}}{\text{Budget} \times \% \text{ complete}} = \dfrac{\text{Actual}}{\% \text{ complete}} = \dfrac{B \times C}{B \times D} = \dfrac{C}{D}$

Example 1: Reasonable Progress

A	B	C	D	E	F	G
Activity	Budget Hours	Actual Hours	% Complete	Value Hours B × D	Forecast Final Hours C/D	Efficiency (CPI) E/C
1	1000	200	20	200	1000	1.00
2	200	100	50	100	200	1.00
3	600	300	40	240	750	0.80
Total	1800	600		540	1950	

Example 1 shows the earned value table for a small project consisting of three activities where there was reasonable progress.

The overall percentage complete of the work can be obtained by adding all the value hours in column E and dividing them by the total budget hours in column B, i.e., E/B.

Thus: Overall percentage complete $= \dfrac{\text{Total Value}}{\text{Total budget}} = \dfrac{E}{B} = \dfrac{540}{1800} = 0.3$ or 30%

The forecast final hours $F = \dfrac{\text{Total actual}}{\text{Overall \%}} = \dfrac{600}{0.3} = 2000 \qquad \dfrac{C}{D}$

As total efficiency of the project (CPI) $= \dfrac{\text{Value}}{\text{Actual}} = \dfrac{540}{600} = 0.9$ or 90% $\qquad \dfrac{E}{C}$

Alternatively, the forecast final hours $F = \dfrac{\text{Budget}}{\text{Efficiency}} = \dfrac{1800}{0.9} = 2000 \qquad \dfrac{B}{G}$

It can be seen that the difference between the calculated final hours of 2000, and the sum of the values of column F of 1950, is only 50 hours or 2.5%, and this tends to be the variation on projects with a large number of activities.

When an analysis is carried out after a period of poor progress as shown in the table of Example 2, the increase in the forecast final hours and the decrease in the efficiency become

Example 2: Very Poor Progress Due to Rework

A	B	C	D	E	F	G
Activity	Budget Hours	Actual Hours	% Complete	Value Hours B × D	Forecast Final Hours C/D	Efficiency (CPI) E/C
1	1000	200	5	50	4000	0.25
2	200	100	10	20	4000	0.20
3	600	300	40	240	750	0.80
Total	1800	600		310	8750	

immediately apparent. An examination of the table shows that this is due to the abysmal efficiencies (column G) of activities 1 and 2.

In this example the overall % complete is:

$$E/B = 310/1800 \quad = 0.17222 \text{ or } 17.222\%$$
$$\text{The efficiency (CPI) is } E/C = 310/600 \quad = .5167 \text{ or } 52\% \text{ approx and}$$
$$\text{The forecast final hours are } C/D = 600/0.17222 = 3484$$
$$\text{or } B/G = 1800/0.5167 = 3484$$

This is still a large overrun, but it is considerably less than the massive 8750 hours produced by adding up the individual forecast final hours in column F.

Clearly such a discrepancy of 5266 hours in Example 2 calls for an examination. The answer lies in the offending activities 1 and 2, which need to be restated so that the actual hours reflect the actual situation on the job. For example, if it is found that activities 1 and 2 required rework to such an extent that the original work was completely wasted and the job had to be started again, it is sensible to restate the actual hours of these activities to reflect this, i.e., all the abortive work is 'written off' and a new assessment of 0% complete is made from the starting point of the rework. There is little virtue in handicapping the final forecast with the gross inefficiency caused by an unforeseen rework problems. Such a restatement is shown in Example 2a.

Comparing Examples 2 and 2a it will be noted that:

1. The total budget hours are the same, i.e., 1800;
2. The total actual hours are now only 350 in 2a because 180 hours have been written off for activity 1A and 70 hours have been written off for activity 2A;
3. The value hours are the same, i.e., 310;
4. The overall % complete is the same;
5. The forecast final hours are now only 1700 because although the 250 aborted hours had to be included, the efficiency of the revised activities 1B and 2B has improved;
6. The overall efficiency is 310/350 = 0.885 or 88.5%.

Example 2a: Very Poor Progress Due to Rework

A Activity	B Budget Hours	C Actual Hours	D % Complete	E Value Hours $B \times D$	F Forecast Final Hours C/D	G Efficiency (CPI) E/C
1A	0	0	100	0	180	0
1B	1000	20	5	50	400	2.5
2A	0	0	100	0	70	0
2B	200	30	10	20	300	0.67
3	600	300	40	240	750	0.80
Total	1800	350		310	1700	

(1A or 2A are the works that have been written off.)

The forecast final hours calculated by dividing the budget hours by the efficiency comes to 1800/0.885 = 2033 hours. This is more than the 1700 hours obtained by adding all the values in column F, but the difference is only because the percentage complete assessment of activities is so diverse.

In practice, such a difference is both common and acceptable because:

1. On medium or large projects, wide variations of % complete assessments tend to follow the law of 'swings and roundabouts' and cancel each other out.
2. In most cases therefore the sensible method of forecasting the final hours is to either:
   (a) divide the budget hours by the efficiency, i.e., B/G or
   (b) divide the actual hours by the % complete, i.e., C/D. Both of course, give the same answer.
3. The column F (forecast final hours) is in most cases not required, but should it be necessary to find the forecast final hours of a specific activity, this can be done at any stage by simply dividing the actual hours of that activity by its percentage complete.
4. It must be remembered that comparing the forecast final hours with the original budget hours is only a reporting function and its use should not be given too much emphasis. A much more important comparison is that between the actual hours and the value hours as this is a powerful and essential control function.

As stated earlier, two of the criteria of the system were the absolute minimum amount of form filling for reporting progress and the accurate assessment of percentage complete of specific activities. The first requirement is met by cutting down the reporting items to three essentials:

1. The activity numbers of the activities worked on in the reporting period (usually one week).
2. The *actual* hours spent on each of these activities, taken from the time cards.
3. The assessment of the percentage complete of each reported activity. This is made by the 'man on the spot'.

The third item is the most likely one to be inaccurate, since any estimate is a mixture of fact and opinion. To reduce this risk (and thus comply with the second criterion, i.e., accuracy) the activities on the network have to be chosen and 'sized' to enable them to be estimated, measured, or assessed in the field, shop, or office by the foreman or supervisor in charge. This is an absolute prerequisite of success, and its importance cannot be over-emphasized.

Individual activities must not be so complex or long (in time) that further breakdown is necessary in the field, nor should they be so small as to cause unnecessary paperwork. For example, the erection of a length of ducting and supports (Figure 32.1) could be split into the activities shown in Figure 32.2 and 32.3.

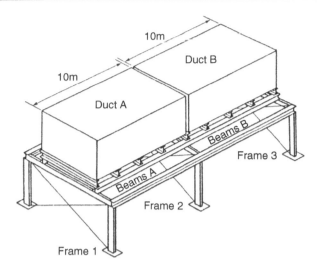

Figure 32.1
Duct support

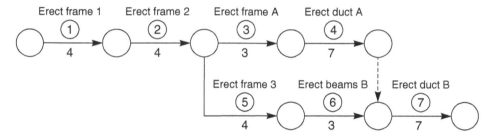

Figure 32.2

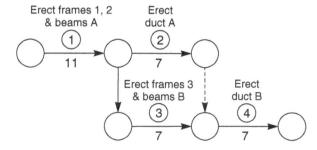

Figure 32.3

Any competent supervisor can see that if the two columns of frame 1 (Activity 1) have been erected and stayed, the activity is about 50% complete. He may be conservative and report 40% or optimistic and report 60%, but this ±20% difference is not important in the light of the total project. When all these individual estimates are summated the discrepancies tend to cancel out. What is important is that the assessment is realistic and checkable. Similarly, if 3 m of the duct between frames 1 and 2 have been erected, it is *about* 30% complete. Again, a margin on each side of this estimate is permissible.

However, if the network were prepared as shown in Figure 32.3 the supervisor may have some difficulty in assessing the percentage complete of activity 1 when he had erected and stayed the columns of frame 1. He now has to mentally compute the manhours to erect and stay two columns in relation to four columns and four beams. The percentage complete could be between 10% and 30%, with an average of 20%. The ± percentage difference is now 50%, which is more than double the difference in the first network. It can be seen therefore that the possibility of error and the amount of effort to make an assessment or both is greater.

Had the size of each activity been *reduced* to each column, beam, or brace, the clerical effort would have been increased and the whole exercise would have been less viable. It is important therefore to consult the men in the field or on the shop floor before drafting the network and fixing the sequence and duration of each activity.

EVA for Civil Engineering Projects

Most civil engineering contracts have an in-built earned value system, because the monthly re-measure is in fact a valuation of the work done. By using the composite rates in the bill of quantities or the schedule of rates, the monetary value of the work done to date can be easily established. This can then be translated into a curve and the value at any time period can then be divided by the corresponding value of the cash flow curve (which is in effect the planned work) to give the approximate % complete (SPI). However these values do not give a true picture of the work actually done, as the rates in the bills include overheads and profit as well as contingency allowance.

In order to compare Actual Costs with Earned Value, it is necessary therefore to take the contingency, profit, and overhead portion out of the unit rates so that only true labour, material, and plant costs remain. This reduction could be between 5% and 10%. Re-measuring of the completed work can then take place as normal in the conventional physical units of m. m^2, m^3, tonne, etc. The measured quantities are then multiplied by the new (reduced) rates to give the useful work done in monetary terms and become in effect the Earned Value.

Planned costs

These cab be taken from the "S" curve of the histogram or cash flow curve. These will include labour, materials and plant costs, but they must again be at the reduced rate. i.e. without overheads and profit)

Budget costs

Labour	Total assessed time for activity x average labour rate
Materials	Estimated purchase price of materials required for this activity
Plant	Where plant is exclusive to the activity, e.g.scaffolding:
	Plant hire cost (external or internal) for the planned period of the activity
	e.g. £/day x anticipated number of days the plant is required.
	Where plant is shared with other activities, e.g.dumper truck:
	An approximate % assessment of usage must be made for each activity

Actual costs

Labour	Time sheet hours x average labour rate
Materials	Invoice cost of material quantities used to date, or unit rate of material x quantity used to date.
Plant	Where plant is exclusive to activity:
	Invoiced plant hire rate, e.g. £/day x number of days worked to date.
	Where plant is shared with other activities:
	Assessed % of plant hire rate. e.g £/day x number of days worked to date.

Site Overheads

Establishment costs and indirect labour costs cannot be part of the EVA system. However, the must be recorded separately. The indirect labour costs must be plotted on the bottom of the EVA set of curves to ensure that they have not been inflated to off-set overruns of direct labour costs.

As with the normal EVA system, at any particular point in time,

Earned Value (EV)	= Measured work in £, $, Euros etc.
Overall % complete	= EV / total original budget
Efficiency (CPI)	= EV / Actual Costs
SPI	= EV / Planned Costs

Where there are no Bills of Quantities, labour costs can still be used for monitoring progress using the EVA system. Material and plant costs can never be used for monitoring progress.

To operate such an EVA system, the budget costs, planned costs and actual costs (and the calculated earned value) must must be broken down into labour, materials and plant for each activity, as each is measured differently. For any particular point in time, it is quite possible to have installed the planned materials materials and expended the associated plant costs and yet have an overrun or under run of labour costs, depending on the effectiveness of supervision, method of working, climatic conditions or a myriad of other factors.

It can be seen therefore, that this breakdown of each activity into three cost items can be very time consuming and for this reason the conventional monthly measurement of completed work based on on the rates in the Bills of Quantities, is the preferred method used by most companies in the civil engineering and building industries."

In case of any queries to this correction, I have attached a word document with how it should look.

Example

A large storage tank consisting of 400 steel plates, which at $150 each, gives a material cost of $60000 (for simplicity, other material costs have been ignored).

The duration for erecting and testing is six weeks.

The labour (mostly welding) manhours are 1200, which at $20/hour, gives a labour cost of $24000.

Again, for simplicity, plant costs (cranes and welding sets) are regarded as site overheads and can be ignored.

All the plates arrive on site on day 1 and have to be paid for at day 28.

Work starts as soon as the plates have been unloaded (by others).

By day 7 (1 week later) 60 plates have been erected and welded.

Therefore at day 7, the percentage complete is $60/400 = 15\%$.

The EV calculations are carried out weekly and the time sheets show that after the first week, the men have booked a total of 200 manhours.

The earned value is therefore 15% of $1200 = 180$, so that the efficiency (CPI) $= 180/200 = 90\%$.

If the men are paid production bonuses, they would not get a bonus for this week as the productivity bonus (as agreed with the unions) only starts at 97% efficiency.

The costs incurred to date are therefore only labour costs and are $200 \times \$20 = \$4000$.

At day 28 (after 4 weeks) 300 plates have been erected.

The % complete is now $300/400 = 75\%$.

The total manhours booked to date are now 850.

The earned value can be seen to be 75% of $1200 = 900$.

As the efficiency (CPI) is now $900/850 = 105\%$, the men get their bonus.

The cost to date is now $850 \times \$20 = \$17000$ plus the *total* material cost of $\$60000 = \$77000$.

Note that *all* the material has to be paid for – not just the material erected.

Alternative Payment Schedule

Supposing the terms of the contract were that the tank contractor had to be paid *weekly* for material and labour. He would therefore be paid as follows:

At the end of week 1:

> Labour: 180 manhours = $3600 (the earned value, not the actual cost)
> Material: 60 plates at $150/plate = $150 \times 60 = \$9000$ (the plates erected)
> The total payment is therefore $\$3600 + 9000 = \$12600$ (ignoring retentions).

It can be seen therefore that materials can be a useful aid in assessing percentage complete, and although in order to obtain the total costs, they must be added to the labour costs, they *cannot* be part of the EV analysis. In other words, with the exception of the individual percentage complete assessment, the earned value, CPI, SPI, anticipated final cost, and anticipated final completion time can only be calculated from the labour data.

The types of work that lend themselves to a similar treatment as the storage tank are:

- Pipework measured in metres
- Cable runs measured in metres
- Insulation measured in metres or sq. metres
- Steelwork measured in tonnes
- Refractory work measured in sq. metres
- Equipment measured in number of pieces, etc.

All labour must of course be measured in manhours or money units.

If plant costs have to be booked to the work package, they can be treated in a similar way to equipment, except that payments (when the plant is hired) are usually made monthly.

Control Graphs and Reports

Apart from the numerical report shown in Figure 33.5, two very useful management control graphs can then be produced:

1. Showing budget hours, actual hours, value hours, and predicted final hours, all against a common time base;
2. Showing percentage planned, percentage complete, and efficiency, against a similar time base.

The actual shape of the curves on these graphs give the project manager an insight into the running of the job, enabling appropriate action to be taken.

Figure 33.1 shows the site returns of manhours of a small project over a nine-month period, and, for convenience, the table of percentage complete and actual and value hours has been drawn on the same page as the resulting curves. In practice, the greater number of activities would not make such a compressed presentation possible.

A number of interesting points are ascertainable from the curves:

1. There was obviously a large increase in site labour between the fifth and sixth months, as is shown by the steep rise of the actual hours curve.
2. This has resulted in increased efficiency.
3. The learning curve given by the estimated final hours has flattened in month 6 making the prediction both consistent and realistic.
4. Month 7 showed a divergence of actual and value hours (indicated also by a loss of efficiency), which was corrected (probably by management action) by month 8.
5. It is possible to predict the month of actual completion by projecting all the curves forward. The month of completion is then given:
   (a) When the value hours curve intersects the budget line; and
   (b) When the actual hours curve intersects the estimated final hours curve.

Project Management, Planning, and Control. http://dx.doi.org/10.1016/B978-0-08-098324-0.00033-0

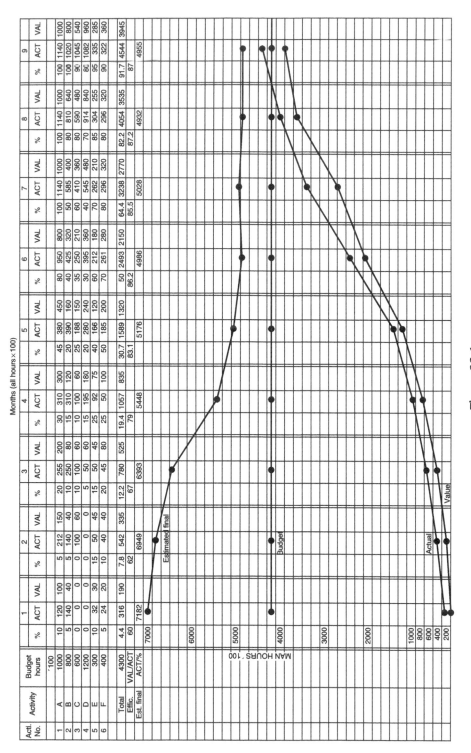

Figure 33.1
Control curves

In this example, one could safely predict completion of the project in month 10.

It will be appreciated that this system lends itself ideally to computerization, giving the project manager the maximum information with the very minimum of site input. The sensitivity of the system is shown by the immediate change in efficiency when the value hours diverged from the actual hours in month 7. This alerts management to investigate and apply corrections.

For maximum benefit the returns and calculations should be carried out weekly. By using the normal weekly time cards very little additional site effort is required to complete the returns, and with the aid of a good computer program the results should be available 24 hours after the returns are received.

An example of the application of a manual EVA analysis is shown in Figures 33.2 to 33.9. The site construction network of a package boiler installation is given in Figure 33.2. Although the project consisted of three boilers, only one network, that of Boiler No. 1 is shown. In this way it was possible to control each boiler construction separately and compare performances. The numbers above the activity description are the activity numbers, while those below are the durations. The reason for using activity numbers for identifying each activity, instead of the more conventional beginning and end event numbers, is that the identifier must always be uniquely associated with the activity description.

If the event numbers (in this case the coordinates of the grid) were used, the identifier could change if the logic were amended or other activities were inserted. In a sense, the activity number is akin to the node number of a precedence diagram, which is always associated with its activity. The use of precedence diagrams and computerized EVA is therefore a natural marriage, and to illustrate this point, a precedence diagram is shown in Figure 33.3.

Once the network has been drawn, the manhours allocated to each activity can be represented graphically on a bar chart. This is shown in Figure 33.4. By adding up the manhours for each week, the totals, cumulative totals, and each week's percentage of the total manhours can be calculated. If these percentages are then plotted as a graph the planned percentage complete curve can be drawn. This is shown in Figure 33.7.

All the work described up to this stage can be carried out before work starts on-site. The only other operation necessary before the construction stage is to complete the left-hand side of the site returns analysis sheet. This is shown in Figure 33.5, which covers only periods 4 to 9 of the project. The columns to be completed at this stage are:

1. The activity number
2. The activity title
3. The budget hours

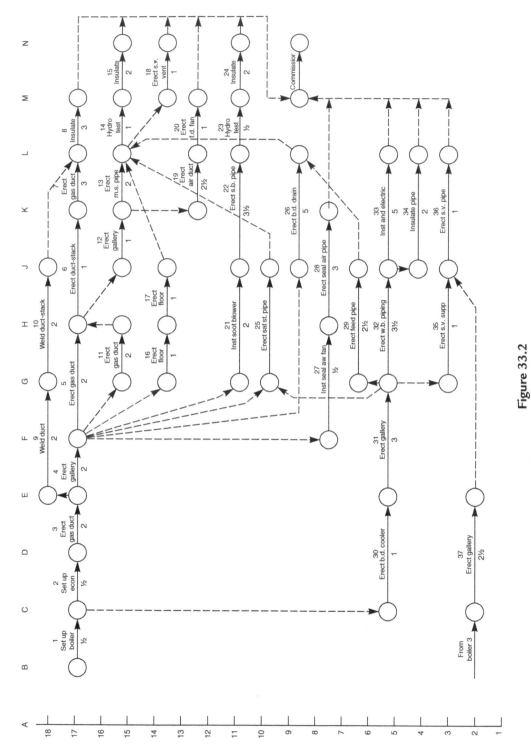

Figure 33.2

Boiler No. 1. Network arrow diagram

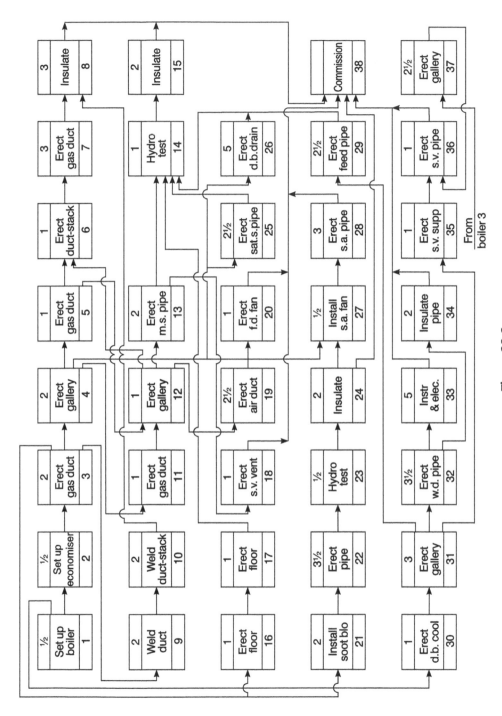

Figure 33.3
Boiler No. 1. Precedence diagram

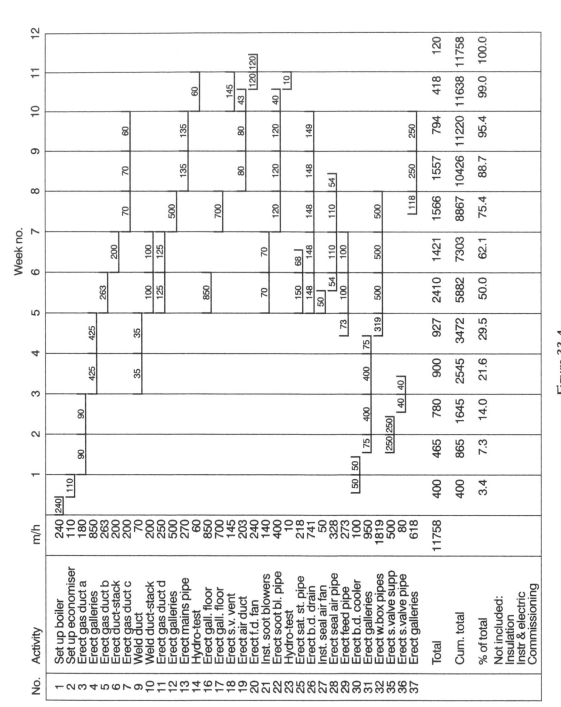

Figure 33.4

Boiler No. 1. Bar chart and manhour loadings

No.	Title	Orig	Rev	P4 W	P4 A	P4 %	P4 V	P5 W	P5 A	P5 %	P5 V	P6 W	P6 A	P6 %	P6 V	P7 W	P7 A	P7 %	P7 V	P8 W	P8 A	P8 %	P8 V	P9 W	P9 A	P9 %	P9 V
1	Set up boiler	240			233	100	240		230	100	240		230	100	240		230	100	240		230	100	240		230	100	240
2	Set up economiser	110			90	100	110		90	100	110		90	100	110		90	100	110		90	100	110		90	100	110
3	Erect gas duct a	180			155	100	180		155	100	180	69	155	100	180		155	100	180		155	100	180		155	100	180
4	Erect galleries	860		352	352	60	510	389	741	80	460	200	810	100	850		810	100	850		810	100	850		810	100	850
5	Erect gas duct b	263										200	200	60	158	55	255	100	263		255	100	263		255	100	263
6	Erect duct-stack	200																		180	180	80	160	5	185	100	200
7	Erect gas duct c	200																		62	62	30	60	88	150	75	150
9	Weld duct	70		32	32	50	35	33	65	100	70	42	65	100	70		65	100	70		65	100	70	70	65	100	70
10	Weld duct-stack	200											92	60	150		70	80	200	110	110	60	120		180	100	200
11	Erect gas duct d	250														18				5	175	100	250		175	100	250
12	Erect galleries	500																		405	405	95	475	5	410	100	500
13	Erect mains pipe	270																						105	105	40	108
14	Hydro-test	60		–																							–
16	Erect gall. floor	850										420	420	60	510	360	780	90	765	5	785	100	850		785	100	850
17	Erect gall. floor	700																		340	340	45	315	310	650	100	700
18	Erect s.v. vent	145		–																							–
19	Erect air duct	203																						75	75	30	81
20	Erect f.d. fan	240		–																							–
21	Inst. soot blowers	140										65	65	55	77	65	130	95	133	5	135	100	140		135	100	140
22	Erect soot bl. pipe	400																		100	100	20	80	110	210	40	160
23	Hydro-test	10		–																							–
25	Erect sat. st. pipe	218										125	125	60	131	85	210	100	218		210	100	218		210	100	218
26	Erect b.d. drain	741										130	130	20	148	132	212	45	333	136	398	60	445	122	520	80	593
27	Inst. seal air fan	50														40	40	100	50		40	100	50		40	100	50
28	Erect seal air pipe	328						52	52	25	68	45	45	20	66	37	82	40	131	128	210	85	279	10	220	90	293
29	Erect feed pipe	273			105	100	100		105	100	100	93	145	60	164	100	245	100	273		245	100	273		245	100	273
30	Erect b.d. cooler	100										25	105	100	100		105	100	100		105	100	100		105	100	100
31	Erect galleries	950		390	810	75	713	30	240	80	760	460	865	100	950		865	100	950		865	100	950		865	100	950
32	Erect w.box pipes	1819						300	300	20	364		760	30	546	445	1205	65	1182	405	1610	80	1455	340	1950	100	1819
35	Erect s.valve supp	500			440	100	500		460	100	500		460	100	500		460	100	500		460	100	500		460	100	580
36	Erect s.valve pipe	80		40	80	100	80		80	100	80		80	100	80		80	100	80		80	100	80	120	80	100	80
37	Erect galleries	618																							120	30	185
	Totals:-	11758		814	2314	21	2468	804	3118		3152	1724	4842	43	5030	1397	6239	56	6628	1881	8120	72	8513	1360	9480	86	10095
						106% 11019				101% 11548				104% 11260				106% 1141				106% 1277				106% 1095

Legend

A – Actual accumulated manhours used to end of period
% – Estimate of percentage completion of activity
V – Value hours = budget x percentage completion
W – Hours worked this period

Figure 33.5
Earned value analysis sheet

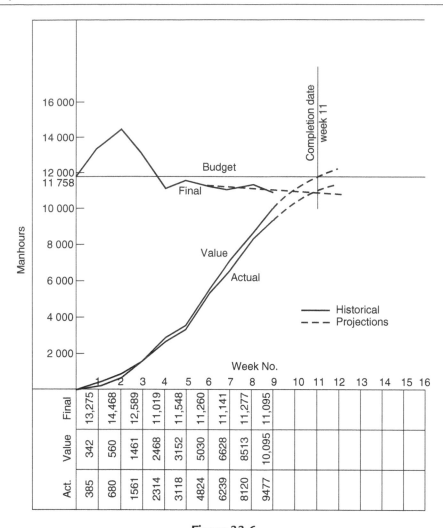

Figure 33.6
Boiler No. 1. Man hour–time curves

Once work has started on-site, the construction manager reports weekly on the progress of each activity worked on during that week. All he has to state is:

1. The activity number
2. The actual hours expended *in that week*
3. The percentage complete of that activity to date

If the computation is carried out manually, the figures are entered on the sheet (Figure 33.5) and the following values calculated weekly:

1. Total manhours expended this week (W column)
2. Total manhours to date (A column)
3. Percentage complete of project (% column)

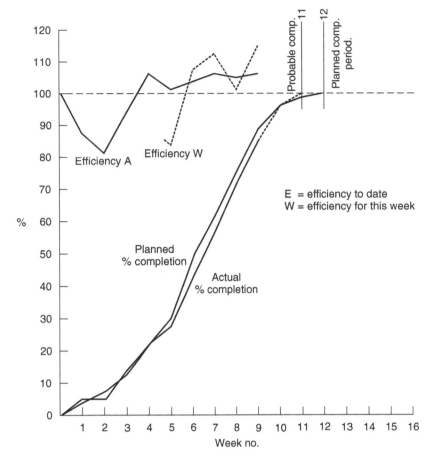

Figure 33.7
Boiler No. 1. Percentage–time curves

4. Total value hours to date (V column)
5. Efficiency
6. Estimated final hours

Alternatively, the site returns can be processed by computer and the resulting printout of part of a project is shown in Figure 33.8. Whether the information is collected manually or electronically, the return can be made on a standard timesheet with the only addition being a % complete column. In other words, no additional forms are required to collect information for EVA. There are in fact only three items of data to be returned to give sufficient information:

1. The activity number of the activity actually being worked on in that time period
2. The actual hours being expended on each activity worked on in that time period
3. The cumulative % complete of each of these activities

Foster Wheeler Power Products Ltd.

Contract No. 2-322-04298
Construction at Suamprogetti

Site manhours and costing system (EVA)
Standard report
Manhours report

Events prec succ	Description	No off unit 0-rate	Hrs/ unit C-rate	Budgets original/ current	Period this accum.	% com	Cimp. value	Est. to compl.	Forecast last rep total	Var. from last rep total	Extra	Remarks
	Setup boiler											
0001-0001-01	Setup boiler	BLR 1	240.00	240.00 240.00	0.00 55.00	100	240	0	55 55	0 185		
	Setup econ											
0001-0002-01	Setup economizer	ECON 1	110.00	110.00 110.00	0.00 52.00	100	110	0	52 52	0 58		
	Erect ducts											
0001-0003-01	Erect ducts blr/econ	DUCT 1	180.00	180.00 180.00	0.00 257.00	100	180	0	257 257	0 −77		
	Erect b/d cooler											
0001-0004-01	Erect b/d cooler	VESSL 1	100.00	100.00 100.00	0.00 128.00	100	100	0	128 128	0 −28		
	Erect galleries											
0001-0005-01	Erect galls for blr	GALLS 1	850.00	850.00 850.00	0.00 651.00	850	850	0	651 651	0 199		
	Erect duct											
0001-006-01	Erect duct chimney dampers	DUCT 1	250.00	250.00 250.00	0.00 169.00	98	245	3	172 172	0 78		
	Erect galleries											

Figure 33.8
Standard E.V. Report printout

All the other information required for computation and reporting (such as activity titles and activity manhour budgets) will already have been inputted and is stored in the computer. A typical modified timesheet is shown in Figure 33.9.

A complete set of printouts produced by a modern project management system are shown in Figures 33.10 to 33.14. It will be noted that the network in precedence format has been produced by the computer, as have the bar chart and curves. In this program the numerical EVA analysis has been combined with the normal critical path analysis from one database, so that both outputs can be printed and updated at the same time on one sheet of paper. The reason the totals of the forecast hours are different from the manual analysis is that the computer calculates the forecast hours for each activity and then adds them up, while in the manual system the total forecast hours are obtained by simply dividing the actual hours by the percentage complete rounded off to the nearest 1%.

As mentioned earlier, if the budget hours, actual hours, value hours, and estimated final hours are plotted as curves on the same graph, their shape and relative positions can be extremely revealing in terms of profitability and progress. For example, it can be seen from Figure 33.6 that the contract was potentially running at a loss during the first three weeks, since the value hours were less than the actual hours. Once the two curves crossed, profitability returned and in fact increased, as indicated by the diverging nature of the value and actual hour curves. This trend is also reflected by the final hours curve dipping below the budget hour line.

The percentage-time curves in Figure 33.7 enable the project manager to compare actual percentage complete with planned percentage complete. This is a better measure of performance than comparing actual hours expended with planned hours expended. There is no virtue in spending the manhours in accordance with a planned rate. What is important is the percentage complete in relation to the plan and whether the hours spent were *useful* hours. Indeed, there should be every incentive to spend *less* hours than planned, provided that the *value* hours are equal or greater than the actual, and the percentage complete is equal or greater than the planned.

The efficiency curve in Figure 33.7 is useful, since any drop is a signal for management action. Curve 'A' is based on the efficiency calculated by dividing the cumulative value hours by the cumulative actual hours for every week. Curve 'W' is the efficiency by dividing the value hours generated in a particular week by the actual hours expended in that week. It can be seen that Curve 'W' (shown only for the periods 5 to 9) is more sensitive to change and is therefore a more dramatic warning device to management.

Finally, by comparing the curves in Figures 33.6 and 33.7 the following conclusions can be drawn:

1. *Value hours exceed actual hours* (Figure 33.6). This indicates that the site is efficiently run.
2. *Final hours are less than budget hours* (Figure 33.6). This implies that the contract will make a profit.

WEEKLY TIMESHEET

Name			Staff no.					Week ending			

Project no.	Activity/ document no.	Mon.	Tues.	Wed.	Thu.	Fri.	Sat.	Sun.	Total	% complete	Remarks

Signed		Date		Approved		Date

Figure 33.9
Weekly time sheet

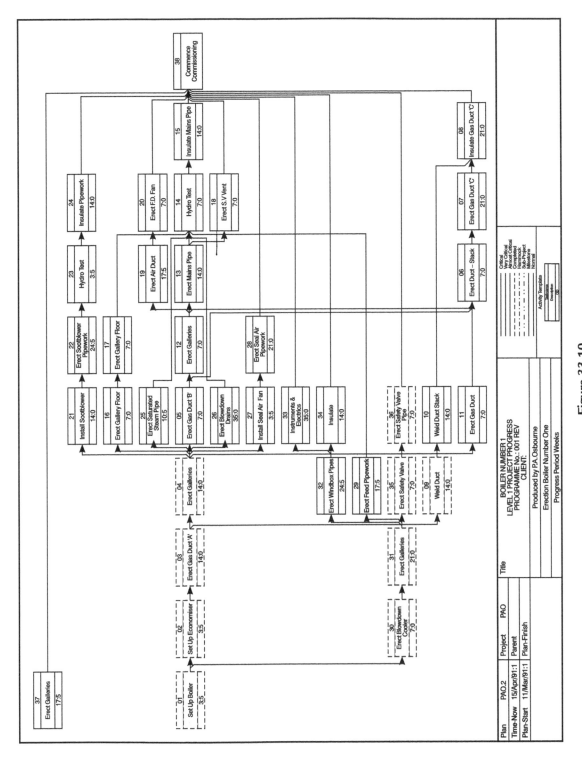

Figure 33.10
AoN diagram of boiler

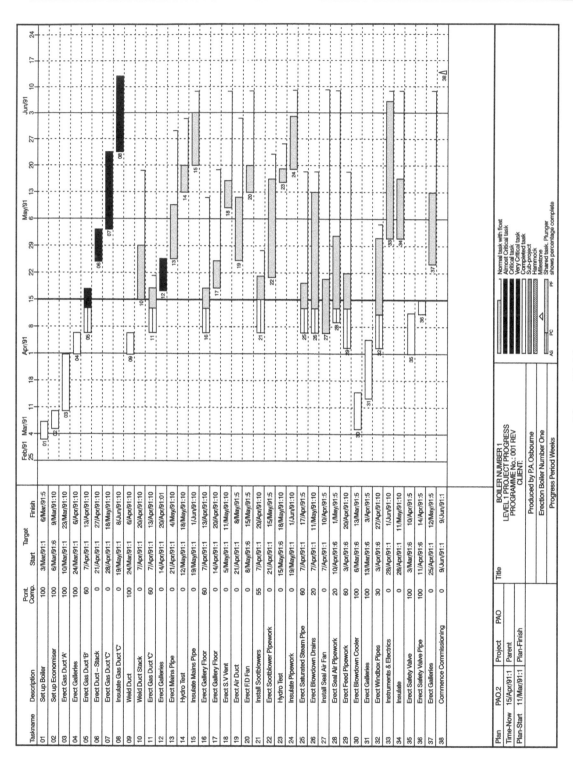

Figure 33.11
Bar chart

Activity Number	Description	Planned to date	Budget hours	Actual hours	Complete %	Value hours	Estimate comp	Forecast hours	Variance +/-	EFF	Orig durn	Rem durn	Early start	Early Finish	Late start	Late finish	Target start	Target finish	Actual start	Actual finish	Float remain	Programme status against target	Action required
01	Set up boiler	240	240	230	100%	240	0	230	10	104	3.5	0.0	3MAR91	6MAR91	3MAR91	6MAR91	3APR91	6MAR91	3MAR91	6MAR91	0.0	1 day (s) slippage	Complete
02	Set up economiser	110	110	90	100%	110	0	90	20	122	3.5	0.0	6MAR91	9MAR91	6MAR91	9MAR91	6APR91	9MAR91	6MAR91	9MAR91	0.0	On target	Complete
03	Erect gas duct 'A'	180	180	155	100%	180	0	155	25	116	14.0	14.0	10MAR91	23MAR91	10MAR91	23MAR91	10APR91	23MAR91	10MAR91	23MAR91	0.0	On target	Complete
04	Erect gas galleries	850	850	810	100%	850	0	810	40	105	14.0	0.0	24MAR91	6APR91	24MAR91	6APR91	24APR91	6APR91	24MAR91	6APR91	0.0	On target	Complete
05	Erect gas duct 'B'	263	263	200	60%	158	133	133	-70	79	7.0	2.8	7APR91	17APR91	7APR91	13APR91	7APR91	13APR91	7APR91	/	-3.8	4 day (s) slippage	Yes
06	Erect duct – stack	0	200	0	0%	0	200	200	0	0	7.0	7.0	24APR91	1MAY91	21APR91	27APR91	21APR91	27APR91	/	/	3.8	4 day (s) slippage	Yes
07	Erect gas duct 'C'	0	200	0	0%	0	200	200	0	0	21.0	21.0	1MAY91	22JUN91	28APR91	18MAY91	28APR91	18MAY91	/	/	3.8	4 day (s) slippage	Yes
08	Insulater gas duct 'C'	0	0	0	0%	0	0	0	0	0	21.0	21.0	22JUN91	12JUN91	19MAY91	8JUN91	19MAY91	8JUN91	/	/	3.8	4 day (s) slippage	Yes
09	Weld duct	70	70	65	100%	70	0	65	5	108	14.0	0.0	24MAY91	6APR91	24MAR91	6APR91	24MAR91	6APR91	24MAR91	6APR91	0.0	On target	Complete
10	Weld duct stack	108	200	200	0%	0	200	200	0	0	14.0	14.0	15APR91	28APR91	5MAY91	18MAY91	7APR91	20APR91	/	/	20.0	8 day (s) slippage	
11	Erect gas duct 'C'	125	250	92	60%	150	61	153	97	163	7.0	2.8	7APR91	17APR91	7APR91	20APR91	7APR91	13APR91	7APR91	/	3.2	4 day (s) slippage	
12	Erect galleries	0	500	0	0%	0	500	500	0	0	7.0	17.5	17APR91	24APR91	14APR91	20APR91	14APR91	20APR91	7APR91	/	-3.8	4 day (s) slippage	Yes
13	Erect mains pipe	0	270	0	0%	0	270	270	0	0	14.0	14.0	24MAY91	8MAY91	5MAY91	18MAY91	21APR91	4MAY91	/	/	10.2	4 day (s) slippage	
14	Hydro test	0	60	0	0%	0	60	60	0	0	7.0	7.0	13MAY91	19MAY91	19MAY91	25MAY91	12MAY91	18MAY91	/	/	6.0	1 day (s) slippage	
15	Insulate mains pipe	0	0	0	0%	0	0	0	0	0	14.0	14.0	20MAY91	2JUN91	26MAY91	11MAY91	19MAY91	1JUN91	/	/	6.0	1 day (s) slippage	Complete
16	Erect galley floor	850	850	420	60%	510	280	700	150	121	2.8	2.8	13APR91	13APR91	13APR91	11MAY91	7APR91	13APR91	7APR91	/	24.2	4 day (s) slippage	Complete
17	Erect galley floor	0	700	0	0%	0	700	700	0	0	7.0	7.0	17APR91	24APR91	12MAY91	18MAY91	14APR91	20APR91	/	/	24.2	4 day (s) slippage	
18	Erect S.V. vent	0	145	0	0%	0	145	145	0	0	7.0	7.0	8MAY91	15MAY91	2JUN91	8JUN91	5MAY91	11MAY91	/	/	24.2	4 day (s) slippage	
19	Erect air duct	0	203	0	0%	0	203	203	0	0	17.5	17.5	24APR91	12MAY91	15MAY91	1JUN91	21APR91	8MAY91	/	/	20.7	4 day (s) slippage	
20	Erect F.D. fan	0	240	0	0%	0	240	240	0	0	7.0	7.0	12MAY91	19MAY91	2JUL91	8JUL91	8MAY91	15MAY91	/	/	20.7	4 day (s) slippage	
21	Install sootblowers	70	140	65	55%	77	53	118	22	118	14.0	6.3	7APR91	21APR91	7APR91	27APR91	7APR91	20APR91	7APR91	/	6.7	0 day (s) slippage	
22	Erect sootblower pipework	0	400	0	0%	0	400	400	0	0	24.5	24.5	21APR91	15MAY91	28APR91	18MAY91	21APR91	11MAY91	/	/	6.7	0 day (s) slippage	
23	Hydro test	0	10	0	0%	0	10	10	0	0	3.5	3.5	15MAY91	19MAY91	22MAY91	25MAY91	15MAY91	18MAY91	/	/	6.7	0 day (s) slippage	
24	Insulate pipework	0	0	0	0%	0	0	0	0	0	14.0	14.0	19MAY91	2JUN91	26MAY91	8JUN91	19MAY91	1JUN91	/	/	6.7	0 day (s) slippage	
25	Erect saturated steam pipe	150	218	125	60%	131	83	208	10	105	10.5	4.2	7APR91	19APR91	7APR91	18MAY91	7APR91	17APR91	7APR91	/	29.8	2 day (s) slippage	
26	Erect blowdown drains	148	741	130	0%	148	520	650	91	114	35.0	28.0	7APR91	12MAY91	7APR91	18MAY91	7APR91	11MAY91	7APR91	/	6.0	1 day (s) slippage	
27	Install seal air fan	50	50	0	0%	0	50	50	0	0	3.5	3.5	7APR91	18APR91	7APR91	8JUN91	7APR91	10APR91	7APR91	/	51.5	8 day (s) slippage	
28	Erect seal air pipework	54	328	45	20%	66	180	225	103	146	21.0	16.8	10APR91	1MAY91	10APR91	8JUN91	10APR91	1MAY91	10APR91	/	38.2	0 day (s) slippage	
29	Erect feed pipework	173	273	145	60%	164	97	242	31	113	17.5	7.0	3APR91	21APR91	3APR91	18MAY91	3APR91	20APR91	3APR91	/	27.0	1 day (s) slippage	
30	Erect blowdown cooler	100	100	105	100%	100	0	105	-5	95	7.0	0.0	6MAR91	13MAR91	6APR91	13MAR91	6MAR91	13MAR91	6MAR91	13MAR91	0.0	1 day (s) slippage	Complete
31	Erect galleries	950	950	865	100%	950	1773	2533	85	110	21.0	21.0	13MAR91	3APR91	13MAR91	3APR91	13MAR91	3APR91	13MAR91	3APR91	0.0	1 day (s) slippage	Complete
32	Erect windbox pipes	819	1819	760	30%	546	0	0	-714	72	24.5	17.2	3APR91	2MAY91	3APR91	4MAY91	3APR91	27APR91	3APR91	/	2.8	1 day (s) slippage	
33	Instruments & electrics	0	0	0	0%	0	0	0	0	0	35.0	35.0	2MAY91	6JUN91	5MAY91	8JUN91	28APR91	1JUN91	/	/	2.8	4 day (s) slippage	
34	Insulate	0	0	0	0%	0	0	0	0	0	14.0	14.0	2MAY91	16MAY91	26MAY91	8JUN91	28APR91	11MAY91	/	/	23.8	4 day (s) slippage	
35	Erect safety valve	500	500	460	100%	500	0	460	40	109	7.0	0.0	3APR91	3APR91	3APR91	10APR91	3APR91	10APR91	3APR91	10APR91	0.0	1 day (s) slippage	Complete
36	Erect safety valve pipe	80	80	80	100%	80	0	80	0	100	7.0	0.0	11APR91	14APR91	11APR91	14APR91	11APR91	14APR91	11APR91	14APR91	0.0	1 day (s) slippage	Complete
37	Erect galleries	0	618	0	0%	0	618	618	0	0	17.5	17.5	25APR91	12MAY91	22MAY91	8JUN91	25APR91	12MAY91	/	/	27.5	On target	
38	Commence commissioning	0	0	0	0%	0	0	0	0	0	0.0	0.0	12JUN91	12JUN91	9JUN91	9JUN91	12MAY91	9JUN91	/	/	-3.8	4 day (s) slippage	Yes
Total		5882	11758	4842	43%	5029	6977	11819	-61	104													

Figure 33.12

Combined CPA and EVA print out

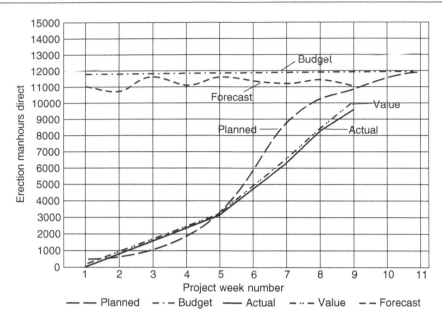

Figure 33.13
Boiler No. 1. Erection manhours

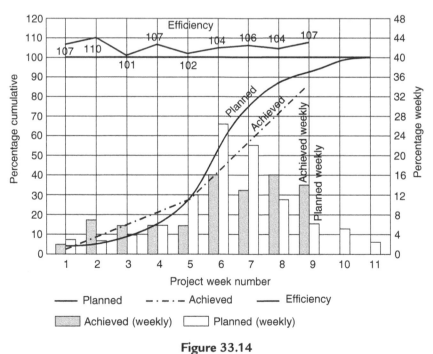

Figure 33.14
Boiler No. 1. Percentage complete and efficiency

3. *The efficiency is over 100% and rising* (Figure 33.7). This bears out conclusion 1.
4. *The actual percentage complete curve* (Figure 33.7), although less than the planned, has for the last four periods been increasing at a greater rate than the planned (i.e., the line is at a steeper angle). Hence the job may well finish *earlier* than planned (probably in Week 11).
5. By projecting the value hour curve forward to meet the budget hour line, it crosses in Week 11 (Figure 33.6).
6. By projecting the actual hour curve to meet the projection of the final hour curve, it intersects in Week 11 (Figure 33.6). Hence Week 11 is the probable completion date.

The computer printout shown in Figure 33.8 is updated weekly by adding the manhours logged against individual activities. However, it is possible to show on the same report the *cost* of both the historical and current manhours. This is achieved by feeding the average manhour rate for the contract into the machine at the beginning of the job and updating it when the rate changes. Hence the new hours will be multiplied by the current rates. A separate report can also be issued to cover the indirect hours such as supervision, inspection, inclement weather, general services, etc.

Since the value hour concept is so important in assessing the labour content of a site or works operation, the following summary showing the computation in non-numerical terms may be of help:

If

B = Budget hours (total)
C = Actual hours (total)
D = % complete
E = Value hours (earned value) (total)
F = Forecast final hours
G = Efficiency % (CPI)

Then:

$$E = B \times D, \quad D\% = E/B \times 100, \quad G\% = E/C \times 100, \quad F = C/D \text{ or } B/G$$

Overall Project Completion

Once the manhours have been 'costed' they can be added to other cost reports of plant, equipment, materials, subcontracts, etc., so that an overall percentage completion of a project can be calculated for valuation purposes on the only true common denominator of a project – money.

The total *value* to date divided by the revised budget × 100 is the percentage complete of a job. The value hour concept is entirely compatible with the conventional valuation of costing such as value of concrete poured, value of goods installed, cost of plant utilized – activities, which can, by themselves, be represented on networks at the planning stage.

Table 33.1 shows how the two main streams of operations, i.e., those categories measured by cost and those measured by manhours can be combined to give an overall picture of the percentage completion in terms of cost and overall cost of a project. While the operations shown relate to a construction project, a similar table can be drawn for a manufacturing process, covering such operations as design, tooling, raw material purchase, machinery, assembly, testing, packing, etc.

Cost of overheads, plant amortization, licences, etc., can, of course, be added like any other commodity. An example giving quantities and cost values of a small job involving all the categories shown in Table 33.1 is presented in Tables 33.2 to 33.4. It can be seen that in order to enable an overall percentage complete to be calculated, all the quantities of the estimate (Table 33.2) have been multiplied by their respective rates – as in fact would be done as part of any budget – to give the estimated costs.

Table 33.3 shows the progress after a 16-week period, but in order to obtain the value hours (and hence the cost value) of Category D it was necessary to break down the manhours into work packages that could be assessed for percentage completion. Thus, in Table 33.4, the pipelines A and B were assessed as 35% and 45% complete, respectively, and the pump and tank connections were found to be 15% and 20% complete, respectively. Once the value hours (3180) were found, they could be multiplied by the average cost per manhour to give a cost value of $14628.

Table 33.5 shows the summary of the four categories. An adjustment should therefore also be made to the value of plant utilization Category C since the two are closely related. The adjusted value total would therefore be as shown in Column V.

With a true value of expenditure to date of $104,048, the percentage completion in terms of cost of the whole site is therefore:

$$\frac{104048}{202000} \times 100 = 51.5$$

It must be stressed that the of cost completed is not the same as the completion of construction work. It is only a valuation method when the material and equipment are valued (and paid for) in their month of arrival or installation.

When the materials or equipment are paid for as they arrive on site (possibly a month before they are actually erected), or when they are supplied 'free issue' by the employer, they must not be part of the value or complete calculation.

Table 33.1

| Basic Method of Measurement | Cost (Money) | | | Manhours |
Method of Measurement Category	Bills of Quantities A	Lump Sum B	Rates C	Rate/Hour D
Type of activity	Earth moving Civil work Painting Insulation Piping supply	Tanks Equipment (compressors, pumps, towers, etc.)	Mechanical plant Cranes Scaffolding Transport	Erection of: piping, electrical work, instrumentation, machinery, steelwork; testing, commissioning
Base for comparison of progress	Total of bills of quantities	Total of equipment items	Plant estimate	Manhour budget
Periodic valuation	Measured quantities	Cost of items delivered	Cost of plant on-site	Value hours = % complete × budget
Method of assessment	Field measurement	Equipment count	Plant count	Physical % complete
Percentage complete for reporting	Measured quantities × rates / Total in bill of quantities	Delivered cost / Total equipment cost	Cost of plant on-site / Plant estimate	Value hours / Manhour budget

Total cost = measured quantity × rates + cost of items delivered + cost of plant on-site + actual hours × rate

$$\text{Total site percentage complete} = \frac{100 \; (\text{cost of A} + \text{cost of B} + \text{adjusted cost of C} + \text{value hours of D} \times \text{average rate})}{\text{Total budget}}$$

Methods of measurement

Table 33.2: Example Showing Effect of Percentage Completion of Different Categories

Estimate Category	Item	Unit	Quantity	Rate ($/hr)	Cost $
A	Concrete pipe	M³	1 000	25	25 000
	6-inch	M	2 000	3	6 000
	Painting	M²	2 500	10	25 000
					56 000
B	Tanks	No	3	20 000	60 000
	Pumps	No	1	8 000	8 000
	Pumps	No	1	14 000	14 000
					82 000
C	Cranes (hire)	Hours	200	6015	12 600
	Welding plant	Hours	400		6000
					18 000
D	Pipe fitters	Hours	4000	4} Av.	16 000
	Welders	Hours	6000	5} 4.6	30 000
			10 000		46 000

Table 33.3: Progress After 16 Weeks

Category	Item	Unit	Quantity	Rate ($/hr)	Cost $
A	Concrete poured	M³	900	25	22 500
	Pipe 6-inch supplied	M	1 000	3	3 000
	Painting	M²	500	10	5 000
					30 500
	% complete: $\dfrac{30500}{56000} \times 100 = 54.46\%$					
B	Tanks delivered	No	2	20 000	40 000
	Pumps A	No	1	8 000	8 000
	Pumps B	No	1		–
					48 000
	% complete: $\dfrac{48000}{82000} \times 100 = 58.53\%$					

It is clearly unrealistic to include materials and equipment in the complete and efficiency calculation as the cost of equipment is not proportional to the cost of installation. For example, a carbon steel tank takes the same time to lift onto its foundations as a stainless steel tank, yet the cost is very different! Indeed, in some instances, an expensive item of equipment may be quicker and cheaper to install than an equivalent cheaper item, simply because the expensive item may be more 'complete' when it arrives on site.

Table 33.4

Category	Item	Unit		Quantity	Rate ($/hr)	Cost $
C	Cranes on-site	Hours		150	60	9 000
	Welding plant	Hours		200	15	3 000
						12 000
		% complete : $\frac{12000}{18000} \times 100 = 66.66$				
D	Pipe fitters	Hours		1 800	4	7 200
	Welders	Hours		2 700	5	13 500
						20 700

Erection work	Budget M/H	Percentage complete	Value hours	Actual hours
Pipeline A	3800	35	1 330	1 550
Pipeline B	2800	45	1 260	1 420
Pump connection	1800	15	270	220
Tank connection	1600	20	320	310
	10000		3 180	3 500

$$\% \text{ complete}: \frac{3180}{10000} \times 100 = 31.80$$

cost value (Av.) = 3180 × 4.6 = \$14628

Table 33.5: Total Cost to Date

I	II	III	IV	V
Category	*Budget*	*Cost*	*Value*	*Adjusted Value*
A	56 000	30 500	30 500	30 500
B	82 000	48 500	48 000	48 000
C	18 000	12 000	12 000	10 920
D	46 000	20 700	14 628	14 628
Total	\$202 000	\$111 200	\$105 128	\$104 048

All the items in the calculations can be stored, updated, and processed by computer, so there is no reason why an accurate, up-to-date, and regular progress report cannot be produced on a weekly basis, where the action takes place – on the site or in the workshop.

Clearly, with such information at one's fingertips, costs can truly be controlled – not merely reported!

It can be seen that the value hours for erection work are only 3180 against an actual manhours usage of 3500. This represents an efficiency of only

$$\frac{3180}{3500} \times 100 = 91\% \text{ approx.}$$

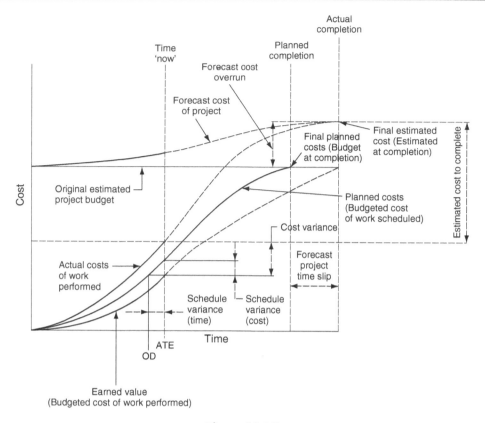

Figure 33.15
Earned Value Chart reproduced from BS 6079 'Guide to Project Management' by permission of British Standards Institution.

An adjustment should therefore also be made to the value of plant utilization, i.e., 12000 × 91% = 10,920. The adjusted value total would therefore be as shown in column V.

The SMAC system described on the previous pages was developed in 1978 by Foster Wheeler Power Products, primarily to find a quicker and more accurate method for assessing the % complete of multi-discipline, multi-contractor construction projects.

However, about 10 years earlier the Department of Defense in the USA developed an almost identical system called Cost, Schedule, Control System (CSCS), which was generally referred to as Earned Value Analysis (EVA). This was mainly geared to the cost control of defence projects within the USA, and apart from UK subcontractors to the American defence contractors, was not disseminated widely in the UK.

While the principles of SMAC and EVA are identical, there developed inevitably a difference in terminology, which has caused considerable confusion to students and practitioners. Figure 33.16 lists these abbreviations and their meaning, and Figure 33.17 shows the

ECTC	is Estimated Cost To Complete
BAC	is Budget At Competition (current) (budget)
BCWS	is Budgeted Cost of Work Scheduled (current) (planned)
BCWP	is Budgeted Cost of Work Performed (actual)
ACWP	is Actual Cost of Work Performed (actual)
OD	is Original Duration planned for the work to date
ATE	is the Actual Time Expended for the work to date
PTPT	is the planned Total Project Time
EAC	is Estimated Cost at Completion
ETPT	is Estimated Project Time
CPI	is Cost Performance Index = BCWP/ACWP = Efficiency
SPI	is Schedule Performance = BCWP/BCWS (cost based) = OD/ATE = % complete

Figure 33.16

comparison between the now accepted EVA 'English' terms (shown in **bold**) and the CSCS jargon (shown in *italics*).

The CSCS also introduced four new parameters for cost efficiency and, for want of a better word, time efficiency:

1. The cost variance: this is the arithmetical difference at any point between the earned value and the actual cost.
2. The schedule variance: this is the arithmetical difference between the earned value and the scheduled (or planned) cost. However, comparing progress in time by subtracting the planned cost from the earned value is somewhat illogical as both are measured in monetary terms. It would make more sense to use parameters measured in time to calculate the time variance. This can be achieved by subtracting the actual duration (ATE) for a particular earned value from the originally planned duration (OD) for that earned value. This is shown clearly in Figure 33.15. It can be seen that if the project is late, the result will be negative. There are therefore two schedule variances (SV):
   (a) SV (cost), which is measured on the cost scale of the graph;
   (b) SV (time), which is measured on the time scale.
3. The cost efficiency is called the cost performance index (CPI) and is
   earned value/actual cost.
4. The 'time efficiency' is called the schedule performance index (SPI) and is
   earned value/scheduled (or planned) cost.
   Again, as with the schedule variance, measuring efficiency in time by dividing the earned value by the planned cost, which are both measured in monetary terms or man-hours, is equally illogical and again it would be more sensible to use parameters measured in time. Therefore by dividing the planned duration for a particular earned value by the actual duration for that earned value a more realistic index can be calculated. All time measurements must be in terms of hours, day numbers, week numbers, etc. – not calendar dates!

EARNED VALUE

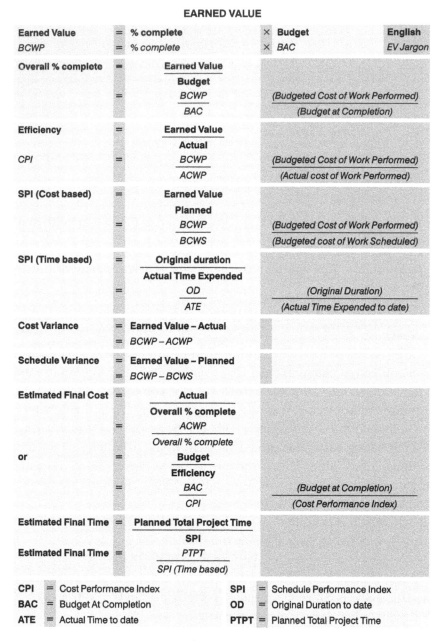

Earned Value	=	% complete		×	**Budget**	**English**
BCWP	=	*% complete*		×	*BAC*	*EV Jargon*

Overall % complete =	$\dfrac{\text{Earned Value}}{\text{Budget}}$	
=	$\dfrac{BCWP}{BAC}$	*(Budgeted Cost of Work Performed)* / *(Budget at Completion)*
Efficiency =	$\dfrac{\text{Earned Value}}{\text{Actual}}$	
CPI =	$\dfrac{BCWP}{ACWP}$	*(Budgeted Cost of Work Performed)* / *(Actual cost of Work Performed)*
SPI (Cost based) =	$\dfrac{\text{Earned Value}}{\text{Planned}}$	
=	$\dfrac{BCWP}{BCWS}$	*(Budgeted Cost of Work Performed)* / *(Budgeted cost of Work Scheduled)*
SPI (Time based) =	$\dfrac{\text{Original duration}}{\text{Actual Time Expended}}$	
=	$\dfrac{OD}{ATE}$	*(Original Duration)* / *(Actual Time Expended to date)*
Cost Variance =	**Earned Value – Actual**	
=	*BCWP – ACWP*	
Schedule Variance =	**Earned Value – Planned**	
=	*BCWP – BCWS*	
Estimated Final Cost =	$\dfrac{\text{Actual}}{\text{Overall % complete}}$	
=	$\dfrac{ACWP}{\text{Overall % complete}}$	
or =	$\dfrac{\text{Budget}}{\text{Efficiency}}$	
=	$\dfrac{BAC}{CPI}$	*(Budget at Completion)* / *(Cost Performance Index)*
Estimated Final Time =	$\dfrac{\text{Planned Total Project Time}}{\text{SPI}}$	
Estimated Final Time =	$\dfrac{PTPT}{\text{SPI (Time based)}}$	

CPI	=	Cost Performance Index	**SPI**	=	Schedule Performance Index
BAC	=	Budget At Completion	**OD**	=	Original Duration to date
ATE	=	Actual Time to date	**PTPT**	=	Planned Total Project Time

Figure 33.17
Comparison of EV terms

There are therefore now two SPIs:
(a) SPI (cost) measured on the cost scale, i.e., earned value/scheduled cost;
(b) SPI (time) measured on the time scale, i.e., planned duration/actual duration.

In practice, the numerical difference between these two quotients is small, so that SPI (cost), which is easier to calculate, is sufficient for most purposes, bearing in mind that the result is still only a prediction based on historical data.

In 1996 the National Security Industrial Association (NISA) of America published their own Earned Value Management System (EVMS), which dropped the terms such as ACWP, BCWP, and BCWS used in CSCS and adopted the simpler terms of Earned Value, Actual, and Schedule instead.

Since then the American Project Management Institute (PMI), the British Association for Project Management (APM), and the British Standards Institution (BSI) have all discarded the CSCS abbreviations and have also adopted the full English terms. In all probability this will be the future universal terminology.

Figure 33.17 clearly shows the earned value terms in both English (in **bold**) and EV jargon (in *italics*).

Earned Schedule

It has long been appreciated that Schedule Performance Index (cost) (SPI_{cost}) based on the cost differences of the earned value and planned curves is somewhat illogical. An index reflecting schedule changes should be based on the time differences of a project. For this reason Schedule Performance Index (time) (SPI_{time}) is a more realistic approach and gives more accurate results, although in practice the numerical differences between SPI_{cost} and SPI_{time} are not very great. SPI time for any point in time or 'time now' can be obtained graphically by dropping a vertical line from the planned curve to the time base line, from the point of the curve where the planned value is equal to the earned value for the selected 'time now'. This SPI_{time} is in effect the time efficiency of the project and has therefore been sometimes referred to 'earned schedule'.

In the same way that Budget Cost / CPI = the final predicted cost,

Estimated Duration/SPI_{time} = final completion time

It goes without saying that all units on the time scale must be in day, week, or month numbers, not in calendar dates!

Integrated Computer Systems

Until 1992, the EVA system was run as a separate computer program in parallel with a conventional CPM system. Now, however, a number of software companies have produced project management programs that fully integrate critical path analysis with earned value analysis. One of the best programs of this type, Primavera P6, is fully described in Chapter 51.

The system can, of course, be used for controlling individual work packages, whether carried out by direct labour or by subcontractors, and by multiplying the total *actual* manhours by the average labour rate, the cost to date is immediately available. The final results should be carefully analysed and can form an excellent base for future estimates.

As previously stated, apart from printing the EVA information and the conventional CPM data, the program also produces a computer drawn network. This is drawn in precedence format.

The information shown on the various reports include:

1. The manhours spent on any activity or group of activities
2. The % complete of any activity
3. The overall % complete of the total project
4. The overall manhours expended
5. The value (useful) hours expended
6. The efficiency of each activity
7. The overall efficiency
8. The estimated final hours for completion
9. The approximate completion date
10. The manhours spent on extra work
11. The relationship between programme and progress
12. The relative performance of subcontractors or internal subareas of work

Procurement

Chapter Outline

Project Management, Planning, and Control. http://dx.doi.org/10.1016/B978-0-08-098324-0.00034-2

Procurement is the term given to the process of acquiring goods or services.

The importance of procurement in a project can be appreciated by inspecting the pie chart (Figure 34.1) from which it can be seen that for a typical capital project, procurement represents over 80% of the contract value.

The main functions involved in the procurement process are:

1. Procurement strategy
2. Approved tender list
3. Pre-tender survey
4. Bidder selection
5. Request for quotation (RfQ)
6. Tender evaluation
7. Purchase order
8. Expediting, monitoring, and inspection
9. Shipping and storage
10. Erection and installation
11. Commissioning and handover

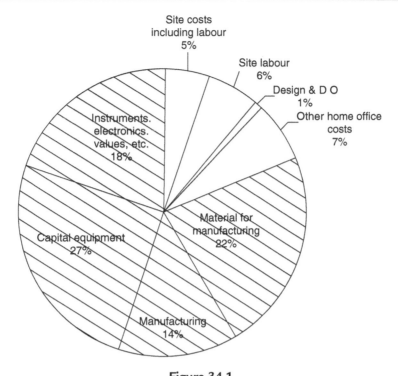

Figure 34.1
Total project investment breakdown. Procurement value = 18 + 27 + 14 + 22 = 81%

These main functions contain a number of operations designed to ensure that the desired goods are correctly described and ordered, are delivered when and where required, and conform to the specified quality and performance criteria.

The functions are described below, and will be followed by a more detailed discussion of the types of contracts and supporting requirements.

Procurement Strategy

Before any major purchasing operation is considered, a purchasing strategy must be drawn up which sets out the criteria to be followed. These include:

- The need and purpose of the proposed purchase
- Should the items be bought or leased?
- Should the items be made in-house?
- Will there be a construction element involved? If there is, it will be necessary to draw up subcontract documentation. If there is no construction or assembly element, a straight purchase order will suffice.
- How many companies will be invited to tender?

- Will there be open or selective tendering? This will often depend on the value and strategic or security restrictions. European Union regulations require contracts (subject to certain conditions) over a certain value to be opened up for bidding to competent suppliers in all member states.
- Are there prohibited areas for purchase or restrictive shipping conditions?
- What are the major risks associated with the purchase at every stage?
- What is the country of jurisdiction for settling disputes and what disputes procedures will be incorporated?
- What contract law is applicable, what language will be used, and in what currency will the goods be paid for?
- How will bids be opened and assessed and by whom? With public authorities tenders are usually opened by a committee.

Supply Chain

Every purchaser must bear in mind that most items of equipment contain components that are outsourced by the manufacturer or subcontractor to specialist companies. These companies may themselves subcontract subcomponents to other specialists, so that the original purchaser is utterly dependent on each subcontractor being technically and commercially rigorous and astute to ensure that the basic criteria of reliability, repeatability, and sustainability are maintained through the whole supply chain.

The technical failure of any component, or the non-compliance with the relevant environmental and health & safety legislation, can have serious repercussions on the reputation and profitability of the purchaser. The problem has been exacerbated by the globalisation of many manufacturing organisations where the inspection of the actual component or manufacturer's premises may be difficult and costly. A monitoring and reporting mechanism must therefore be set up right through the supply chain to ensure that technical and environmental standards are maintained.

This issue has to be confronted by every manufacturing industry whether it be construction, engineering, electronics, pharmaceuticals, textiles, toys, etc.

Approved Tender List

When the procurement strategy has been agreed, a list of approved (or client-nominated) tenderers can be drawn up.

On many large infrastructure or power plant projects it is beneficial to obtain the advice of specialist suppliers or contractors during the design stage. This will then inevitably lead to the possibility of reducing the number of competitive bids or even contravening of (in Europe) of EU procurement laws. In practice, many such projects will be constructed by consortia comprising civil, mechanical, and electrical contractors, and manufacturers, so that this risk has been transferred from the client/operator to the consortium.

Most major companies operate a register of the vendors who have carried out work on earlier contracts and have reached the required level of performance. These are generally referred to as 'approved vendor lists' and should only contain the names of those vendors who have been surveyed and whose capacities have been clearly established.

The use of computers enables such lists to be very sophisticated and almost certainly to contain information on company capacity and performance together with details of all the work the contractor has carried out and the level of his performance in each case.

Lists are normally prepared on a commodity basis, but they are only of real use if they are constantly updated.

A typical entry against a vendor on such a list includes:

- Company name and address
- Telephone, telex, and telefax no.
- Company annual turnover
- List of main products
- Value of last order placed
- Performance rating on:
  - adherence to price
  - adherence to delivery
  - adherence to quality requirements
- Ability to provide documents on time
- Co-operation during design stage
- Responses to emergency situations

Many organizations also keep a second list, comprising companies that have expressed their wish to be considered for certain areas of work, but are not on the approved vendors list. When the opportunity arises for such companies to be considered, they should be sent the pre-qualification questionnaires shown in Figure 34.2 and, if necessary, be visited by an inspector.

In this way, the list will be a dynamic document onto which new companies are added and from which unsatisfactory companies are removed.

Pre-Tender Survey

The approved vendor list is a useful tool, but it cannot be so comprehensive as to cover every commodity and service in every country of the world. For new commodities or new markets,

FOSTER WHEELER POWER PRODUCTS LTD
P.O. Box 160, Greater London House, Hampstead Road, London NW1 7ON

PRE-QUALIFICATION QUESTIONNAIRE

This questionnaire shall be completed by Vendors of direct materials.

1.0 General

1.1 Full legal Company name ....

1.2 Legal status of Vendor - whether a Private or Public Limited Liability Company, Partnership, Consortium, etc.

1.3 Registered office and/or other legal address ....
....

Telephone No: ....
Telex No: .... Fax No: ....
Telegraphic Address ....
Contact (English speaking) ....
Position Held by Contact ....
Vendor Type Manufacture/Distributor/Stockist/Agent

1.4 Full address of all Branch Offices ....
....
....

1.5 Full name & address of all Subsidiary & Associate Companies ....
....
....

1.6 Name & address of Parent or Holding Company ....

Telephone No: .... Telex ....
Contact .... Position
Annual Turnover

Where there is insufficient space for a full reply to a question Please attach additional sheets.

PRE-QUALIFICATION QUESTIONNAIRE

2.0 Description of Goods Supplied

2.1 Type of Goods & Ranges ....
....
....
....

2.2 Location & capacity of manufacturing & servicing facilities
....
....
....

2.3 List of design codes to which previous work has been produced
....
....

2.4 Licenses & monograms held ....

2.5 Test Facilities ....
....

2.6 Details of components normally sub-contracted to others
....
....

2.7 Quality Assurance Details in accordance with attached Quality Assurance Approval letter (2 sheets) and form A.

3.0 Financial

3.1 Levels of turnover of Vendor in equivalent pounds sterling:

Figure 34.2

PRE-QUALIFICATION QUESTIONNAIRE

a) Levels of turnover of Company for the last four financial years:-

198_
198_
198_
198_

3.2 Total value of orders waiting manufacture or in progress, in equivalent pounds sterling:

3.3 Companies expecting to supply goods totalling more than £25,000 sterling shall provide evidence of their financial standing. This shall include a statement of the situation for the current year to date (unaudited) and of the two previous years.

Please provide the following information:

a) Are shares quoted on Public Stock Exchange .... If so at which Exchange ....

b) Copy of Accounts of Company and of Group.

c) Liquidity Ratio i.e. Trade amounts to available cash.

d) Average Working Capital of Company and of Group.

3.4 Bank References

4.0 Work History

Client	Goods Supplied	Value	Date of Delivery

5.0 Number of Employees - Company Only

a) Administrative Staff   -
b) Graduate Engineers   -
c) Draughting Personnel   -
d) Quality Assurance Personnel   -
e) Inspection Personnel   -
f) Procurement Personnel   -
g) Qualified Welders   -
h) Machine Shop Personnel   -
i) General Works Staff   -

6.0 Comments

Date Completed ....   By ....

Signed ....

PRE-QUALIFICATION QUESTIONNAIRE (REVIEW)   FWPP Ref:
(NOT TO BE ISSUED TO VENDOR)

Vendor Name ....
Location ....
Date Questionnaire sent to Vendor ....
returned ....

Recommendation of Review:

Additional Contacts with Vendor:

Name 1) .... Position 1) ....
Name 2) .... Position 2) ....

Reviewed by Project Procurement Manager & Manager of Quality Assurance or his nominee

Recommendation

1) Enter Vendor on FWPP list   YES/NO
2) Reject   YES/NO

Project Procurement Manager   Signature ....
                                 Date ....

Manager of Quality Assurance or nominee
                                 Signature ....
                                 Date ....

Recommendation Agreed by Procurement Manager   YES/NO
                                 Signature ....
                                 Date ....

Figure 34.2 cont'd

therefore, a pre-tender survey is necessary. This is particularly important when a contractor is about to undertake work in a new country with whose laws and business practices he is not familiar.

The summary should be as comprehensive as possible and the buyer should draw information from every available source. The subject is discussed more fully later in this chapter under 'Overseas bidder selection', but the principles are equally applicable to surveys carried out in the UK or USA.

Bidder Selection

Before an enquiry can be issued, a list of bidders has to be compiled. Although the names of qualified companies will almost certainly be taken from the purchaser's approved vendor list, the actual selection of the prospective bidder for a particular enquiry requires careful consideration.

During the preparation stage of the equipment requisition, a number of suitable companies may well have been suggested to the buyer by the project manager, the engineering department, the construction department, and, of course, the client. While all this, often unsolicited, advice must be given serious consideration, the final choice is the responsibility of the procurement manager. In most cases, however, the client's and project manager's suggestions will be included.

The number of companies invited to bid depends on the product, the market conditions, and the 'imposed' names by the client. In no case, however, should the number of invited bidders be less than three, or more than ten. In practice, a bid list of six companies gives a reasonable spread of prices, but if too many of the invited companies refuse to bid, further names can be added to ensure a return of at least three valid tenders.

To ensure that the minimum number of good bids is obtained, some purchasers tend to favour a long bidders list. This practice is costly and time consuming, especially if the support documentation, such as specifications and drawings, is voluminous. A far better method is to telephone prospective bidders before the enquiry is sent out and, after describing the bid package in broad terms, obtain from the companies in question their assurance that they will indeed submit a bid when the documents are received. In this way the list can be kept to a reasonable size. If a vendor subsequently refuses to bid he runs the risk of being crossed off the next bid list.

While the approved vendor list is an excellent starting point for bidder selection, further research is necessary to ensure that the selected companies are able to meet the quality and programme requirements. The prospective vendor must be experienced in manufacturing the item in question, have the capacity to produce the quantity in time, be financially stable, and meet all the necessary technical and contractual requirements.

The buyer who keeps in touch with the market conditions can often obtain useful information from colleagues in other companies, from technical journals, and even the daily press on the workload or financial status of a particular supplier. An announcement in the financial columns that a company has obtained an order for 10,000 pumps to be delivered over two years should trigger off an investigation into the company's total capacity, before it is asked to bid for more pumps. Loading up a good supplier until he is overstretched and becomes a bad supplier is not in the best interest of either party.

When a purchaser lets it be known that a certain enquiry will be issued, the buyer will undoubtedly be contacted by many companies eager to be invited to bid. The information given by the salesman of one company must be compared with the, often contradictory, information from the opposition before a realistic decision can be taken. Only by being familiar with the product and the market forces at the time of the enquiry, can the buyer make objective judgements. On no account should vendors be included on a bid list to keep salesmen at bay or just to make up numbers.

For items not usually ordered by a purchaser, a different approach is necessary. In all probability no approved vendors list exists for such a product and unless someone in the procurement department, or in one of the other technical departments, knows of suitable suppliers, a certain amount of research has to be carried out by the buyer.

The most obvious sources of suitable vendors are the various technical trade directories and technical indexes available on microfilm or microfiche. In addition, a telephone call to the relevant trade association will often yield a good crop of prospective suppliers. Whatever the source, and whoever is chosen for the preliminary list, it will be prudent to send a questionnaire to all the selected companies, requesting the following details:

* Full company name and address (postal and website)
* Telephone, fax no., and e-mail address
* Company annual turnover
* Name of bank
* Names of three references
* Confirmation that the specified product can be supplied
* What proportion will be sub-contracted
* Name of parent company – if any

It is clear from the above that these questions must be asked well in advance of the enquiry date. It is necessary, therefore, for the buyer to scan the requisition schedule as soon as it is issued to extract such materials and items for which no vendors list has been compiled. In other words, the buyer will have to perform one of the most important functions of project management – forward planning!

Request for Quotation (RfQ)

The procurement manager will first of all:

* Produce a final bid list of competent tenderers
* Decide on the minimum and maximum number of bids to be invited
* Ensure that the specification and technical description of goods or services to be purchased are complete
* Decide on the type of delivery and insurances required (Ex works, FAS, FOB, etc.)
* Agree to the programme with delivery dates for information and goods with the project manager
* Decide on the most appropriate general conditions of purchase or contract
* Draft the special conditions of contract (if required)
* Decide the type and number of document requirements (manuals)
* Agree with project manager on the amount of liquidated damages required against late delivery
* Draft and Issue the Request for Quotation letter (RfQ)
* Enclose with the RfQ the list of drawings and other data required for submitting bids

It is important to include any special requirements over and above the usual conditions of contract with the enquiry documents. Failure to do so might generate claims for extras once the contract is under way. Such additional requirements may include special delivery needs and restrictions, access restrictions, monitoring procedures such as EVA, inspection facilities and currency restrictions, etc. Discounts and bonus payments are best discussed at the interviews with the preferred supplier/contractor after the tenders have been opened.

The following information must be requested *from* tenderers in the RfQ:

* Price and delivery as specified
* Discounts and terms of payment required
* Date of issue of advance date technical data, layout drawings, setting plans, etc.
* Production and delivery programme
* Expediting schedule for sub-orders
* Spares lists and quotation for spares
* Guarantees and warranties offered
* Alternative proposal for possible consideration

Tender Evaluation

The procurement manager must next:

- Decide on bid opening procedure and open bids
- Receive and log the bids as they arrive
- Assess previous experience and check reference contracts or sites
- Check financial stability of bidders
- Check list of past clients
- Set up bidders meetings
- Interview site or installation manager or foreman (if applicable)
- Discuss quantity discounts
- Negotiate early or other payment discounts
- Obtain parent company guarantees when the contract is with a subsidiary company
- Agree the maintenance (guarantee) period
- Discuss and agree bid, performance, maintenance, and advance payment bonds
- Produce bid summary (bid tabulation) (Figure 34.3)
- Carry out technical evaluation
- Carry out commercial evaluation

The following are the main items that have to be compared when assessing competing bids:

- Basic cost
- Extras
- Delivery and shipping cost
- Insurance
- Cost of testing and inspection
- Cost of documentation
- Cost of recommended spares
- Discounts
- Delivery period
- Terms of payment
- Retentions guarantees
- Compliance with purchase conditions

All the vendors' prices must, of course, be compared with the estimator's budgets, which will also appear on the sheet.

If the value of the bids (or at least one bid) is within the budget, the bid evaluations can proceed as described below. If, on the other hand, all the bids are higher than the budget,

COMMERCIAL SUMMARY OF BIDS

CONTACT _____ CLIENT _____ LOCATION _____ REQUISITION NO. _____

ITEM NO.	DESCRIPTION	BIDDERS						REMARKS
		1 BID DATE	2 BID DATE	3 BID DATE	4 BID DATE	5 BID DATE		

GUARANTEE REQUIRED
OBTAINED
GROSS PRICE
NEGOTIATED REDUCTION
ORDER VALUE
CURRENCY CONVERSION RATE
TERMS OF PAYMENT

TECHNICALLY APPROVED SUPPLIER/S

RECOMMENDED SUPPLIER/S

PRICE FIXED/SUBJECT TO ECALATION
RULING DATE
DELIVERY ITEMS
BUDGET PRICE
REQUIRED DELIVERY
PROMISED DELIVERY
ACCEPTANTCE OF GEN. COND.
BID VALIDITY DATE

REASON FOR RECOMMENDATION

APPROVALS	ORIGINATOR	PROC.	PROJ MGR.	DIRECTOR	CLIENT	PAGE
	DATE	DATE	DATE	DATE	DATE	OF

Figure 34.3
Bid summary

a meeting has to be convened with the project manager at which one of four decisions has to be taken, depending largely on the overall project programme:

1. If there is time, the enquiry can be reissued to a new list of vendors, spreading the area to include overseas suppliers, provided foreign suppliers are not excluded by the terms of the contract.
2. Review the original design or design standards to see whether savings can be made. While such savings may be difficult to make with equipment items, which have to meet specified technical requirements, it may be possible to effect cost savings in the type of finish, or the materials of construction. For example, a tank that was originally specified as galvanized may be acceptable with a good paint finish.
3. If there is no time to reissue the enquiry it may be necessary to call in the two lowest bidders and either negotiate their prices down to the budget level or ask them to rebid within a few days on the basis of a few quickly ascertainable design changes.
4. If it is discovered that the original budget estimate was too low, the lowest bid (if technically and commercially acceptable) will have to be submitted for approval by the project manager.

These commercial comparisons must be carried out for every enquiry. However, an additional technical assessment must be produced when the material and equipment is other than a general commodity item, which only has to comply with a material specification.

Although the purpose of this evaluation process is to find the cheapest bidder who complies with the specification and contractual requirements, the technical capability of the bidder as well as their track record of timely completion and past relationships (good or bad) with other stakeholders should also be taken into consideration. This may of course be a problem in itself with government contracts, where there can always be accusations of favouritism or even nepotism if the least expensive bidder has not been selected.

Purchase Order

When the selected bid has been agreed, the purchase order or contract document must be issued with all the same attachments, which were part of the request for quotation. The only additional documents are those containing the terms and conditions agreed at the bidder's meeting or other written agreements made during the bid evaluation process. These will include:

- The procedures for payments as set out in contract
- The stages for issuing interim and final acceptance certificates

Any changes to the specification or drawings, etc., after the date of the purchase order will be issued as a variation order and controlled using established configuration management procedures.

As the contract proceeds, the procurement manager must:

- Set priorities from programme and revisions to programme if necessary
- Set out bonded areas and marking when advance payments are considered
- Arrange to carry out regular expediting and check expediting reports
- Carry out stage inspection and specified tests and check inspection and test reports
- Carry out route survey (if required)
- Issue final delivery instruction for documents and goods
- Issue packing instructions
- Issue shipping information, sailing times
- Advise on shipping restrictions (conference lines requirements, etc.)
- Advise on current site conditions for storage
- Agree and finalise close out procedures
- Obtain details of after sales service

Expediting, Monitoring, and Inspection

Between the time of issuing the purchase order or contract documents and the actual delivery of the goods or services, the progress of the production (manufacture) of the goods must be monitored if quality and delivery dates are to be maintained.

The contract documentation will have (or should have) included the appropriate portion of the overall project programme. From this the contractor or supplier will be required to produce his own construction or manufacturing programme. This should be issued to the main contractor or client within two or three weeks of receipt of order and will be used to monitor progress.

It is beneficial if an expediter or inspector visits the works or offices of the supplier within two weeks of contract award to ensure that:

1. the contract documents have been received and understood;
2. the contractor's or supplier's own programme has been started or is ready to be issued;
3. the supplier or contractor has all the information he or she needs.

At regular intervals an expediter will then visit the supplier and check that the supplier's own programme is being met and in particular that sub-orders for materials and components have been placed, bearing in mind the lead times for these items. Any slippage to either the purchaser's or supplier's programme must be reported immediately to the project manager so that appropriate action can be taken. Where time is of the essence, a number of options are available to bring the delivery of the goods back on schedule. These are:

1. The purchaser's programme may have sufficient float (permissible delay) in the delivery string, which may make further action unnecessary, but pressure to deliver to the revised date must now be applied.

2. If there is no float available, the supplier must be urged to work overtime and/or weekends to speed up manufacture. It may be worthwhile for the purchaser to pay for these premium manhours in full or in part, as even if liquidated damages are imposed and obtained, the financial cost of delay, apart from loss of prestige or reputation, is often very much greater than the value of any liquidated damages received.
3. If there is no liquidated damages clause in the purchase agreement or contract but delivery by a stated date is a fundamental requirement, the purchaser may threaten the supplier with legal proceedings, breach of contract, or any other device, but while this may punish the supplier, it will not deliver the goods. For this reason, early warning of slippage is essential.

Because of the danger from the imposition of damages at large, due to possible delayed delivery, prudent suppliers, and especially contractors, should request that a liquidated damages clause be inserted into the contract, even if this has not been originally provided.

Similarly an inspector will visit the supplier early on to ensure that the necessary materials and certificates of conformity are either on order or available for inspection.

This expediting and inspection process should continue until all the items are delivered or the specified services are commenced.

Shipping and Storage

To ensure that the materials or equipment are delivered on time and at the correct location, checks should be carried out to ensure that, in the case of overseas procurement, there are no shipping, landing, or customs problems. Overland deliveries of large components may require a route survey to ensure that bridges are high or strong enough, roads wide enough, and the necessary local authority permits and police escorts are in place.

Special attention must be given to contracts when dealing with the construction industry. Here the timing of deliveries has to be carefully calculated to ensure that there is sufficient unloading and storage space, that adequate cranage is available, that there is no interference with other site users, and that adequate temporary protective materials are ready.

In the case of urban construction, interference with traffic or the general public must be minimized.

Erection and Installation

Once on site, fabrication, erection, and installation work should be regularly monitored using earned value techniques, which are related to the construction programme. Only in this way can the efficiency of a subcontractor be assessed and predictions as to final cost and completion be made.

As the cost of delays to completion can be many times greater than the losses to production or usage of the facility being constructed, the imposition of liquidated damages (even if possible) will not recover the losses and are at best a deterrent. A far better safeguard against late completion is the insistence that realistic and updated construction programmes are produced, regularly (preferably weekly) monitored, and, with the aid of network analysis and EVA techniques, kept on target. Although this may require additional resources and incentives, it may well be the best option.

As most construction contracts will have been placed on the basis of an agreed set of conditions of contracts (whether general or special) it will be necessary to check that the conditions are fully met, in particular those relating to quality, stage completion, and health and safety. It is also necessary to ensure that adequate drawings and commissioning procedures are ready when that stage of the contract commences.

Throughout the manufacturing or construction process, documents such as operating and maintenance procedures, spares lists, and as-built drawings must be collated and indexed to enable them to be handed over in a complete state when the official handover takes place.

Commissioning and Handover

Before a plant is handed over for operation, it must be checked and tested. This process is called commissioning and involves both the contractor and operator of the completed facility.

Careful planning is necessary especially if the plant is part of an existing facility that has to be kept fully operational with minimal disruption during the commissioning process. Integration with existing systems (especially when computers are involved) has to be seamless, and this may require the existing and new systems to be operated concurrently until all the teething problems have been resolved.

On certain types of plants, equipment will have to be operated 'cold', i.e., without the operating fluids (gases, liquids, or even solids) being passed through the system. Only when this stage has been successfully completed can 'hot' commissioning commence with the required media being processed in the final operating condition (temperature, viscosity, pressure voltage, etc.) and at the specified rates of usage (flow, wattage, velocity, etc.).

It is often convenient to involve the operating personnel in the commissioning process so that they can become familiar with the new systems and operating procedures.

After the various tests and pilot runs have been completed and the operating criteria and KPIs have been met to the satisfaction of the client, the new facility can be formally handed over for operation. This involves handing over all the stipulated documents, such as built drawings, operating and maintenance instructions, lubrication schedules, and spares lists, as well as commissioning records, test certificates of equipment, materials and operators, certificates of origins and compliance, guarantees, and warranties.

A close-out report will have to be written, which records the major events and problems encountered during the manufacturing, construction, and commissioning stages. This will be indexed and filed to enable future project managers to learn from past experience.

Types of Contracts

Contracts consist of three main types:

1. Lump sum contracts
2. Remeasured contracts
3. Reimbursable contracts

It is possible to have a sub-contract that is a combination of two or even all three types. For example, a mechanical erection sub-contract could have design content that is a lump sum fixed fee, a remeasured piping erection portion, and a reimbursable section for heavy lifts in which the client wants to be deeply involved. The main differences between the three types are as follows.

Lump Sum Contracts

A prerequisite to a lump sum (a fixed price) contract, with or without an escalation clause, is a complete set of specifications and construction drawings. These documents will enable the sub-contractor to obtain a clear picture of the works, assess the scope and quality requirements, and produce a tender that is not inflated by unnecessary risk allowances or hedged with numerous qualifications.

The trend in the USA is to have all the designs and drawings complete before tenders are invited and, provided such usual risk items as sub-soil, climatic, and seismic conditions are clearly given, a good competitive price can be expected.

Most lump sum contract documents should, however, contain a schedule of rates, so that variations can be quickly and amicably costed and agreed. Clearly, these variations should be kept to a minimum and must not exceed reasonable limits, since the rates used by the subcontractor are based on the tender drawings and quantities and a major change could affect his manhour distribution, supervision level, and site organization. A common rough limit accepted as reasonable is a value of variations of 15% of the contract value.

It must be stated that a variation can be a decrease as well as an increase in scope, and although the quantities may be reduced by the client, the reduction in price will not be proportional to that reduction. Indeed, when a sub-contract includes a design element, a reduction in hardware (say the elimination of a small pump), may increase the contract value due to costly drawing changes and cancellation charges.

Remeasured Contracts

Most civil engineering sub-contracts in the UK are let as remeasured contracts. In other words, the work is measured and costed (usually monthly) as it is performed in accordance with a priced bill of quantities agreed between the purchaser and the sub-contractor. The documents that are required for the tender are:

1. Specifications
2. General arrangement drawings
3. Bills of quantities

The bills of quantities are usually prepared by a quantity surveyor employed by the purchaser and are in fact only approximate, since the only drawings available for producing the bills are the general arrangement drawings and a few details or sketches prepared by the designers. Obviously, the more details that are available, the more accurate the bills of quantities are, but since one of the objectives of this type of contract is to invite tenders as quickly as possible after the basic design stage, full details are rarely available for the quantity surveyor. The sub-contractor prices the items in the bills of quantities, taking into account the information given in the drawings, the specification, the preambles in the bills, and, of course, the location and conditions of the proposed site.

Although the items in the bills of quantities are often described in great detail, they are included for costing purposes only, and do not constitute a specification. In the same way, the quantities given in the bill are for costing purposes only and are no guarantee that they will be the actual quantities required.

As with lump sum contracts, variations, which inevitably occur, must not exceed reasonable limits in either direction, since these could invalidate the unit rates inserted by the sub-contractor. For example, if the bills of quantities call for 10000 m^3 of excavation of a depth of 2 m, and it subsequently transpires that only 1000 m^3 have to be excavated, the sub-contractor is entitled to demand a rerating of this item on the grounds that a different excavator would have to be employed, which costs more per cubic metre than the machine envisaged at time of tender.

On remeasured sub-contracts, it is common practice to carry out a monthly valuation on site as a basis for progress payments to the sub-contractor. These valuations consist of three parts:

1. Value of materials on site, but not yet incorporated in the works;
2. Value of work executed and measured in accordance with the method of measurement stated in the bills of quantities;
3. Assessed value of preliminary items and provisional sums set out in the bills.

The value of materials, which have been paid for in a previous month (when they were delivered, but not yet incorporated in the works), is deducted from the measured works in the subsequent month (by which time they were incorporated) since the billed rates will include the cost of materials as well as labour and plant.

At the end of the contract period, the final account will require a complete reconciliation of the cost and values, so that any overpayments or underpayments will be balanced out.

Reimbursable Contracts

When a client (or purchaser) wishes to place a contract as early as possible, but is not in a position to supply adequate drawings or specifications, or when the scope has not been fully defined, a reimbursable contract is the most convenient vehicle.

In its simplest form, the contractor (or sub-contractor) will supply all the materials, equipment, plant, and labour as and when required, and will invoice the client at cost, plus an agreed percentage to cover overheads and profit. To ensure 'fair play', the client has the right to audit the contractor's books, check his invoices, and labour returns, etc., but he has little control over his method of working or efficiency. Indeed, since the contractor will earn a percentage on every hour worked, he has little incentive in either minimizing his manhour expenditure or finishing the job early.

To overcome the obvious deficiencies of a straight reimbursable (or cost-plus) contract, a number of variations have been devised over the years to give the contractor an incentive to be efficient and/or finish the job on time.

Most cost reimbursable contracts have two main components:

1. A fee component that can cover design costs, site and project management costs, overheads, and profits;
2. A prime cost component covering equipment, materials, consumables, plant, site labour, sub-contracts, and site establishment.

In some cases, the site establishment may be in the fee component or the design costs may be in the prime cost portion. The very flexibility of a reimbursable contract permits the most convenient permutation to be adopted.

By agreeing to have the fee portion 'fixed' the contractor has an incentive to finish the job as quickly as possible, since his fee and profits are recoverable over a shorter period and he can then release his resources for another contract. Furthermore, since he only recovers his prime cost expenditure at cost, he has absolutely no advantage in extending the contract – indeed, his reputation will hardly be enhanced if he finishes late and costs the client more money.

If the scope of work is increased by the client, the contractor will usually be entitled to an increase of the fixed fee by an agreed percentage, but frequently such an increase only comes into effect if the scope charge exceeds by 10% of the original contract value.

The factors to be considered when deciding on the constitution of the fixed fee and prime cost components are:

1. Time available for design;
2. Extent of the client's involvement in planning and design;
3. Extent of the client's involvement in site supervision and inspection;
4. Need to permit operations of adjacent premises to continue during construction. This is particularly important in extensions for factories, hotels, hospitals, or process plants;
5. Financial interest of the client in the contractor's business or vice versa;
6. Location of site in relation to the main area of equipment manufacture;
7. Importance of finishing by a specified date, e.g., weather windows for offshore operations, or committed sales of product;
8. Method of sharing savings or other incentives. It can be seen that if both parties are in agreement a contract can be tailored to suit the specific requirements, but the client still has only an estimate of the final cost.

Target Contracts

To counter some of the disadvantages of the straight reimbursable contracts, target contracts have been devised. In these contracts, an estimate of the works has been prepared by the employer (or his consultants) and agreed with a selected contractor. The prime cost is then frozen as a target cost the contractor must not exceed. The fee component is fixed, but it can also be calculated to be variable in such a way that the contractor has an incentive to complete the works below target or on programme or both.

Again, there are numerous variations of this theme, but the following are the more common methods of operation:

1. If the final measured prime costs are less than the target value, the difference is shared between the parties in a previously agreed proportion. If the final costs exceed the target value, the contractor pays the difference in full. The fee portion remains fixed.
2. As in 1, but the fee portion increases by an agreed percentage as the prime cost portion decreases. This gives the contractor a double incentive to complete the contract below the target value and thus increase the fee and ensure savings.
3. As in 1, but the fee portion increases by an agreed percentage for every week of completion prior to the contractual completion date.
4. The employer pays the final prime cost or the target cost, whichever is lower, but the difference, if any, is not shared with the contractor. The fee, however, is increased if there is a saving of prime cost or time or both.

If the contractor is responsible for purchasing equipment as part of the prime cost portion, the procurement costs such as purchasing, expediting, and inspection may be reimbursable at an hourly rate (subject to an annual review), but all discounts, including bulk and prompt payment discounts, must be credited to the employer. The contractor still has an incentive to obtain the best possible prices, since the prime cost will be lower.

Design, Build, and Operate Contracts (PPP & PFI)

These types of contracts have been developed since 1992 to reduce the financial burden on the public purse. This process is known as Private Public Partnership (PPP). A subset of PPP is Private Finance Initiative (PFI). These new types of contracts have a number of variations as described below, but their main purpose is for the private sector to take most if not all of the financial risk and employ their specialist knowledge and commercial experience, not always available in the public sector. Generally a special service company, known as a Special Service Vehicle (SPV) is formed, consisting of the contractors (building, maintenance, and operating) and the financiers (bank or other finance house) to construct, maintain, and operate the asset for a contractually agreed period. This company then signs the contract with the public authority (central or local) and arranges any leases with the actual operating organisation such as a prison or hospital.

The big difference between PFI and more conventional contracts is that the contractor finances the whole project from his own resources and then recoups the cost and profit from the operation revenues or from levies charged on the public sector organization that awarded the contract. The following three examples explain the different types.

Free Standing

A typical example of this type of PFI contract is where the public authority drafts a performance specification for a new bridge and asks a number of large contractors to submit costed schemes. The successful contractor then designs, builds, and maintains the bridge for a specified period. During this operating period, the cost and profit are recovered by the contractor/operator from tolls charged for the crossing. The contractor carries the risk that the volume of traffic will not reach the anticipated levels to yield the required revenues, possibly because the end user, the general public, considers the tolls to be too high.

Levies on the Public Sector

An example of such a contract is where the government requires a new prison. The successful contractor designs and builds the prison to the client's specification and operates it in accordance with the standards laid down by the prison service. As the end users, the prisoners, can hardly be expected to pay rent, an operating levy is charged on the prison service to cover building, financing, and operating costs plus profit.

Joint Ventures

Joint venture PFI projects are often used for road construction where there are no toll charges on the road users. Here the design and construction costs are shared by an agreed amount by the public and private sector and the contractor recovers his costs by a fee based on a benefit/revenue formula agreed in advance.

In all these types of contracts, the contractor is not chosen on the basis of the cheapest price, but on the viability of his business case, design concept, experience, technical expertise, track record, and financial backing. In many cases the contractor leads a consortium of specialist contractors, design consultants, and financial institutions.

Basic Requirements for Success

Whichever formula is agreed upon, it is clear that in any reimbursable contract, the employer retains a measure of control and hence a considerable responsibility. If the employer is also involved in the design process, the release of process information, construction drawings, and operating procedures must form part of the programme and should be marked as a series of key dates that must be kept if adherence to the programme is imperative.

The success of a contract (or sub-contract) depends, however, on more factors than a well-drawn-up set of contract conditions, and the following points must not be overlooked:

1. Good cost control of prime cost items
2. Good site management
3. Careful planning and programming
4. Punctual release of information to site
5. Timely deliveries of equipment and materials
6. Elimination of late design changes
7. Good labour relations
8. Good relationship between contracting parties

The main types of contracts are summarized in Figure 34.9.

Bonds

A bond is a guarantee given by third party, usually a bank or insurance company, that specified payments will be made by the supplier or contractor to the client if certain stipulated requirements have not been met. There are four main types of bonds a client may require to be lodged before a contract is signed. These are shown in order of submission:

1. Bid bond
2. Advance payment bond

3. Performance bond
4. Retention bond

Any of these bonds can be either conditional or on-demand. Conditional bonds, which are usually issued by an insurance company or similar financial institution, carry a single charge that is independent of the time the bond is in force and can only be called if certain predetermined conditions have been met. Such a condition might be that the supplier or contractor has to agree that the bond is called (i.e., that the money is paid to the client), or that the client or purchaser must prove loss to the satisfaction of the issuing house due to the default of the supplier. While such a bond may be very advantageous to the supplier, it is often regarded as unacceptable to a purchaser, since the collection and submission of evidence of default or proof of loss can be a time-consuming business.

The on-demand bond, on the other hand, has no such restrictions. As the name implies, it enables the purchaser to call in the bond as soon as and when he or she believes that a default by the supplier has occurred. Such a bond is normally issued by a recognized bank and will be paid without question and without the need for justification as soon as the demand for payment is made by the purchaser. Clearly the main element of such a bond is trust. Both the bank and the supplier trust the purchaser to be reasonable and honourable not to call the bond until the contractual terms permit it. These bonds cost more than a conditional bond and are only for a fixed duration, usually a particular stage of the project. They can, however, be extended for a further period for an additional fee (see Figure 34.4).

Apart from the benefit of speedier payment should there be a default by the supplier, another advantage to the purchaser of such a bond is that, as the cost of the bond depends on the bank's perception of the risk and the supplier's financial rating, a measure of the supplier's standing can be obtained. A low bond fee usually means that the supplier is regarded by the bank as reliable and financially stable.

Bid Bond

On major contracts many overseas clients require a bid bond to be submitted with the tender documents. The purpose of this bond, which is usually an on-demand bond, is to discourage the tenderer from withdrawing his bid after submission. This can be of considerable potential danger to a tenderer who discovers, after the bids have been dispatched, that there was an error in his tender price, or that other contractual requirements have been overlooked. Unless his price was originally higher than that of his competitors, the unfortunate tenderer has to decide whether to proceed with a potentially loss-making contract or to forfeit his bond.

The client will undoubtedly argue that the main purpose of the bond is to eliminate frivolous bids and ensure that those bids submitted are not only serious but also firm.

```
Foster Wheeler Power Products Limited.,
P.O. Box 160,
Greater London House,
Hampstead Road,
LONDON  NW1 7QN
```

BANK GUARANTEE DRAFT

We understand that you have entered into contract number ....

with .... for ....

at an agreed price of ....

In this connection, we hereby give you our guarantee in the sum of ....

.... such guarantee being effective in the

event that .... fail in their contractual

obligations in respect of this contract.

Claims under this guarantee are to be received by ....

at .... no later than ....

Such claims will be payable by us upon your first demand accompanied by your

statement that .... have

failed in their contractual obligations in respect of the contract specified

above.

Authorised signatories of Bank.

Figure 34.4
Bank guarantee draft

However, there can be considerable financial disadvantages to a tenderer since a bid bond, if issued by his bank, is equivalent to an overdraft, so that the working capital can be greatly reduced for a considerable period. When one considers that it can take between three months and a year to know whether a large contract has been won or lost, the loss in financing facilities and interest charges for the bond can be so great as to deter all but the largest contractors from tendering.

Advance Payment Bond

There are circumstances when a seller requires payments to be made before the goods are delivered. This arrangement is frequently required to finance expensive raw materials. The purchaser may also wish to make advance payments to reserve a place in the manufacturing queue or, as in the case of public authorities, to meet an expenditure deadline.

Until the goods are delivered, however, the purchaser has little or no guarantee that the advance payments will not be completely lost, should the supplier go into liquidation or the directors disappear to South America. To eliminate this risk, the purchaser requires the supplier to deposit with him a bond, usually underwritten by a bank, which guarantees a refund should any of the above misfortunes of the above type occur. The bond usually has a time limit that is often geared to a physical stage of the contract, such as the receipt by the purchaser of preliminary drawings or the arrival of raw materials. The latter stage is often accompanied by a certificate of ownership which vests the proprietorial rights with the purchaser.

Such a certificate is often supplemented by labels, which are affixed to the equipment or materials, declaring that the items marked are the property of the purchaser. This enables the purchaser to recover his goods (for which he has after all paid) should the vendor go into liquidation. The wording of such a notice should be vetted by the purchaser's legal advisers to ensure that the goods can, indeed, be recovered without further court action.

Where bulk materials have to be protected in this way, it is usual to fence off a 'bonded' area and erect notice boards at a number of locations. A typical notice of transfer of ownership is shown in Figure 34.5.

While an advance payment bond will usually be required for progress payments for work carried out off site, it is not normally required for work on site, since the completed works are the immediate property of the purchaser and could be finished by another sub-contractor in the case of bankruptcy or default (see Figure 34.6).

Performance Bond

This type of bond is more usually associated with sub-contracts and is an underwritten guarantee by a bank or other financial institution that the sub-contractor will perform his contract and complete the works as specified.

Even if the sub-contractor is paid by progress payments, the purchaser may still suffer considerable loss and frustration if the works are not completed due to the sub-contractor withdrawing from site.

The performance bond should be of sufficient value to cover the cost of finding and negotiating with a new sub-contractor and paying for the additional costs that the new sub-contractor may incur. There may, of course, be the additional costs of delays in completing the project, which are often far greater than the difference in price of two sub-contractors.

To be typed on Company Headed Paper

TRANSFER OF OWNERSHIP CERTIFICATE

In pursuance of invoice(s)   , dated   presented
by (insert full Company Name & Address) in respect of Foster Wheeler Power
Products Limited Purchase Order No: dated   and in
accordance with the terms and conditions thereof it is hereby warranted that
the materials, components and equipment used and all employees, agents,
sub-contractors and sub-suppliers employed by (insert full Company Name) for
the purposes of manufacturing and delivering the Goods and/or rendering the
services have been properly and fully paid in respect of the Goods listed
below whether completed or otherwise and are marked, identified and set aside
to become the absolute property of Foster Wheeler Power Products Limited
wherever they may be situated. Transfer of ownership of the Goods set aside
as aforesaid under the above numbered Purchase Order shall become effective to
the extent payment(s) are made in respect of the Goods listed below.

(Insert full Company Name) hereby agrees that the transfer of ownership shall
be without prejudice to any other warranty provided in accordance with the
Purchase Order.

List of Goods

. . . .   (To be signed by a Director of
                                                         your Company).
                                                         _____

Figure 34.5
Transfer of ownership certificate

Usually, the value of a performance bond is between 2 ½% and 5% of the contract value, which covers most contingencies.

Once the certificate of substantial completion has been issued, the performance bond is returned to the sub-contractor. Alternatively, the bond can be extended to cover the maintenance period and thus takes the place of the retention bond (see following section), provided, of course, that the percentage of the contract value is the same for both bonds (see Figure 34.7).

Retention Bond (or Maintenance Bond)

Many purchase orders and most sub-contracts require a retention fund to be established during the life of the manufacturing or construction stage. The purpose of a retention bond is to release the monies held by the purchaser at the end of the construction period and yet give the purchaser the available finance to effect any necessary repairs or replacements if the sub-contractor

TO BE ON BANK LETTERHEADED PAPER

ADVANCE PAYMENT BOND
BANK GUARANTEE
DRAFT
———————

We understand that you have entered into a contract with PIPEFABRO LTD.

for ECONOMISER TUBES at an agreed price of £30,000.

and we are informed that in this connection a Bank Guarantee for £3000

being 10% of the contract value is required.

In consideration of ALBAN POWER CO. paying the sum of £3000.

as advanced payment for goods to be delivered, we hereby give you

our Guarantee in the sum of £3000, such Guarantee being

effective in the event that PIPEFABRO LTD. fail in their contractual

obligations in respect of this contract.

Claims under this Guarantee are to be received by NAT.CITY BANK

at 12 HOWE STREET, LEEDS, no later than 1 p.m. 19TH OCTOBER, 1988,

Such claims will be payable by us upon your first demand accompanied

by your statement that PIPEFABRO LTD. have failed in their

contractual obligations in respect of the contract specified above.

This Guarantee should be returned to us on expiry.

Our maximum liability under this Guarantee is limited to the sum of

£3000.

This Guarantee shall be construed and shall take effect in all respects

in accordance with English Law.

Authorised Signatories of Bank.

Figure 34.6
Advance payment bond draft

or vendor fails to fulfil his contractual obligations during the maintenance period. The value of the bond is exactly equal to the value of the retention fund (usually between 2 ½% and 10% of the contract value), and is issued by either a bank or an insurance company.

When the maintenance period has expired and the final certificate of acceptance has been issued, the retention bond is returned by the purchaser to the sub-contractor, who in turn returns it to his bank (see Figure 34.8).

<u>Performance Bond</u>

Bond No. ....   Amount £..............

Know all men by these presents

That we,

As Principal, and the .... (hereinafter called the "Principal")
a corporation duly organised under the laws of England, (hereinafter called the
"Surety"), as Surety, are held and firmly bound unto ....
(hereinafter called the "Obligee"), in the sum of .... Pounds
(£..............), for payment of which sum well and truly to be made, we,
the said Principal and the said Surety, bind ourselves, our heirs, executors,
administrators, successors and assigns, jointly snd several, firmly by these
presents.

THE CONDITION OF THIS OBLIGATION IS SUCH, that whereas the Principal
entered into a certain Contract with the Obligee, dated ....19... for

In accordance with the terms and conditions of said contract, which is hereby
referred to and made a part hereof as if fully set forth herein;

NOW THEREFORE, THE CONDITION OF THIS OBLIGATION SUCH, that if
the above bounden Principal shall well and truly keep, do and perform each
and every, all snd singular, the matters and things in said contract set forth
and specified to be by said Principal kept, done and performed, at the times
and in the manner in said contract specified, or shall pay over, make good
and reimburse to the above named Obligee, all loss and damage by which said
Obligee may sustain by reason of failure or default on the part of said Principal
so to do, then this obligation shall be null and void; otherwise shall remain in
full force and effect.

Sealed with our seals and dated this .... day of ....
A.D. nineteen hundred and ....

                             ....
                                   Principal
                            By.........................
                            By.........................
                           And......................

Figure 34.7
Performance bond

Letter of Intent

If protracted negotiations created a situation where it is vital to issue an order quickly to meet the overall project programme, it may be necessary to issue a letter or fax of intent. Formal purchase orders, especially if extensive amendments have to be incorporated, can take days if not weeks to type, copy, and distribute. A device must thus be found to give the vendor a formal instruction to proceed to enable the agreed delivery period to be maintained.

The letter of intent fulfils this function, but unless properly drafted it can turn out to be a very dangerous document indeed. Invariably the buyer tends to be brief, restricting the letter or fax to essentials only. The danger lies in the fact that by being too brief, he may underdefine the contract, leaving the position open for an unscrupulous or genuinely confused vendor to lodge claims for extras. To make matters worse, instructions to proceed may have to be given before a number of apparently minor contractual points have been fully agreed, and while the buyer may try to build a safeguard into the letter by a clause, such as: 'This authority is given subject to final agreement being reached on the outstanding matters already noted', he has not, in fact, protected anybody.

The following examples show how a letter or fax of intent should and should **not** be drafted.

Good letter

Following our Invitation to Bid and your quotation No. 2687 of ...... together with all subsequent documentation, please accept this Fax as your instruction to proceed with the works.

This Authority is given subject to final agreement being reached on the outstanding matters already noted.

Bad letter

This fax gives the vendor the right to start work and incur costs which can be recovered by him even if the final negotiations break down and the formal contract is not issued. A fax of intent should be drafted on the following lines:

Following an Invitation to Bid of the ...... and your quotation No. 2687 of ...... together with Amendments Nos. 1, 2, 3 and 4, and Minutes of Meeting of ......, and ......, please proceed with the design portion of the works and the preparation of sub-order requisitions to a max. value of £2000 to maintain a contract completion of ......

The firm order for the remainder of the contract of the agreed value of £59 090 (subject to adjustment) will be issued if the outstanding matters, i.e., amount of liquidated damages and cost of extended drive shafts are agreed by the ......

MAINTENANCE RETENTION BOND

Bond No: ....   Amount £.............

KNOW ALL MEN BY THESE PRESENTS,

That we,

                                        (hereinafter called the "Principal"),
as   Principal,   and   a
corporation duly organised under the laws of the   , and duly
licensed to transact business in the State of   (hereinafter
called the "Surety"), as Surety, are held and firmly bound unto

                                        (hereinafter called the "Obligee"),
in the sum of   Pounds
(£..............), for the payment of which sum well and truly to be made, we,
the said Principal and the said Surety, bind ourselves, our heirs, executors,
administrators, successors and assigns, jointly and severally, firmly by these
presents.

Sealed with our seals and dated this   day of   ,
A.D. nineteen hundred and   .

WHEREAS, the said Principal has heretofore entered into a contract with said
Obligee dated   , 19   , for

                                                          and;

WHEREAS, the said Principal is required to guarantee the
installed under said contract, against defects in materials or workmanship,
which may develop during the period.

NOW, THEREFORE, THE CONDITION OF THIS OBLIGATION IS SUCH, that if said
Principal shall faithfully carry out and perform the said guarantee, and
shall, on due notice, repair and make good at its own expense any and all
defects in materials or workmanship in the said work which may develop during
the period

or shall pay over, make good and reimburse to the said Obligee all loss and
damage which said Obligee May sustain by reason of failure or default of said
Principal so to do, then this obligation shall be null and void; otherwise
shall remain in full force and effect.

                              ....
                              Principal

                              ....
                              BY

Figure 34.8
Retention bond

Summary of main types of contract

	Lump sum	Remeasured	Reimbursable
Documents required	Schedule of rates for variations	Bill of quantities	Little definition
Design requirement	Full design	Almost full design	Basic design
	Full specification Full set of drawings	Full specification Almost full drawings	Part specification Basic drawings
Price	Fixed or subject to escalation	Fixed or subject to escalation	Preliminary, subject to escalation
Client's involvement	Minimum	Negotiations of star rates and extras	Monitoring of manhours. Auditing of costs of materials
Supervision	Quality only	Quality and variations to contract	Close quality control and variation
Advantages	Price known	Drawings need not be complete	Can start early on site
Disadvantages	Drawings must be complete	Costs could rise as design is changed	Final cost could be very high. Contractor has little incentive to reduce costs

Target contract
Fee for contractor is fixed
Prime cost (materials and labour) is frozen
If final prime cost is less than target, saving is shared
If final prime cost exceeds target, contractor pays the difference
Contactor's fee increases as prime cost decreases
Contractor's fee increases if overall project time is reduced

Figure 34.9
Summary of main types of contract

This fax of intent is undoubtedly longer, but it contains all the essential information and tells the vendor what his limits of expenditure are before the final order is placed. The vendor also knows the scope of supply (including all the agreed amendments) and the date by which the equipment has to be delivered. By releasing the vendor to commence the design and sub-order preparation, the delivery date will not be jeopardized, provided, of course, that the stated outstanding issues are resolved.

The vendor realizes that he may, in fact, still lose the order if he does not come to terms with the purchaser, and this gives him an incentive to complete the deal.

The fax also states what the contract sum (subject to the negotiated adjustments) will be and what the items are that are subject to adjustment.

Clearly, the best procedure is to be in a position to issue the formal purchase order as soon as the negotiations have been completed. This can be done provided the buyer works up to the preparation of the purchase order during the negotiation phase. As clauses or specification details are amended and agreed, they are added to the draft purchase order document so that when the final meeting has taken place, any last-minute extra paragraphs can be added and the price and delivery boxes filled in. It should then be possible to send the final draft to the typing section within 24 hours.

A further advantage of following the above procedure is that the buyer is aware of, and can make quick reference to, the current status of the discussions with the vendor so that he can brief other members of the organization at short notice.

Sub-Contracts

Definition of Sub-Contracts

The difference between a sub-contract and a purchase order is that the sub-contract has a site labour content. The extent of this content can vary from one operative to hundreds of men. The important point is that the presence of the man on site requires documents to be included in the enquiry and contract package that set out the site conditions for labour and advise the sub-contractor of the limitations and restrictions on the site. While this distinction is undoubtedly true, there are numerous cases where the decision between issuing a relatively simple purchase order or a full set of sub-contract documents is not quite as straightforward as it would appear.

For example, if an order is placed for a gas turbine and it is required that the manufacturers send a commissioning engineer to site to supervise setting up and commissioning, does this constitute a site labour content or not? Similarly, if a control panel vendor prefers to complete the wiring of a panel on site (possibly due to programming requirements) and has then to send two or three technicians to site, can this be classed as a sub-contract?

There are undoubtedly good reasons why, if at all possible, the issuing of a full set of subcontract documents should be avoided. The cost of collating and issuing what is often a very thick set of contractual requirements, site conditions, specifications, safety regulations, etc., is obviously greater than the few pages that constitute a normal purchase order. Furthermore, the vendor has to read and digest all these instructions and warnings and may well be inclined to increase his price to cover for conditions that may not even relate to his type of work. On the other hand, if a vendor brings a man onto the site who performs similar work to other site operatives but is paid more, or belongs to an unacceptable trade union (or no union), or works longer hours, or enjoys unspecified conditions better than the other men, the effect on site labour relations may be catastrophic. The cost of even half a day's strike is infinitely greater than a bundle of contract documents.

It can be seen, therefore, that there is a grey area that can only be resolved in the light of actual site conditions known at the time, plus a knowledge of the scope of work to be carried out by the vendor's site personnel. The following guidelines may be of some assistance in deciding the demarcation between a purchase order and a sub-contract, but the final decision must reflect the specific labour content and site conditions.

Typical sub-contracts

- Demolition
- Site clearance and fencing
- Civil engineering
- Steel erection
- Building work and decorating
- Mechanical erection and piping
- Electrical and instrumentation installation
- Insulation application
- Painting
- Specialist tray erection
- Specialist telecommunication installations
- Specialist tank erection
- Specialist boiler or heater erection
- Water treatment
- Effluent treatment
- Site refractory works
- Site cleaning (including office cleaning)
- Security and night watchmen
- Radiography and other non-destructive testing (NDT)

Sub-Contract Documents

The documentation required for a sub-contract can be roughly classified into three main groups:

1. Commercial conditions
2. Technical specification
3. Site requirements

Although all three types of documents are interrelated, they cover very different aspects of the contract and are therefore prepared by different departments in the purchaser's organization.

The commercial conditions are usually standardized for a particular contract or industry, and if not actually written by the commercial or legal department, are certainly vetted and agreed by them.

The technical specification may be prepared by the relevant technical department and includes the necessary technical description, material and work specifications, standards, drawings, data sheets, etc.

The site requirements originate from the construction department or client and set out the site conditions, labour restrictions, safety and welfare requirements, and programme (sometimes called the schedule).

The sub-contract manager's function is to pull these three sets of documents together and produce one combined set of papers that tell the sub-contractor exactly what he must do, how, where, and when.

Commercial Conditions – General

The conditions of sub-contract, like the general or main conditions of contract, are most effective if they follow a standardized and familiar form. Most civil engineers are conversant with the Institution of Civil Engineers' (ICE) General Conditions of Contract and NEC3, and every mechanical engineer should at least have a knowledge of MF/1/ as published by the Institution of Mechanical Engineers (I.Mech.E.) and the Institution of Electrical Engineers (IEE). In 1993 the ICE published the New Engineering Contract (NEC), called the Engineering & Construction Contract (ECC). This has since been updated and is now known as NEC3. The NEC family of contracts now covers contract conditions for main contractors, subcontractors, professional services, supply and adjudicators. A table of the more important standard conditions of contract, which frequently form the basis of the sub-contract conditions, is given in Figure 34.10, but it is not imperative that any of these standard conditions be used. Many large companies, such as oil companies, chemical manufacturers, or nationalized industries, have their own conditions of contract. In turn, many of the contractors, whether civil or mechanical, have their own conditions of sub-contract. Generally, the terms and clauses of all these conditions are fairly similar, since if they were unreasonably onerous, contractors would either not quote or would load their tenders accordingly. However, there are differences in a number of clauses a prospective tenderer would be well advised to heed. Such differences are often incorporated by the purchaser in the light of actual unfortunate experiences he has no intention of repeating. One can well imagine the commercial officer writing these conditions and applying the adage that the difference between a wise man and a fool is that a wise man learns from his experience.

The alternative to using standard conditions, whether issued by established institutions or by the purchaser's organization, is to write tailor-made general conditions for a particular project. This is usually only viable when the project is very large and when a multitude of sub-contracts is envisaged. There are considerable advantages for the purchaser or main contractor in tailoring the conditions to a particular project, since in this way the same base

STANDARD CONDITIONS OF CONTRACT

NEW ENGINEERING CONTRACT (NEC3) CONDITIONS OF CONTRACT AND FORMS OF TENDER, AGREEMENT AND BOND FOR USE IN CONNECTION WITH WORKS OF CIVIL ENGINEERING CONSTRUCTION:	Institution of Civil Engineering Association of Consulting Engineers Federation of Civil Engineering Contractors
JCT 80 STANDARD FORM OF BUILDING CONTRACT:	Joint Contracts Tribunal (JCT) Royal Institute of British Architects National Federation of Building Trades Employees Royal Institution of Chartered Surveyors
MODEL FORM OF GENERAL CONDITIONS OF CONTRACT (INCLUDING FORMS _ OF AGREEMENT AND GUARANTEE)	Institution of Mechanical Engineering Institution of Electrical Engineers Association of Consulting Engineers
GENERAL CONDITIONS OF CONTRACT FOR STRUCTURAL ENGINEERING WORKS:	Institution of Structural Engineers
MODEL CONDITIONS OF CONTRACT FOR PLANT (INCLUDING ERECTION)	EB (ELECTRICITY BOARD) B E A M A (British Electrical and Allied Manufacturers Association)
CONDITIONS OF CONTRACT (INTERNATIONAL) FOR WORKS OF CIVIL ENGINEERING CONSTRUCTION	F I D E C Fédération Internationale des Ingénieres - Conseils
MODEL FORM OF CONDITIONS OF CONTRACT FOR PROCESS PLANTS (SUITABLE FOR LUMP SUM CONTRACTS IN THE U.K.)	Institution of Chemical Engineers
MODEL CONDITIONS OF CONTRACT FOR REPAIR, MODIFICATION AND REHABILITATION OF BOILERS AND ASSOCIATED PLANT. (CONDITONS RMR)	GB(GENERATING BOARD) WTBA (WATERTUBE BOILERMAKERS ASSOCIATION)
GENERAL CONDITIONS OF GOVERNMENT CONTRACT FOR BUILDING AND CIVIL ENGINEERING WORKS (GC/WORKS 1)	H.M. GOVERNMENT

Figure 34.10
Standard conditions of contract

documents can be used for every discipline. In other words, instead of the civil contractor being governed by the ICE conditions, the piping erection contractor by model form 'A', and the insulation contractor by the Thermal Insulation Contractors Association (TICA) conditions, all the sub-contractors must work to the same general conditions written especially for the project. To ensure that the various disciplines can work to one set of conditions, great care must be taken in their compilation. Since most of the clauses must be applicable to all the sub-contracts, they should be of a general nature. Clauses specific to a particular discipline or trade are collected together in what are known as 'special conditions of sub-contract'. These are described later.

Obviously, such a comprehensive set of conditions will contain clauses that are not relevant to some of the disciplines. This problem is overcome by either incorporating a list of non-relevant clauses in the accompanying special conditions, or relying on the common sense of all parties to ignore clauses that are not usually applicable by custom and practice. For example, a clause relating to underground hazards (usually in a civil contract) would be irrelevant in an insulation contract.

The advantages of a common set of general conditions are:

1. There is no confusion at the issuing stage as to which conditions of contract must be used for a specific sub-contract.
2. The site sub-contract administrator becomes conversant with the terms of the contract and will thus find it easier to administer them.
3. There is no risk of contradiction between certain terms that may have a different interpretation in different standard conditions. A typical example is Clause 24 in the I.Mech.E. model form 'A' of general conditions of contract. This clause lists industrial disputes as a reason for granting an extension of time. The corresponding clause in the ICE conditions (Clause 4A) does not list this particular occurrence as a valid claim for extension of time. Clearly, it is highly desirable that such an important factor as industrial disputes has the same implications for all contractors on a particular site.

Special Conditions

As mentioned earlier, one way of advising the tenderer that certain clauses in the general conditions are not applicable to his particular contract is to list all those non-applicable conditions in a special conditions of contract that form part of the package.

Where the general conditions have *not* been tailor-made for a contract, the special conditions contain all those clauses peculiar to a particular site, especially the labour relations procedures. In theory, general conditions of contract apply to any site in the UK (overseas sites usually require separate conditions), so that particular items such as site establishment requirements, utility facilities, security, site car parking, site agreement notifications, and other special clauses must be drawn to the attention of the tenderer in a separate document. Because of the specific nature of these clauses, special conditions of contract usually precede the general conditions in the hierarchy of importance. In other words, a modification or qualification in the special conditions takes precedence over the unqualified clause in the general conditions. Other clauses in the special conditions are terms of payment and, of course, the form of agreement.

Technical Specification

The technical portion of the sub-contract document consists of six main sections:

1. Description of work
2. Specification and test requirements
3. Bills of quantities (if applicable)
4. List of drawings
5. List of reports to be submitted and details of cost codes
6. Payments schedule (if related to work packages)

Some organizations also include the planning schedule and insurance requirements in this section, but these two items are more logically part of the site requirements and will be dealt with later.

Description of Work

Again, this section can be divided into two parts:

1. Description of the site and a general statement of the objectives relating to the project as a whole;
2. Description of that portion of the work relating to the sub-contract in question.

Thus, sub-section 1 would state the purpose of the project (e.g., to produce 1000 tonnes of cement per day using the dry process, etc.). Sub-section 2 would describe (in the case of a civil sub-contract) which structures are in concrete, which are steel with cladding, the extent of roads, pavings, and sewers, and the soil conditions likely to be encountered.

Needless to say, more detailed technical descriptions will appear on the drawings, in the technical specifications and in the bills of quantities, giving, in effect, the scope of the subcontract.

Liquidated Damages (or Ascertainable Liquidated Damages)

Liquidated damages have been defined by Lord Dunedin in a court case in 1913 as 'a genuine covenanted pre-estimate of damages', and as such is the compensation payment by a vendor to a purchaser when the goods were not delivered by the contract date. In cases of sub-contracts, liquidated damages can be imposed if the contract is not completed by the agreed date.

Liquidated damages are not penalties. They are primarily designed to cover the losses suffered by a purchaser because the goods or services were not available to him by the agreed date. As the amount of liquidated damages was agreed by both parties in advance, the purchaser does not have to prove he has lost money. The fact that the goods are late is sufficient reason for claiming the damages.

Over the years, however, liquidated damages have been assessed in quite an arbitrary way that bears no relationship to the losses suffered. Usually, they are calculated as a percentage of the contract value and vary with the number of days or weeks for which the goods have been delayed.

In most cases the Courts will uphold such a clause, provided the actual amount of liquidated damages is less than the amount that could have been realistically shown to have been the loss. It is argued that both parties knew at the time of signing the contract that the loss would probably be greater, but agreed to the lower figure. If, on the other hand, the amount is greater than the real loss and the vendor could demonstrate to the Courts that the purchaser was, in fact, imposing a penalty, then the clause would not be enforceable.

A normal figure used for assessing liquidated damages is ½% per week of delay with a maximum of 2 ½%. This means that the vendor's maximum liability becomes operative after a five weeks' delay and is limited to 2 ½% of the contract value. If the purchaser does not really need the goods, even after five weeks' delay, he can still claim his 2 ½%, which is, in effect, pure profit. On the other hand, if, because of the delay of one item of equipment, the whole plant remains inoperative, his losses could be enormous. The receipt of a miserable 2 ½% of the value of one relatively small item is insignificant.

It can thus be seen that the real purpose of liquidated damages is to encourage the vendor to deliver on time, since a loss of 2 ½% represents a large proportion of his profit. It is quite naïve to suggest that the vendor should pay the true value of a loss that could be suffered by a purchaser, which could be many times greater than the cost of the goods in question.

If no liquidated damages clause is included in the purchase order, the purchaser may claim damages at large, and may, indeed, recover the full, or a substantial proportion of the full amount of his loss, due to the goods being delayed. For this reason many vendors actually request that a liquidated damages clause is inserted, so that their liability is limited to the agreed amount.

For large sub-contracts, it is prudent to produce some form of calculation for assessing the amount of liquidated damages, since if they are challenged, they must be shown to be reasonable. There are a number of ways these can be assessed:

1. If the whole plant has been prevented from producing the desired product, the loss of net profit per week of production can be used as a basis;
2. If the works are non-profit earning, such as a road or reservoir, the additional weekly interest payment on the capital cost is a realistic starting point;
3. If the delayed items hold up work by another sub-contractor, the waiting time for plant and additional site overheads are considered as real losses. To these could be added the standby time of labour if it cannot be redirected to other work.

Liquidated damages may be imposed on the total contract or on sections. This means that late delivery of layout or even final drawings could be subject to liquidated damages. The amount of these damages could easily be calculated as the manhours of waiting time by engineers being held up for information.

After all these calculations have been produced, the total value of the damages must be compared with the contract value of the goods. If the amount is high in relation to the contract value, it must be reduced to a figure that a vendor can accept. At the end of the day, if the purchaser requires the goods, he must find a vendor who is prepared to supply them.

Insurance

Normally a purchaser–contractor requires his goods to be fully insured from the point of manufacture up to the stage when the client has taken over the whole project. In practice, this insurance is effected in a number of stages, which vary with the terms of the main contract between the contractor and his client. The more usual methods adopted are as follows:

1. The manufacturer insures the goods from the time they leave his works to the time they are off-loaded on site. The insurance cover for this stage ceases when the contractor's crane lifts the goods off the transport. The contractor's all-risk insurance policy now covers the goods until they are actually taken over by the client.
2. The manufacturer insures the goods in 1 above – the *client's* overall site insurance policy covers the goods as soon as they are lifted off the transport. In such circumstances the goods will be paid for at the next payment stage and will become the property of the client, although they may, in fact, not yet be erected or installed. Depending on the terms of the conditions of purchase, the goods will have become the property of the purchaser when delivered to site or paid for, whichever was earlier.

For large capital projects, the second method is the more common for the following reasons:

1. A large site may involve a number of contractors, all of whom have to insure their works. The cost of this insurance will, if provided by the different contractors, have to be paid eventually by the client as part of the contract sum. By taking out his own insurance for the total value of all the various contractors' works the client will be able to negotiate far better terms with a large insurance company than if the different works were insured individually.
2. Most contractors require payment for materials delivered or erected in accordance with agreed terms of payment, which form part of the contract. When these payments are made, that portion of the finished works becomes the property of the client. It is reasonable, therefore, for the client to be responsible for the insurance also.

It can be seen that if the contractor's insurance were to cover the goods from receipt on site to the date of payment, a whole series of insurance changeover dates would have to be agreed. The additional administrative problems would be both time consuming and costly.

3. In many cases, the new works will be constructed on a site close to, or even integrated in an existing operational plant owned or run by the client. Any damage to the existing plant, due to an accident on the new plant, can be covered by the same insurance policy.

4. The project, though large in itself, may only be a part of an even bigger project, e.g., an onshore oil terminal may be part of a major development of an offshore oil field involving a number of oil rigs. In such a situation, the client will negotiate a massive insurance policy, perhaps with a consortium of insurers, at a really attractive rate.

Needless to say, the goods will only be covered by the client's policy once they have arrived on the job site. If the goods have to be stored temporarily in an off-site warehouse, the contractor will have to arrange for insurance, even if the goods have been paid for in the form of advance payments.

The exact stages at which the insurance risk passes from the seller to the buyer depends on the conditions of purchase of the purchaser and the shipping terms. For a more detailed explanation see the section on Incoterms, which discusses the shipping responsibilities used internationally by all the trading nations.

Discounts

During the pre-order discussions with the prospective supplier, the buyer must try to reduce the price as much as possible. This can be achieved by asking the supplier to give a price reduction in the form of a discount. These, often considerable, reductions can take the form of:

1. Negotiated discounts and hidden discounts
2. Bulk purchase discounts
3. Annual order discounts
4. Prompt payment discounts
5. Discount for retention bond

Negotiated Discounts

There comes a stage during most negotiations when all the technical points have been resolved and all the commercial conditions agreed. However, the final price can still be unresolved since the very technical and commercial points discussed have probably affected the original bid. This is the time for the buyer to bring up the question of discounts. The arguments put forward could be:

1. The technical requirements are now to a different specification requiring less material, etc.;
2. The commercial conditions are now less onerous.

Both these changes could warrant a price reduction. If, on the other hand, the opposite is the case, i.e., the specification is higher or the conditions harsher, a 'hidden discount' can be obtained by insisting that the price remain as tendered. To clinch the deal, the vendor may well agree to this at this stage. A salesman would be very loath to return to his Head Office without an order, having got so far in the negotiations.

It must be remembered that there is no such thing as a fixed profit percentage. Most salesmen are allowed to negotiate between prescribed limits, and it is the buyer's job to take advantage of these margins. When the bid analysis is prepared, the discounts obtained should be shown separately so that the bid price can be checked against the original tender documents. This is especially important if the bid price is made up of a number of individual prices that have to be compared with those of competitors.

Bulk Purchase Discounts

When large quantities of a particular material have to be purchased, the vendor, in order to make the offer more attractive, may offer a bulk purchase discount on the basis that some of the economies of scale can be shared with the purchaser. If such a discount is not volunteered by the vendor, it can still be suggested by the buyer.

Annual Order Discounts

A vendor may offer (or be persuaded to offer) a discount if the purchaser buys goods whose total value over a year exceeds a pre-determined amount. This will encourage a purchaser to order all similar items of equipment, say electric motors, from the same vendor. The items may be of different size or specification, but will still be obtained from the same supplier. At the end of the year, a percentage of the total value of all orders is paid back to the purchaser as a discount.

Prompt Payment Discount

Although the conditions of sale may stipulate payment within 30 days of the date of the invoice (assuming the item has been received by the purchaser in good condition), many companies tend not to pay their bills unless the vendor has issued repeated requests or even threatened legal action. To encourage the prompt payment of invoices, an additional discount is frequently offered. The value of this is usually only a few per cent and reflects the financing charges the vendor may have to pay due to late receipt of cash.

Discount for Retention Bond

Most contracts or sub-contracts contain a retention clause, which requires a percentage of the contract value to be retained by the purchaser for periods of between six and twelve months.

To improve the vendor's cash flow, a retention bond can often be accepted by the purchaser, which guarantees the retention value, but this will deprive the purchaser of the use of these monies during the retention period. To compensate the purchaser for this, a vendor may offer a discount, which in effect is a proportion of the interest charges the vendor would have to pay for borrowing the retention sum from a bank. A usual procedure is to split the interest charges 50:50 between the purchaser and the vendor. In this way both parties gain by the transaction.

It can be seen that discounts can frequently be obtained from a supplier, especially if it is a buyer's market. In most reimbursable cost contracts, all discounts except prompt payment discounts must be passed on to the client for whom the goods or services have been purchased. For this reason, all negotiations including the discounts offered must be open and properly documented so that they can stand up to any subsequent audit.

Counter-Trade

Despite the name, this is not meant to refer to trade carried out over a shop counter, although this use of the term is commonly applied to goods collected from a wholesaler's premises. In the case of international business, the term refers to the payment for goods or services by something other than money. In other words, it is akin to good old-fashioned barter.

The difference between barter and counter-trade is that in barter, one type of goods or services are exchanged for another without money being involved, while in counter-trade, the goods supplied by the buyer are delivered to a third party who sells them (usually at a profit) for the benefit of the seller who then receives cash.

A simple example illustrates how the system works: a potential client in a developing country may need to extend his production facilities. His business may be expanding and highly profitable, but because of government restrictions the company has no access to hard currency. It is in the country's national interest to encourage industrial growth at home, but not to increase its national debt by borrowing dollars or pounds. A new approach is needed and one solution is to resort to counter-trade. If, for example, the country is rich in some natural resource, such as coal, this may be the most convenient commodity to trade-off against the proposed factory extension. The expanding company will buy the coal from the mine in local currency. The UK supplier will provide the production facility expansion and receive an appropriate quantity of coal as payment.

Incoterms

World trade inevitably requires goods to be shipped from one country to another. Raw materials must be transported from the less developed countries to the developed ones, from which finished goods are sent in the opposite direction. Both movements have to be packed, insured, transported, cleared through customs, and unloaded at their point of destination, and in order to standardize the different conditions required by the trading partners, INCOTERMS (Figure 34.11), were developed. These trade terms cover 14 main variations and encompass the spectrum of cost and risk of shipments from 'ex works' where the buyer has all the risk and pays all the costs, to 'delivered duty paid' where the seller contracts to cover delivery costs and insurance.

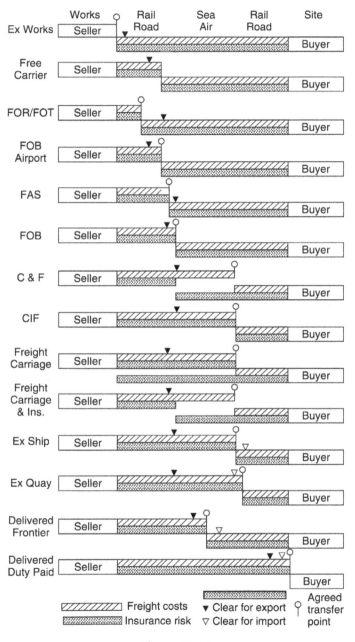

Figure 34.11
Incoterms

Ex works

'Ex works' means that the seller's only responsibility is to make the goods available at his premises (i.e., works or factory). In particular, he is not responsible for loading the goods in the vehicle provided by the buyer, unless otherwise agreed. The buyer bears the full cost and risk involved in bringing the goods from there to the desired destination. This term thus represents the minimum obligation for the seller.

Free Carrier (Named Point)

This term has been designed to meet the requirements of modern transport, particularly such 'multi-modal' transport as container or 'roll-on roll-off' traffic by trailers and ferries. It is based on the same main principle as FOB except that the seller fulfils his obligations when he delivers the goods into the custody of the carrier at the named point. If no precise point can be mentioned at the time of the contract of sale, the parties should refer to the place or range where the carrier should take the goods into his charge. The risk of loss of or damage to the goods is transferred from seller to buyer at the time and not at the ship's rail. 'Carrier' means any person by whom or in whose name a contract of carriage by road, rail, air, sea, or a combination of modes, has been made. When the seller has to furnish a bill of lading, way bill, or carrier's accept, he duly fulfils this obligation by presenting such a document issued by a person so defined.

FOR/FOT

FOR and FOT mean 'free on rail' and 'free on truck'. These terms are synonymous since the word 'truck' relates to the railway wagons. They should only be used when the goods are to be carried by rail.

FOB Airport

FOB airport is based on the same main principle as the ordinary FOB term. The seller fulfils his obligations by delivering the goods to the air carrier at the airport of departure. The risk of loss of or damage to the goods is transferred from the seller to the buyer when the goods have been so delivered.

FAS

FAS means 'free alongside ship'. Under this term the seller's obligations are fulfilled when the goods have been placed alongside the ship on the quay or in lighters. This means that the buyer has to bear all costs and risks of loss of or damage to the goods from that moment. It should be noted that, unlike FOB, this term requires the buyer to clear the goods for export.

FOB

FOB means 'free on board'. The goods are placed on board a ship by the seller at a port of shipment named in the sales contract. The risk of loss of or damage to the goods is transferred from the seller to the buyer when the goods pass the ship's rail.

C & F

C & F means 'cost and freight'. The seller must pay the cost and freight necessary to bring the goods to the named destination, but the risk of loss of or damage to the goods, as well as of any cost increases, is transferred from the seller to the buyer when the goods pass the ship's rail in the port of shipment.

CIF

CIF means 'cost, insurance and freight'. This term is basically the same as C & F but with the addition that the seller has to procure marine insurance against the risk of loss of or damage to the goods during carriage. The seller contracts with the insurer and pays the insurance premium.

Freight Carriage – Paid to ...

Like C & F, 'freight or carriage – paid to ...' means that the seller pays the freight for the carriage of the goods to the named destination. However, the risk of loss of or damage to the goods, as well as of any cost increases, is transferred from the seller to the buyer when the goods have been delivered into the custody of the first carrier and not at the ship's rail. It can be used for all modes of transport including multi-modal operations and container or roll-on or roll-off traffic by trailers and ferries. When the seller has to furnish a bill of lading, waybill, or carrier's receipt, he duly fufils this obligation by presenting such a document issued by the person with whom he has contracted for carriage to the named destination.

Freight Carriage – and Insurance Paid to ...

This term is the same as 'freight or carriage paid to ...' but with the addition that the seller has to procure transport insurance against the risk of loss of or damage to the goods during the carriage. The seller contracts with the insurer and pays the insurance premium.

Ex ship

'Ex ship' means that the seller shall make the goods available to the buyer on board the ship at the destination named in the sales contract. The seller has to bear the full cost and risk involved in bringing the goods there.

Ex quay

'Ex quay' means that the seller makes the goods available to the buyer on the quay (wharf) at the destination named in the sales contract. The seller has to bear the full cost and risk involved in bringing the goods there.

There are two 'ex quay' contracts in use, namely 'ex quay (duty paid)', and 'ex quay (duties on buyer's account)' in which the liability to clear the goods for import are to be met by the buyer instead of by the seller.

Parties are recommended to use the full description of these terms always, namely 'ex quay (duty paid)', and 'ex quay (duties on buyer's account)', or uncertainty may arise as to who is to be responsible for the liability to clear the goods for import.

Delivered at Frontier

'Delivered at frontier' means that the seller's obligations are fulfilled when the goods have arrived at the frontier – but before 'the customs border' of the country named in the sales contract. The term is primarily intended to be used when goods are to be carried by rail or road, but it may be used irrespective of the mode of transport.

Delivered Duty Paid

While the term 'ex works' signifies the seller's minimum obligation, the term 'delivered duty paid', when followed by words naming the buyer's premises, denotes the other extreme – the seller's maximum obligation. The term 'delivered duty paid' may be used irrespective of the mode of transport.

If the parties wish the seller to clear the goods for import but that some of the costs payable upon the import of the goods should be excluded – such as VAT and/or other similar taxes – this should be made clear by adding words to this effect (e.g., 'exclusive of VAT and/or taxes').

CHAPTER 35

Value Management

In a constantly changing environment, methods and procedures must be constantly challenged and updated to meet the needs and aspirations of one or more of the stakeholders of a project. This need for constant improvement was succinctly expressed by the first Henry Ford when he said he could not afford to be without the latest improvement of a machine.

Value management and its subset, *value engineering*, aim to maximize the performance of an organization from the board room to the shop floor. Value management is mainly concerned with the strategic question of 'what' should or could be done to improve performance, while value engineering concentrates more on the tactical issues of 'how' these changes should be done.

Value can be defined as a ratio of function/cost, so in its simplest terms, the aim is to increase the functionality or usefulness of a product while reducing its overall cost. It is the constant search for reducing costs across all the disciplines and management structures of an organization without sacrificing quality or performance that makes value management and value engineering such an essential and rewarding requirement.

The first hurdle to overcome in encouraging a value management culture is inertia. The inherent conservatism of 'if it ain't broke, don't fix it' must be replaced with 'how can a good thing be made better?'. New materials, better techniques, faster machines, more sophisticated programs, and more effective methods are constantly being developed and in a competitive global economy, it is the organization that can harness these developments and adapt them to its own products or services that will survive.

The search and questioning must therefore start at the top. Once the strategy has been established, the process can be delegated. The implementation, which could cover every department and may include prototyping, modelling, and testing, must then be monitored and checked to ensure that the exercise has indeed increased the function/cost ratio. This process is called *value analysis*.

The objectives should be one or more of the following: eliminating waste, saving fuel, reducing harmful emissions, reducing costs, speeding production, improving deliveries, improving performance, improving design, streamlining procedures, cutting overheads, increasing functionality, and increasing marketability. All this requires one to 'think value' and challenge past practices, even if they were successful.

Project Management, Planning, and Control. http://dx.doi.org/10.1016/B978-0-08-098324-0.00035-4
341

In an endeavour to discover what areas of the business should be subjected to value analysis, brainstorming sessions or regular review meetings can be organized, but while such meetings are fundamentally unstructured, they require a good facilitator to prevent them from straying too far off the intended route.

Value analysis can be carried out at any stage of the project as can be seen from the simplified life cycle diagram of Figure 35.1. For the first two phases it is still at the 'What' stage and can be called value planning while during the implementation phase it is now at the 'How' stage and is known as value engineering. The diagram has been drawn to show value management during the project phases, i.e., before handover. However, value management can be equally useful when carried out during the operation and demolition phases in order to reduce the cost or manufacturing time of a product, or simplify the dismantling operations, especially when, as with nuclear power stations, the decommissioning phase can be a huge project in its own right.

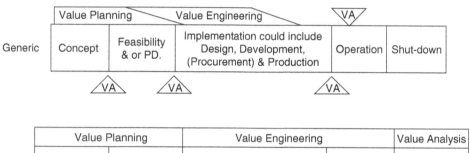

Figure 35.1
VM and the project life cycle

In addition to brainstorming, a number of techniques have been developed to systemize or structure the value engineering process of which one of the best known ones is FAST, or *function analysis system technique*. This technique follows the following defined stages:

1. Collect and collate all the information available about the product to be studied from all the relevant departments, clients, customers, and suppliers.
2. Carry out a functional analysis using the 'Verb and Noun' technique.
    This breaks down the product into its components and the function (verb) of each component is defined. The appropriate noun can then be added to enable a cost value to be ascertained. This is explained in the following example:

It has been decided to analyse a prefabricated double glazed widow unit. The functions in terms of verbs and nouns are:

Verbs	**Nouns**
Transmit	Light, Glass
Eliminate	Draughts, Seals
Maintain	Heat, Double glazing
Facilitate	Cleaning access, Reversibility
Secure	Handles, Locking catches

Each function and component can now be given a cost value and its percentage of the total cost calculated.

3. Find alternative solutions. For example, it may be possible to reduce the thickness of the glass but still maintain the heat loss characteristics by increasing the air gap between the panes. It may also be cheaper to incorporate the lock in the handles instead of as a separate fitting.
4. Evaluation. The suggested changes are now costed and analysed for a possible saving and the function/cost ratio compared with the original design.
5. Acceptance. The proposed changes must now be approved by management in terms of additional capital expenditure, marketability, sales potential, customer response, etc.
6. Implementation. This is the production and distribution stage.
7. Audit. This is carried out after the product has been on the market for a predetermined time and will confirm (or otherwise) that the exercise has indeed given the perceived additional value or function/cost ratio. If the results were negative, the process may have to be repeated.

Value management is not only meeting the established success criteria or KPIs but improving them by periodic reviews. Having previously carried out a stakeholder analysis and identifying their needs, it should be possible to meet these requirements even if the costs have been reduced. Indeed customer satisfaction may well be improved and environmental damage reduced, resulting in a win–win situation for all the parties.

CHAPTER 36

Health and Safety and Environment

Chapter Outline

In the light of some spectacular company collapses following serious lapses and shortcomings in safety, health and safety is now on the very top of the project management agenda. Apart from the pain and suffering caused to employees and the public by accidents attributable to lax maintenance of safety standards, the inability to provide high standards of safety and a healthy environment is just bad business. Good reputations built up over years can be destroyed in a day due to one serious accident caused by negligence or lack of attention to safety standards. In addition, under the Corporate Manslaughter Act 2007, Directors of companies can now be held responsible for fatalities caused by contraventions of H&S regulations.

It is for this reason that the British Standards Institution's 'Guide to project management in the construction industry' BS 6079 Part 4: 2006 has placed the 'S' for safety in the centre of the project management triangle, indicating that a project manager can juggle the priorities between cost, time, and performance, but he must never compromise safety.

Health and Safety was given a legal standing with the British Health and Safety at Work Act 1974. This creates a legal framework for employers to ensure that a working environment is maintained in which accidents and unhealthy and hazardous practices are kept to a minimum.

Subsequent legislation included:

- Management of Health and Safety Regulations 1992
- Control of Asbestos at Work Regulations 1987
- Noise at Work 1989
- Workplace (H, S & Welfare) Regulation 1992
- Personal Protective Equipment Regulation 1992
- Manual Handling Operations Regulations 1992
- Fatal Accidents Act 1976
- Corporate Manslaughter Act 2007

As well as a raft of European Community directives such as EC Directive 90/270/EEC.

Project Management, Planning, and Control. http://dx.doi.org/10.1016/B978-0-08-098324-0.00036-6

The Act set up a Health and Safety Executive that has wide powers to allow its inspectors to enter premises and issue improvement or prohibition notices as well as instigating prosecutions where an unsafe environment has been identified. The Act also gives legal responsibilities to employers, employees, self-employed persons, designers, manufacturers, suppliers, and persons generally in control of premises where work is performed.

Each of these groups has been identified as to their responsibilities for health, safety, and welfare, which broadly are as follows.

Employer

- Provide and maintain safe equipment
- Provide safe and healthy systems
- Provide a safe and healthy workplace
- Ensure safe handling and storage of chemicals and toxic substances
- Draw up a health and safety policy statement
- Provide information, training, instruction, and supervision relating to safety issues

Employee

- Co-operate with employer
- Take care of one's own health and safety
- Look after the health and safety of others
- Do not misuse safety equipment
- Do not interfere with safety devices

Self-employed

- Must not put other people at risk by their method of working

Designers, manufacturers, suppliers, and installers

- Must use safe substances
- Must ensure designs are safe
- Must ensure testing and construction operations are safe
- Must provide information, instructions, and procedures for safe operation and use

People in control of premises

- Ensure the premises are safe and healthy

To ensure that the requirements of the *Health and Safety at Work Regulations* are met, employers are required to manage the introduction and operation of safety measures by:

- Setting up planning, control, and monitoring procedures
- Training and appointing competent persons

- Establishing emergency procedures
- Carrying out regular risk assessments
- Auditing and reviewing procedures
- Disseminating health and safety information

The Act has been given teeth by the formation of an enforcement authority called the *Health and Safety Executive (HSE)*, which appoints inspectors with wide powers to conduct investigations, enter premises and sites, take photographs and samples, issue, where necessary, improvement or prohibition notices, and even initiate prosecution.

However, a well-run organization will make sure that visits from the HSE are not required. The watchword should always be: prevention is better than cure. Accidents do not just happen. They are caused by poor maintenance, inappropriate equipment, unsafe practices, negligence, carelessness, ignorance, and any number of human frailties.

Because accidents are caused, they can be prevented, but this requires a conscious effort to identify and assess the risks that can occur and then ensure that any possible ensuring accident or hazard is avoided. Such risk assessment is a legal requirement and does assist in increasing the awareness of health and safety and reducing the high costs of accidents.

The most common forms of accidents in commercial, manufacturing, and construction or even domestic premises are caused by:

- Equipment failure
- Fire
- Electricity
- Hazardous substances
- Unhealthy conditions
- Poor design
- Unsafe operating practices
- Noise and lighting

Each of the above factors can be examined to find what hazards or shortcomings could cause an accident.

Equipment failure

- Poor maintenance
- Sharp edges to components
- Points of entrapment or entanglement
- Ejection of finished products
- High temperatures of exposed surfaces
- Ill-fitting or insecure guards
- No safety features (overload or pressure relief devices)

- Badly sited emergency stop buttons
- Lack of operating manuals or procedures
- Lack of operator training
- Tiredness of operator

Fire

- Overheated equipment
- Sparks from electrical equipment
- Naked flames
- Hot surfaces
- Hot liquids
- Combustible liquids
- Combustible rubbish
- Explosive gases
- Smoking
- Blocked vents
- Deliberate sabotage (arson)
- Lack of fire extinguishers

Electricity

- Poor insulation
- Bad earthing
- Overrated fuses
- Lack of overload protectors
- Underrated cables or switches
- Unprotected circuits
- No automatic circuit breakers
- No warning signs
- Trailing cables
- Unqualified operators or installers
- Poor maintenance
- Lack of testing facilities
- Dirty equipment

Hazardous substances

- Badly sealed containers
- Corroded containers
- Poor storage
- Unlockable enclosures
- Lack of sign-out procedures

- Poor ventilation
- Bad housekeeping, dirt, spillage
- Inadequate protective clothing
- Lack of emergency neutralising stations
- Badly designed handling equipment
- Lack of staff training

Unhealthy conditions

- Dirty work areas
- Dusty work areas
- Lack of ventilation
- Fumes or dusty atmosphere
- Smoke
- Noxious smells
- Poor lighting
- Lack of protective equipment
- Excessive heat or cold
- Vibration of handles
- Slippery floors, etc.

Poor design

- No safety features
- Awkward operating position
- Lack of guards
- Poor ergonomic design
- Poor sight lines
- Vibration
- Poor maintenance points
- Poor operating instructions
- Awkward filling points

Unsafe operating practices

- No permit system
- Inadequate lifting equipment
- Inadequate handling equipment
- Poor protective clothing
- Untied ladders
- No obligatory rules for hard hats, boots, etc.
- Inadequate fencing
- Poor warning notices

- Poor supervision
- Poor reporting procedures
- No hazard warning lights, etc.
- Blocked or inadequate emergency exits
- No emergency procedures
- No safety officer
- Poor evacuation notices
- Long working hours

Noise and lighting

- Excessive noise
- High-pitched noise
- Vibration and reverberation
- Inadequate noise enclosures
- Inadequate silencers
- No ear protectors
- Poorly designed baffles
- Poor lighting
- Glare
- Intermittent light flashes
- Poor visibility
- Haze or mist

All the above hazards can be identified and either eliminated or mitigated. Clearly such a risk assessment must be carried out at regular intervals, say every six months, as conditions change, and new practices may be incorporated as the project develops.

Apart from the direct effect of accidents and health-related illness on the individual who may suffer great physical and mental pain, the consequences are far reaching. The following list gives some indication of the implications:

- Cost of medical care
- Cost of repair or replacement
- Absence of injured party
- Cost of fines and penalties
- Cost of compensation claims
- Loss of customer confidence
- Loss of public image
- Loss of market due to disruption of supply
- Loss of production
- Damages for delays

- Loss of morale
- Higher insurance premiums
- Legal costs
- Possible loss of liberty (imprisonment)
- Closure costs

CDM Regulations

A special set of regulations came into force in 1994 to cover work in the construction industry, which has a poor safety record. These regulations are called the *Construction (Design and Management) Regulation 1994* (CDM). These are concerned with the management of health and safety and apply to construction projects including not only the client and contractors but also the designers, associated professional advisers, and, of course, the site worker. The duties of the five main parties covered by these regulations are:

1 Client

(a) Ensure adequate resources are available to ensure the project can be carried out safely.
(b) Appoint only competent designers, contractors, and planning supervisors.
(c) Provide planning supervisor with relevant health and safety information.
(d) Ensure health and safety plan has been prepared before start of construction.
(e) Ensure health and safety file is available for inspection at end of project.

These duties do not apply to domestic work where the client is the householder.

2 Designer

(a) Design structures that are safe and incorporate safe construction methods.
(b) Minimize risk to health and safety while structures are being built and maintained.
(c) Provide adequate information on possible risks.
(d) Safe designs to be inherent in drawings, specifications, and other documents.
(e) Reduce risks at source and avoid risks to health and safety where practicable.
(f) Co-operate with planning supervisor and other designers.

3 Planning supervisor

(a) Co-ordinate the health and safety aspects of the design and planning phases.
(b) Help draw up the health and safety plan.
(c) Keep the health and safety file.
(d) Ensure designers co-operate with each other and comply with health and safety needs.
(e) Notify the project to the HSE.
(f) Give advice on health and safety to clients, designers, and contractors.

4 Main contractor

(a) Take into account health and safety issues during tender preparation.
(b) Develop and implement site health and safety plan.
(c) Co-ordinate activities of subcontractors to comply with health and safety legislation.
(d) Provide information and training for health and safety.
(e) Consult with employees and self-employed persons on health and safety.
(f) Ensure subcontractors are adequately resourced for the work in their domain.
(g) Ensure workers on site are adequately trained.
(h) Ensure workers are informed and consulted on health and safety.
(j) Monitor health and safety performance.
(k) Ensure only authorized persons are allowed on site.
(l) Display the HSE notification of the project.
(m) Exchange information on health and safety with the planning supervisor.
(n) Ensure subcontractors are aware of risks on site.

5 Subcontractors and self-employed

(a) Co-operate with main contractor on health and safety issues.
(b) Provide information on health and safety to main contractor and employees.
(c) Provide information on health and safety risks and mitigation methods.

The CDM regulations apply to:

(a) *Notifiable* construction work, i.e., if it lasts for more than 30 days.
(b) Work that will involve more than 500 person days of work (approx. 4000 manhours).
(c) Non-notifiable work which involves 5 or more persons being on site at any one time.
(d) Demolition work regardless of the time taken or the number of workers.
(e) Design work regardless of the time taken or the number of workers on site.
(f) Residential property where business is also carried out.

The regulations do not apply to:

(a) Domestic dwellings.
(b) Very minor works.

However, the requirement on designers still applies and the project must be submitted to the HSE.

Health and Safety Plan

The *health and safety plan* consists of two stages:

1. The pre-tender health and safety plan.
2. The construction phase health and safety plan.

1 Pre-tender health and safety plan

This is drawn up by the employer under the direction of the planning supervisor. Its main purpose is to set a pattern for the construction phase plan and should include the following items:

(a) A general description of the work to be carried out.
(b) A programme and key milestones for the project.
(c) A table of risks envisaged at this stage and their effects on workers and staff.
(d) Information to be submitted by the contractor to demonstrate his capabilities regarding resources and management.
(e) Information to be submitted by the contractor regarding the preparation of the health and safety plan for the construction phase and welfare arrangements.

2 Construction phase health and safety plan

This plan is prepared by the main contractor and has to be submitted to the planning supervisor for approval before work can start on site. Its main constituents are:

(a) Health and safety arrangements for all persons on site or who may be affected by the construction work.
(b) Managing and monitoring the health and safety of construction work.
(c) Detailed arrangements of the site welfare facilities.
(d) Evidence of arrangements for keeping the health and safety file.

Health and Safety File

The *health and safety file* is a record of events on site relating to health and safety and in particular the risks encountered and their mitigations as well as possible risks still to be anticipated. The file must be handed to the client at the end of the contract to enable him to manage and deal with possible risks during the carrying out of subsequent renovations or repairs.

The planning supervisor is responsible for ensuring that the file is compiled properly as the project proceeds and that it is handed to the client at the end of the project.

The client must make the file available to all persons involved in future designs of similar structures or those concerned with alterations, additions, maintenance, or demolition of the structure.

Warning Signs

Standard warning signs have been developed to draw attention to either prohibit certain actions, take certain safety precautions, or warn of particular environmental hazards.

These signs have been colour and shape coded to indicate quickly what the type of warning is.

Thus:

* Hazard signs have a yellow background in a triangle (Figure 36.1).
* Prohibition signs are a red circle with a red diagonal (Figure 36.2).
* Mandatory signs have a blue background, usually in a round disc (Figure 36.3).
* Safe condition signs have green background in a square (Figure 36.4).

The following samples of signs, taken from BS 5499-10-2006 are the most common ones in use, but the selection is not exhaustive.

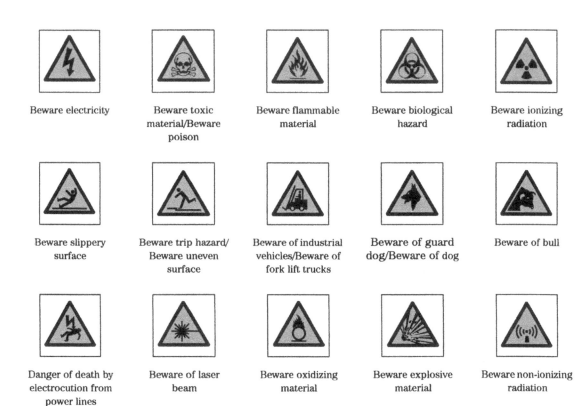

Beware electricity	Beware toxic material/Beware poison	Beware flammable material	Beware biological hazard	Beware ionizing radiation
Beware slippery surface	Beware trip hazard/ Beware uneven surface	Beware of industrial vehicles/Beware of fork lift trucks	Beware of guard dog/Beware of dog	Beware of bull
Danger of death by electrocution from power lines	Beware of laser beam	Beware oxidizing material	Beware explosive material	Beware non-ionizing radiation

Figure 36.1
Hazard signs (yellow background)

No smoking

No access for
pedestrians/No
pedestrians allowed

Do not run

No children allowed

No dogs

Not drinking water

No eating or
drinking

No radios/No
playing of radios

No use of cameras/
No photography

Switch off mobile
phones/Do not use
mobile phones

No industrial
vehicles/No fork lift
trucks

No naked flames

Do not use ladder

Do not wear metal
studded footwear

No admittance for
people with
pacemakers

Figure 36.2
Prohibition signs (red circles)

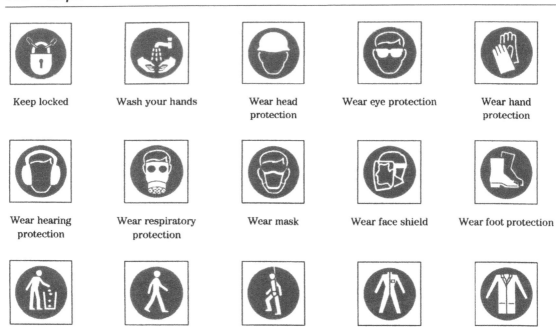

Figure 36.3
Mandatory signs (blue background)

First aid

Emergency call point

Location of emergency telephone

Location of evacuation assembly point

Break to obtain access

Stretcher

Location of first aid shower/Location of emergency shower

Location of first aid eyewash/Location of emergency eyewash

First aid call point

Location of first aid telephone

Drinking water

Location of hand washing facility

Turn clockwise to open

Turn anticlockwise to open

Push to open/Push bar to open

Figure 36.4
Safe condition signs (green background)

Information Management

Chapter Outline

Information, together with communication, are the very lifeblood of project management. From the very beginning of a project, information is required to enable someone to prepare a cost and time estimate, and it is the accuracy and ease of acquisition of this information that determine the quality of the estimate.

The success of a project depends greatly on the smooth and timely acquisition, preparation, exchange, dissemination, storage, and retrieval of information, and to enable all these functions to be carried out efficiently, an information system enshrined in an *information policy plan* is an essential ingredient of the project management plan.

As with many procedures, a policy document issued by management is the starting point of an information system. If issued at a corporate level, such an information policy document ensures not only that certain defined procedures are followed for a particular project, but that every project carried out by the organization follows the same procedures and uses the same systems.

The following list indicates the most important topics to be set out in an information policy plan:

1 Objectives and purpose for having an information management plan
2 Types of documents to be covered by the plan

3 Authority for producing certain documents

4 Methods of distribution of information

5 Methods for storing information and virus protection

6 Methods for retrieving information and acquisition/modification permits

7 Methods for acknowledging receipt of information

8 Security arrangements for information, especially classified documents

9 Disaster recovery systems

10 Configuration control for different types of information

11 Distribution schedule for different documents

12 Standards to be followed

13 Legal requirements regarding time period for information retention

14 Foreseeable risks associated with information

This *information management plan* sets out the basic principles, but the actual details of some of the topics must then be tailored to a particular project. For example, the document distribution schedule, which sets out which document is produced by whom and who receives it, must clearly be project-specific. Similarly, the method of distribution depends on the availability of an IT system compatible with the types of information to be disseminated and the types and styles of documents produced.

Objectives and Purpose

The purpose of explaining the objectives of the policy document is to convince the readers that it is not bureaucratic red tape, but an essential aid to a smooth-running project. There is no doubt that there are numerous projects being run that do not have such a document, but in these cases the procedures are either part of another document or are well known and understood by all parties due to company custom and practice.

Types of Documents

The types of documents covered include correspondence with clients and suppliers, specifications, data sheets, drawings, technical and financial reports, minutes of meetings, records of telephone conversations, and other selected data. All these data will be subject to configuration management to ensure correct distribution and storage.

Authority

Certain types of documents can only be issued by specified personnel. This covers mainly financial and commercial documents such as purchase orders, invoices, and cheques.

Distribution of Information

Distribution of information can be done electronically or by hard copy. While in most cases the sender has the option of which one to use, certain documents may only be sent by a specified method. For example, in some organizations legal documents may be faxed but must also be followed up with hard copies by mail. Generally nowadays, most data are available electronically and can be accessed by selected stakeholders using appropriate access codes or passwords.

Storing Information and Virus Protection

Most data can be stored electronically either by the sender or the receiver, but where special measures have to be taken, suitable instructions must be given. Data must be protected against viruses and hackers and must be arranged and filed in a structured manner according to a predetermined hierarchy based on departmental or operational structures such as work break-down structures or work areas. In some cases hard copies of documents and drawings will have to be filed physically where electronic means are not available, as, for example, on remote construction sites. Important documents such as building leases, official purchase orders, and contracts require storage space that must be both easily accessible and safe from natural disasters and theft. Special fire and waterproof storage facilities may have to be installed.

Retrieving Information and Acquisition/Modification Permits

Retrieving data electronically will generally not be a problem provided a good configuration management system and effective indexing and identification methods are in place. In many cases only certain personnel will be able to access the file and of these only a proportion will have the authority to make changes. Appropriate software will have to be installed to convert data from external sources, in the form of e-mails, faxes, spreadsheets, or other computerized data transmission processes, into a format compatible with the database in use. When handling hard copies, the method of filing documents in a central filing system must be firmly established and an order of search agreed upon. For example, the filing could be by suppliers' names (alphabetically), by product type, by order number, by requisition number, or by date. Again, some organizations have a corporate system while others file by project. Whatever system is used, an enormous amount of time will be wasted searching for documents if the filing system is badly designed and, equally importantly, not kept up to date regularly.

Acknowledging Receipt of Information

In many cases it may be necessary to ensure that the information sent has been received and read. The policy of acknowledging certain documents must therefore be set down. A recipient may glance at a letter or scan an e-mail without appreciating its importance. To ensure that the receiver has understood the message, a request must be made to reply either electronically or, in the case of a hard copy, by asking for the return of an attached 'confirmation of receipt' slip. This is particularly important where documents such as drawings go to a number of different recipients. Only by counting the return slips can the sender be sure all the documents have been received.

Security Arrangements

Classified or commercially sensitive documents usually have a restricted circulation. Special measures must therefore be put in place to ensure the documents do not fall into the wrong hands. In some government departments all desks must be cleared every evening and all documents locked away. Electronic data in this category requires special passwords that may have to be changed periodically. Documents no longer required must be shredded and where necessary incinerated. It has been known that some private investigators have retrieved the wastepaper bags from the outside of offices and reassembled the paper shreds!

Disaster Recovery Systems

In light of both major natural disasters such as earthquakes and hurricanes and terrorist attacks, disaster recovery, also known as business continuity, is now a real necessity. Arrangements must be made to download important data regularly (usually daily) on disk, tape, or even film and store them in a location far enough removed from the office to ensure that they can be retrieved if the base data has been destroyed. It will also be necessary to make arrangements for the replacement of any hardware and software systems that may have been destroyed or corrupted.

Configuration Control

Configuration management is an integral part of information management. Version control, change control, and distribution control are vital to ensure that everybody works to the latest issue and is aware of the latest decisions, instructions, or actions. The subject is discussed in more detail in the chapter on configuration management, but the information policy plan must draw attention to the configuration management procedures and systems being employed.

Distribution Schedule

The distribution of documents can be controlled either by a central computerized data distribution system activated by the originator of the document or by a special department charged with operating the agreed configuration management system. One of the key sections of the project management plan is the document distribution schedule, which sets out in tabular form who originates a document and who receives it. If hard copies have to be sent, especially in the case of drawings, the number of copies for each recipient must also be stated. If this schedule is lodged with the project office or the distribution clerk, the right persons will receive the right number of copies of latest version of a document at the right time. While most document distribution will be done electronically, hard copies may still be required for remote locations or unsophisticated contractors, suppliers, or even clients.

Standards to Be Followed

Standard procedures relating to information management must be followed wherever possible. These standards may be company standards or guidelines, codes of practice, or recommendations issued by national or international institutions. The British Standards Institution has issued guides or codes of practice for configuration management, project management, risk management, and design management, all of which impact on information management. An International Standard BS ISO 15489-I-2001 'Information and Documentation – Records Management – Part 1 – General' is of particular importance.

Legal Requirements

Every project is constrained by the laws, statutory instruments, and other legal requirements of the host country. It is important that stakeholders are aware of these constraints and it is part of the information management process to disseminate these standards to all the appropriate personnel. The Freedom of Information Act and Data Protection Act are just two such legal statutes that have to be observed. There may also be legal requirements for storing documents (usually for about seven years) before they can be destroyed.

Forseeable Risks

All projects carry a certain amount of risk and it is vital that warnings of these risks are disseminated in a timely fashion to all the relevant stakeholders. Apart from issuing the usual risk register, which lists the perceived risks and the appropriate mitigation strategies, warnings of unexpected or serious risks, imminent political upheaval, or potential climatic disasters must be issued immediately to enable effective countermeasures to be taken. This requires a preplanned and rehearsed set of procedures to be set up that can be implemented very rapidly. The appointment of an information risk manager will be part of such a procedure.

An important part of information management is issuing reports. A project manager has to receive reports from other members of the organization to enable him to assess the status of the project and in turn produce his own reports to higher management. Systems must be set up to ensure that these reports are produced accurately, regularly, and timely. The usual reports required by a project manager cover progress, cost, quality, exceptions, risks, earned value, trends, variances, and procurement or production status. These data will then be condensed into the regular (usually monthly) progress report to the programme manager, sponsor, or client as the case may be. Templates and standard formats greatly assist in the production of these reports and modern technology enables much of these data to be converted into graphs, charts, and diagrams, but presentation can never be a substitute for accuracy.

Communication

Chapter Outline

While it is vital that a project manager has a good information system, without an equally good communication system such information would not be available when it is needed.

Generally, all external communications of a contractual nature, especially changes in scope or costs, must be channelled via the project manager or his office. This applies particularly to communications with the client and suppliers/subcontractors. The danger of not doing this is that an apparent small change agreed between technical experts could have considerable financial or program (or even political) repercussions due to the experts not being aware of the whole picture.

Unless a project manager decides to do everything himself, which should certainly not be the case, he has to communicate his ideas, plans, and instructions to others. This requires communication, whether by mouth, in writing, by mail, electronically, or by carrier pigeon.

Communications can be formal and informal, and while contractual, organizational, and technical information should always follow the formal route, communication between team members is often most effective when carried out informally. There are many occasions when a project manager has the opportunity to meet his or her team members, client, or other stakeholders, all of which will enable him or her to discuss problems, obtain information, elicit opinions, and build up trusting relationships which are essential for good project management.

Management by walkabout is an accepted method of informal communication, which not only enables an exchange of information and ideas to take place, often in a relaxed atmosphere, but also has the advantage of seeing what is actually going on as well as setting the framework for establishing personal relationships.

Probably more errors occur in a project due to bad communications than any other cause. Ideas and instructions are often misunderstood, misinterpreted, misheard, or just plainly

Project Management, Planning, and Control. http://dx.doi.org/10.1016/B978-0-08-098324-0.00038-X

ignored for one reason or another; in other words the communication system has broken down. All communication involves a sender and a receiver. The sender has a responsibility to ensure that the message is clear and unambiguous and the receiver has to make sure that it is correctly understood, interpreted, confirmed, and acted upon.

There are a number of reasons why and how failures in communication can occur. The most common of these, generally known as communication barriers, are:

- Cultural differences
- Language differences
- Pronunciation
- Translation errors
- Technical jargon
- Geographical separation of locations
- Equipment or transmission failure
- Misunderstanding
- Attitude due to personality clash
- Perception problems due to distrust
- Selective listening due to dislike of sender
- Assumptions and prejudice
- Hidden agendas
- Poor leadership causing unclear instructions
- Unclear objectives
- Poor document distribution system
- Poor document retention system or archiving
- Poor working environment such as background noise
- Unnecessarily long messages
- Information overload such as too many e-mails
- Withholding of information
- Poor memory or knowledge retention.

Clearly some of these barriers are closely related. In the following further explanation some of these have been collected and will be discussed in more detail together with the techniques which can be used to overcome these communication problems.

Cultural Differences, Language Differences, Pronunciation, Translation, and Technical Jargon

Problems may arise because different cultures have different customs, etiquette, and trading practices. In some instances where two countries use the same language, a particular word may have a totally different meaning. This occurs not only between England and America but for example also between Germany and Austria, who are, as some cynics might say, all 'divided by a common language'.

For example a lift in England would be an elevator in the USA and a water tap would be a faucet. Most project managers will be familiar with the English term *planned* being called *scheduled* in the USA. In addition, regional accents and variations in pronunciation can cause misunderstandings in verbal communications. The solution is simple. Always speak clearly and confirm the salient points in writing.

Forms of address may be fairly informal in some countries like the UK or USA, but unless one knows the other party well, the formal personal pronoun *sie* or *vous* must be used in Germany or France. The incorrect form of address could easily cause offence. It is advisable therefore to seek guidance or attend a short course before visiting a country where such rules apply.

Incorrect translations are not only a source of amusement but can be a real danger. To overcome such errors, the translator should always be a native speaker of the language the text is translated *into*. This will enable the correct word for a particular context to be chosen and the right nuances to be expressed.

Most disciplines or industries have their own technical jargon which can cause difficulties or misunderstanding when the recipient is from a different environment or culture. There may be a reluctance for the receiver to admit to his/her ignorance of the terms used, which can cause errors or delays in the execution of an instruction. The sender should therefore refrain from using jargon or colloquialisms, but by the same token, it is up to the receiver to request that any unfamiliar term is explained as it is mentioned.

Geographical Separation, Location Equipment, or Transmission Failure

Where stakeholders of a project are located in different offices or sites, good electronic transmission equipment is essential. The necessary equipment must be correctly installed, regularly checked, and properly maintained. Generally it is worthwhile to install the latest updates, especially if these increase the speed of transmission, even if they do not reduce the often high operating and often high transmission costs. Where persons in countries with different time zones have to be contacted, care must be taken to take these into consideration. A person from whom one wants a favour will not be very co-operative if woken up at 4 o'clock in the morning!

Misunderstanding, Attitude, Perception, Selective Listening, Assumptions, Hidden Agendas

Senders and receivers of communications are human beings and are therefore prone to prejudice, bias, tiredness, and other failings, often related to their mood or health at the time.

Misunderstandings can occur due to bad hearing or eyesight or because there was not sufficient time to properly read and digest the message. Cases have been known where, because the receiver did not like or trust the sender, the transmitted information was perceived as being unimportant or not relevant and was therefore not be acted upon with the urgency it actually required. The receiver may believe the sender to have a hidden agenda or indeed have his/her own agenda and may therefore deliberately not co-operate with a request. To avoid these pitfalls, all parties must be told in no uncertain manner that the project has priority over their personal opinions. It also helps to arrange for occasional face-to-face meetings to take place.

It is not unusual for the receiver to make assumptions which were not intended. For example, the sender may request a colleague to book some seats to a theatre. The receiver may assume the sender wants the best seats when the opposite may be true. The fault here lies with the sender who was not specific in his request.

Poor Leadership, Unclear Instructions, Unclear Objectives, Unnecessarily Long Messages, Withholding of Information

Instructions whether verbal or written must be clear and unambiguous. They should also be as short as possible as the receiver's as well as the sender's time is often costly. Winston Churchill required all important documents to be condensed onto 'one sheet of foolscap paper' (approximately the size of an A4 sheet). Time is money and the higher one is in the hierarchy, the more expensive time becomes. As with instructions, objectives must also be set out clearly and unequivocally. It is often advantageous to add simple sketches to written communications. These are often more explicit than long descriptions.

When information has to be communicated to a number of recipients, it may not be advisable to tell everybody everything. For example an instruction to a technical department may not include the cost of certain quoted components. Some information is often only disseminated on a 'need to know' basis. The sender therefore has the responsibility to decide which parts of the documentation are required by each receiver. Clearly particular care has to be taken with sensitive or classified information which may be subject to commercial distribution restrictions or even the Official Secrets Act.

It can be seen that while there are many potential communication barriers, they can all be overcome by good communication planning and sensitive project management.

An example of how ones attitude can be affected by receiving good or bad communications is clearly shown by the following scenario.

You are standing on a railway platform waiting for a late train, and you hear the usual bland announcement which simply says:

'This train will be 40 minutes late. We are sorry for the inconvenience this has caused.'

This will probably make you angry and blame the train operators for incompetence.

If, on the other hand, the announcement says: 'This train will be delayed by 40 minutes due to a young girl falling onto the line at the last station', you will be mollified and probably quite sympathetic to the operators who now have the problem of dealing with a very unhappy occurrence.

The difference is that in the second announcement you were given an explanation.

Team Building and Motivation

Chapter Outline

Large or complex projects usually require many different skills that cannot be found in one person. For this reason, teams have to be formed whose members are able to bring their various areas of expertise and experience together to fulfill the needs of the project and meet the set criteria. The project manager is usually the team leader and it may be his responsibility to select the members of the team, although in many instances he may be told by senior management or the HR department who will be allocated to the team. If the project is run as a matrix-type organization, the different specialist team members will almost certainly be selected by the relevant functional department manager, so that the project manager has to accept whoever has been allocated.

There are considerable advantages in operating as a team, which need not require all the members to be fully allocated to the project all the time. Nevertheless, the project manager must create an atmosphere of co-operation and enthusiasm whether the members are permanent or not.

Project Management, Planning, and Control. http://dx.doi.org/10.1016/B978-0-08-098324-0.00039-1

The main advantages of teams are:

*   Teams engender a spirit that encourages motivation and co-operation
*   Different but complementary skills and expertise can be brought to bear on the project
*   Problems can be resolved by utilizing the combined experience of the team members
*   New ideas can be 'bounced' between team members to create a working hypothesis
*   Members gain an insight into the workings of other disciplines within the team
*   Working together forms close relationships which encourage mutual assistance
*   Lines of communications are short
*   The team leader is often able to make decisions without external interference

The following characteristics are some of the manifestations of a successful team:

*   Mutual trust
*   A sense of belonging
*   Good team spirit
*   Firm but fair leadership
*   Mutual support
*   Loyalty to the project
*   Open communications
*   Co-operation and participation
*   Pride in belonging to the team
*   Good mix of talents and skills
*   Confidence in success
*   Willingness to overcome problems
*   Clear goals and objectives
*   Enthusiasm to get the job done
*   Good teams tend to receive good support from top management and sponsors. They are often held up as examples of good project management during discussions with existing and potential clients.

Clearly, if too many of the above characteristics are absent, the team will be ineffective. Merely bringing a number of people together with the object of meeting a common objective does not make a team. The difference between a group and a team is that the team has a common set of objectives and is able to co-operate and perform as a unified entity throughout the period of the project. However, to create such a team requires a conscious effort by the project manager to integrate and motivate them and instill an esprit de corps to create an efficient unit, whether they are in industry, in the armed services, or on the playing field.

Team Development

Building a team takes time and its size and constituency may change over the life of the project to reflect the different phases. Team development has been researched by Tuckman who found that a team has to undergo four stages before it can be said to operate as a successful entity. These stages are:

1 Forming
2 Storming
3 Norming
4 Performing

To these could be added a fifth stage termed mourning, which occurs when the project is completed and the team is being disbanded.

Forming

As the word implies, this is the stage when the different team members first come together. While some may know each other from previous projects, others will be new and unsure, not only of themselves, but also of what they will be required to do. There will be an inevitable conflict between the self-interest of the team member and the requirements of the project, which may impose pressures caused by deadlines and cost restraints.

Clearly at this stage the project manager will have to 'sell' the project to the team and explain what role each member will play. There may well be objections from some people who feel that their skills are not being given full rein or conversely that they do not consider themselves to be well suited for a particular position. The project manager must listen to and discuss such problems, bearing in mind that the final decision rests with him and once decided, must be adhered to. There is no virtue in forcing a square peg into a round hole.

Storming

Once the team has been nominally formed and the main roles allocated, the storming stage will start. Here the personalities and aspirations of the individuals will become apparent. The more dominant types may wish to increase their sphere of influence or their limits of authority, while the less aggressive types may feel they are being sidelined. There will be some jockeying for position and some attempts to write their own terms of reference and it is at this stage that the conflict management skills of the project manager are most needed. It is vital that the project manager asserts his/her authority and ensures that the self-interests of the individual become subservient to the needs of the project.

Norming

When the storming is over, the project should run smoothly into the norming stage. Here all the team members have settled down and have accepted their roles and responsibilities, although the project manager may use a more participative approach and do some 'fine tuning'. The important thing is to ensure that the team are happy to work together, are fully aware of the project objectives and the required regulations and standards, and are motivated to succeed.

Performing

At this stage the team can now be considered a properly integrated working entity with every member confident of his/her role. All the energy will be focused on the well-being of the project rather than the individual. Communications are well established and morale is high. The project manager can now concentrate on the work at hand but must still exercise a degree of maintenance on the team. The organization should now run as 'on well-oiled casters' with everyone being fully aware of the three main project criteria: cost, time, and quality/performance.

Mourning

There is an inevitable anticlimax when a project has come to an end. Members of a project team probably feel what soldiers feel at the end of a war. There is a mixture of relief, satisfaction, and apprehension of what is to follow. Unless there is another similar project ready to be started, the team will probably be disbanded. Some people will return to their base discipline departments, some will leave on their own accord, and some will be made redundant. There is a sense of sadness when friendships break up and relationships built up over many months, based on respect and mutual co-operation, suddenly cease.

The project manager now has to take on the mantle of a personnel officer and keep the team spirit alive right up to the end. There is always a risk, on large long-running projects, that as the end of the project approaches, some people will leave before final completion to ensure further employment without a break. It may then be necessary for the organization to offer termination bonuses to key staff to persuade them to stay on so as to ensure there are sufficient resources to finish the job.

The Belbin Team Types

While the main requirement of a team member must be his or her expertise or experience in his or her particular field, in the ideal team, not only the technical skills but also the characteristics of the team members should complement each other. A study of team characteristics was carried out by Meredith Belbin after nine years of research by the Industrial Training Research Unit in Cambridge. At the end of the study, Belbin identified nine main types that are needed to a greater or lesser extent to make up the ideal team.

Unfortunately in practice it is highly unlikely that the persons with the right skills and the ideal personal characteristics will be sitting on a bench waiting to be chosen. More often than not, the project manager has to take whatever staff has been assigned by top management or functional managers. However, the benefit of the Belbin characteristics can still be obtained by recognizing what Bebin 'type' each team member is and then exploiting his or her strengths (and recognizing the weaknesses) to the benefit of the project. In any case, most people are a mix of Belbin characteristics, but some will no doubt be more dominant than others.

The nine Belbin characteristics are as follows:

- Plant
- Resource investigator
- Co-ordinator
- Shaper
- Monitor/evaluator
- Team worker
- Implementor
- Completer/finisher
- Specialist

The strengths and weaknesses of each of these characteristics are as follows.

Plant

Such persons are creative, innovative, imaginative, self-sufficient, and relish solving difficult problems often using new ideas and fresh approaches. Their unorthodox behaviour may make them awkward to work with and their dislike of criticism, discipline, and protocol may make them difficult to control.

Resource Investigator

These persons are very communicative, probably extroverted, show curiosity in new ideas, and are enthusiastic in responding to new challenges. Once the initial challenge or fascination is over, their interest tends to wane.

Co-ordinator

Co-ordinators are self-controlled, stable, calm, self-confident, can clarify goals and objectives, and are good at delegating and maximizing people's potentials. When given the opportunity they tend to hold the stage.

Shaper

These persons are outgoing, dynamic, and thrive on pressure. Drive and courage to shape events, overcome difficulties, and a desire to challenge inertia or complacency are part of their character. They may therefore be anxious, impatient, and easily irritated by delays and blockages.

Monitor/Evaluator

These people are sober, prudent, and are able to evaluate the options. They have a good sense of judgement, are analytical, and can make critical and accurate appraisals. They could be easily judgmental and their tactless criticism may be destructive.

Team Worker

Such persons are co-operative, sensitive, socially orientated, and help to build a good team. They are often only noticed when they are absent. They may have difficulties in making decisions and tend to follow the crowd.

Implementer

Disciplined and reliable, conservative and practical, such persons turn ideas into actions systematically and efficiently. They could be inflexible and averse to new unconventional methods.

Completer/Finisher

Such people are painstaking, conscientious, and self-controlled perfectionists with a strong sense of urgency. They are good at checking and seeking out errors and omissions. They tend to be over-concerned with minor faults and find it hard to give in.

Specialist

Specialists supply skills that are in short supply. They tend to be single-minded, self-reliant, and dedicated to their profession. Their independence is not easily controlled, especially if they know they are difficult to replace. Being absorbed in their speciality, they may at times have difficulty in seeing the larger picture.

Motivation

The simplest dictionary definition of motivation is 'the desire to do'. The strength or degree depends on the individual's character and the reason or cause of the desire. In many cases the individual may be self-motivated, due to an inner conviction that a particular action or behaviour is necessary for personal, political, or religious reasons, but in a project context, it may be necessary for an external stimulant to be applied. It is undoubtedly the function of a project manager to motivate all the members of the project team and convince them that the project is important and worthwhile. The raison d'etre and perceived benefit, be they political, economic, social, or commercial, must be explained in simple but clear terms so that each team member appreciates the importance of his/her role in the project. In a wartime scenario, motivation can well be a question of survival and is often the result of national pride or convincing propaganda, but such clear objectives are seldom the case in a normal peacetime project, which means that the project manager has to provide the necessary motivation, encouragement, and enthusiasm.

There is little doubt that a large part of the success of the 2012 London Olympics was due to the collaborative approach, the team spirit of the design/construction team, and the motivation to complete the project on time, for what was regarded by everybody as a project of national pride and international importance.

Apart from the initial indoctrination and subsequent pep talks, a project manager can reinforce the message by the conventional management practices of giving credit where it is due, showing appreciation of good performance, and offering help where an individual shows signs of stress or appears to be struggling, mentally or physically.

A good example of the effect of motivating people by explaining the objective of a project or even a work package is shown in the following little story.

A man walking along a street notices that a bricklayer building a wall is very lethargic, clearly not enthusiastic, and looks generally unhappy about his work.

'Why are you so unhappy?' he asks the workman.

'I have just been told to build this wall. Just placing one brick on another is monotonous and boring' was the reply.

The man walked further up the street and met another bricklayer clearly building the other end of the same wall. This workman, on the other hand, worked quickly, was clearly interested in the work, and whistled a happy tune while he laid the bricks

'Why are you so happy?' he asked the man

The man looked up with shining eyes and said proudly: 'I am building a cathedral'.

Maslow's Hierarchy of Needs

A. H. Maslow carried out research into why people work and why some are more enthusiastic than others. He discovered that in general there was, what he called, a *hierarchy of human needs*, which had to be satisfied in an ascending order. These can be conveniently demonstrated as a series of steps in a flight of stairs where a person has to climb one step before proceeding to the next (see Figure 39.1).

The five levels on Maslow's needs are: *physiological, security and safety, social, esteem, and self-actualization.* Maslow argued that the first needs are the ones that enable the human body to perform its functions, i.e., air for the lungs, food and water for the digestive system, exercise for

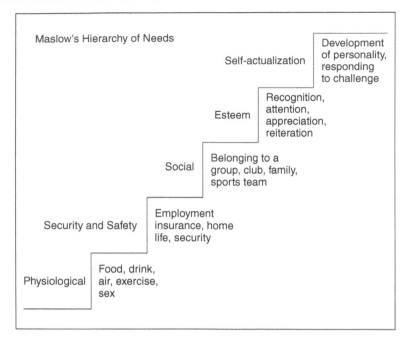

Figure 39.1
Maslow's hierarchy of needs

the muscles, and of course, sex for the continuation of the species. Once these needs have been met, the next requirement is *shelter, security* in employment, and a *safe* environment. This is then followed by *social acceptance* in the society one frequents such as at work, clubs, or pubs, and of course, the family. The next step is *self-esteem*, which is the need to be appreciated and respected. Praise, attention, recognition, and a general sense of being wanted, all generate self-confidence and well-being. The last aspiration is *self-actualization*. This is the need to maximize all of one's potentials, utilize fully one's abilities, and be able to meet new challenges.

As in all theories, there are exceptions. The proverbial starving artist in his garret is more concerned about his esteem and self-actualization than his security or even social acceptance. Similarly the ideals of missionaries take precedence over the desire for physical comfort. However, for the majority of wage or salary earners, the theory is valid and must be of benefit to those wishing to understand and endeavouring to fulfill the needs of people in their charge.

Herzberg's Motivational Hygiene Theory

Herzberg has tried to simplify the motivational factors by suggesting two types:

* Hygiene factors
* Motivators

Hygiene Factors

- Physiological needs
- Security
- Safety
- Social

Motivators

- Recognition of achievement
- Interesting work
- Responsibility
- Job freedom
- Pleasant working conditions
- Advancement and growth prospects

The hygiene factors represent the first three steps of Maslow's needs, i.e., physiological needs, security and safety, and social. The motivators are then esteem and self-actualization. From a management point of view, the first three can almost be taken for granted, as without reasonable pay or security, staff will not stay. To obtain the maximum commitment from an employee (or even oneself) motivators such as recognition of achievement, interesting or challenging work, responsibility, job freedom, pleasant working conditions, and possibility of advancement and growth must be present.

In general people like doing what they are good at and what gives them satisfaction. At the same time they tend to shun what they are less able at or what bores them. It is of benefit to the organization therefore to reinforce these behaviours, once they have been identified.

Leadership

Chapter Outline

Leadership can be defined as the ability to inspire, persuade, or influence others to follow a course of action or behaviour towards a defined goal. In a political context this can be for good or evil, but in a project environment it can generally be assumed that good leadership is a highly desirable attribute of a project manager.

Leadership is not the same as management. Leadership is about motivating, influencing, and setting examples to teams and individuals, while management is concerned with the administrative and organizational facets of a project or company. It can be seen therefore that a good project manager should be able to combine his leadership and management skills for the benefit of the project.

Whether leadership is attributed to birth, environment, or training is still a subject for debate but the attributes required by a leader are the same. The following list gives some of the more essential characteristics to be expected from a good leader. To dispel the impression that there is a priority of qualities, they are given in alphabetical order.

Adaptability	Ability to change to new environment or client's needs
Attitude	Positive can-do outlook, optimism despite setbacks
Charisma	Presence and power to attract attention and influence people
Cognitive ability	Ability to weigh up options, give clear instructions
Commitment	Will to succeed and achieve set goals
Common sense	Ability not to be hoodwinked by irrational suggestions or solutions
Creativity	Able to do some innovative or lateral thinking
Drive	Energy, willpower, and determination to push forward
Fairness	Fair and considerate attitude to human needs and staff problems
Flexibility	Willingness to modify ideas and procedures to new circumstances
Honesty	Trustworthy, reliable, will not tolerate cover-ups
Integrity	Ability to make sound moral judgements, approachable, principled

Project Management, Planning, and Control. http://dx.doi.org/10.1016/B978-0-08-098324-0.00040-8

Intelligence	Clear thinking and ability to understand conflicting arguments
Open-mindedness	Open to new ideas and suggestions even if unconventional
Prudence	Ability to weigh up and take risks without being reckless
Self-confidence	Trust in own decisions and abilities without being self-righteous
Technical knowledge	Understanding of technical needs of the project and deliverables

While these 'paragonial' attributes (apart from being charismatic) sadly do not seem to be necessary in a politician, they are desirable in a project leader and in fact many good project managers do possess these qualities which, in practice, result in the following abilities:

- Good communication skills, such as giving clear unambiguous instructions and listening to others before making decisions.
- Inspiring the team by setting out clearly the aims and objectives and stressing the importance of the project to the organization or indeed, where this is the case, the country.
- Fostering a climate in which new suggestions and ideas are encouraged and giving due credit when and where these can be implemented.
- Allocating to the selected members of the team, the roles and tasks to suit the skills, abilities, and personal characteristics of each member irrespective of race, creed, colour, sex, or orientation.
- Gaining the confidence and respect of the team members by resolving personnel issues fairly, promptly, and sympathetically.

Situational Leadership

Situational leadership simply means that the management style has to be adapted to suit the actual situation the leader finds himself or herself in.

According to Hersey and Blanchard, who made a study of this subject as far back as 1960, managers or leaders must change their management style according to the level of maturity of the individual or group. Maturity can be defined as an amalgam of education, ability, confidence, and willingness to take responsibility. Depending on this level of maturity, a leader must then decide, when allocating a specific task, whether to give firm, clear instructions without inviting questions or delegating the performance of the task, giving the follower a virtual free hand. These are the extreme outer (opposite) points of a behavioural curve. In between these two extremes lies the bulk of management behaviour. For convenience, the level of maturity can be split into four categories:

Category 1
Low skill, low confidence, low motivation

Category 2
Medium skill, fair confidence, fair ability, good motivation

Category 3
Good skill, fair confidence, good ability, high motivation

Category 4
High skill, high confidence, high ability, high motivation

The degree of direction or support given to the follower will depend on the leader's perception of the follower's maturity, but always in relation to a specific task. Clearly a person can be more confident about one task or another, depending largely on his or her level of experience of that task, but situational leadership theory can only be applied to the situation (task) to be performed at this particular time.

The simplest way to illustrate situational behaviour is to look at the way tasks are allocated in the army.

High task, low support

A sergeant will give clear direction to a category 1 recruit, which he or she will not expect to be questioned on. There will be little technical or emotional support – just plain orders to perform the task.

High task, high support

A captain will give an order to the sergeant (category 2) but will listen to any questions or even suggestions the sergeant may make, as this follower may have considerable experience.

Low task, high support

A colonel will suggest a course of action with a major but will also discuss any fears or problems that may arise before deciding on the exact tactics.

Low task, low support

A commander in chief will outline his strategy to his general staff, listen to their views, and will then let them get on with implementing the tasks without further interference.

Clearly in every case the leader must continue to monitor the performance of any follower or group, but this will vary with the degree of confidence the leader has in the follower. At the lowest level it could be a check every half hour. At the highest level it could be a monthly report.

It is not possible to apply mathematical models to managing people who are not only diverse from one another, but can also change themselves day by day depending on their emotional or physical situation at the time.

Figure 40.1 shows the four maturity categories set against the behaviour grid. It also superimposes a development curve that indicates the progression of behavioural change from the lowest to the highest assuming that the follower's maturity develops over the period of the project.

The leader can help to develop the maturity of the follower by gradually reducing the task behaviour, which means explaining the reasons for instructions and increasing the support by praising or rewarding achievements as soon as they occur. There should be a high degree of encouragement by openly discussing mistakes without direct criticism or apportioning blame. Phased monitoring and a well-structured feedback mechanism will highlight a problem before it gets out of hand, but probably the most important point to be hammered home is the conviction that the leader and follower are on the same side, have the same common interest to reduce the effect of errors, and must therefore work together to resolve problems as soon as they become apparent.

It is fear of criticism that inhibits the early disclosure of problems or mistakes, which tend to get worse unless confronted and rectified as soon as possible. Even senior managers risk instant dismissal if they deliberately submit incorrect information or unduly withhold an unpalatable financial position from the board of directors.

Leaders who are confident of their abilities and able to practise the low-task, low-support style will be able to delegate without completely abdicating their own accountability.

Delegation means transferring both the responsibility and authority to another person, but still retaining the right to monitor the performance as and when required. Clearly if this monitoring takes place too frequently or too obtrusively, the confidence of the follower is soon undermined. Generally speaking a monthly report is a reasonable method of retaining overall control, provided of course that the report is up to date, honest, and technically correct.

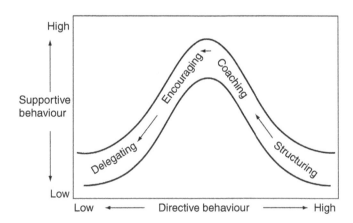

Figure 40.1
Situational leadership *(Source: Hersey and Blanchard)*

Professionalism and Ethics

All the major professional institutions expect their members to observe rules of conduct that have been designed to ensure that the standards of ethical behaviour set out in the charter and by laws of the institution are adhered to.

The rules set out the duties of professional members towards their employers and clients, their profession and its institution, their fellow members, the general public, and the environment. These rules apply equally to professionals in full- or part-time employment and to those acting as professional consultants or advisers on a fee basis for and behalf of private or public clients. Contravention of this code could result in disciplinary proceedings and possible suspension or even expulsion from the institution.

Project management is a relatively new profession when compared with the established professions such as law, medicine, architecture, accountancy, civil engineering, and surveying.

A standard code of conduct was therefore produced by the Association for Project Management (APM), which sets out the required standards of professional behaviour expected from a project manager. These duties and responsibilities can be divided into three main categories as follows.

A further discussion on ethics is given in chapter 43 (Governance)

Responsibilities to Clients and Employers

*   Ensure that terms of engagement and scope are agreed by both parties
*   Act responsibly and honestly in all matters
*   Accept responsibility for own actions
*   Declare possible conflicts of interest
*   Treat all data and information as confidential
*   Act in the best interest of the client or employer
*   Where required, provide adequate professional indemnity insurance
*   Desist from subcontracting work without the client's consent

Responsibilities to the Project

*   Neither give nor accept gifts or inducements, other than of nominal value, from individuals or organizations associated with the project
*   Forecast and report realistic values in terms of cost, time and performance, and quality
*   Ensure the relevant health and safety regulations are enforced
*   Monitor and control all tasks
*   Ensure sufficiency and efficient use of resources as and when required

- Take steps to anticipate and prevent contractual disputes
- Act fairly and equitably in resolving disputes if called upon to do so

Responsibilities to the Profession of Project Management

- Only accept assignments for which he/she considers himself/herself to be competent
- Participate in continual professional development
- Encourage further education and professional development of staff
- Refuse to act as project manager in place of another professional member without instructions from the client and prior notification to the other project manager
- Always act in a manner that will not damage the standing and reputation of the profession, or the relevant professional institution

Normally one of the conditions of membership of a professional institution is that one accepts the rules of conduct without question and that any decision by the disciplinary committee is final.

Negotiation

However well a project is managed, it is inevitable that sooner or later a disagreement will arise between two persons or parties, be they different stakeholders or members of the same project team. If this disagreement escalates to become a formal dispute, a number of dispute resolutions exist (see Chapter 42) that have been designed to resolve the problem. However, it is far better, and certainly cheaper, if the disagreement, which may be financial, technical, or organizational, can be resolved by negotiation.

Negotiation can be defined as an attempt to reach a result by discussion acceptable to both parties. This does not mean that either or both parties are particularly happy with the outcome, but whatever compromise has been agreed, business or relationships between the parties can continue.

The ideal negotiation will end in a win–win situation, where both parties are satisfied that their main goals have been met, even at the expense of some minor concessions. More often, however, one party has not been able to achieve the desired result and may well leave the negotiating table aggrieved, but the fact that work can continue and the dispute does not escalate to higher level, or that a commercial deal is struck rather than a complete breakdown of a business relationship, indicates that the negotiation has been successful.

Once it has been agreed by both parties to enter into negotiation, both parties should follow a series of phases or stages to achieve maximum benefit from the negotiation.

Phase 1: Preparation

As with claims or legal proceedings, negotiations will have a greater chance of succeeding if the arguments are backed by good documentation. The preparation phase consists largely of collecting and collating these documents and distilling them into a concise set of data suitable for discussion. These data could be technical data, test results, and commercial forecasts, and could include precedents of previous discussions. There is really no limit as to what this back-up documentation should be, but the very act of reading these data and condensing them into a few pages will give the negotiator a clear picture of what the issues are.

Phase 2: Planning

It is pointless even considering a negotiating process if there is no intention to compromise. The degree of compromise and the limits of concessions that can be accepted have to be established in this phase. There is usually a threshold, below (or above), that must be respected, and the upper and lower limits in terms of time, delivery, money, and payment arrangements as well as the different levels of compromise for each area must be established in advance. There must be a clear appreciation of what concessions can be accepted and at what stage one must either concede or walk away.

Generally the party that has the most to gain from a negotiated settlement is automatically in the weaker position. In addition factors such as financial strength, future business relationships, possible publicity (good or bad), time pressures, and legal restraints must all be taken into consideration.

The location of the negotiations must be given some consideration as it may be necessary to call in advisers or experts at some stage. There is some psychological advantage in having the negotiation on one's home ground and for this reason the other party may insist on a neutral venue such as a hotel or conference centre.

Phase 3: Introductions

Negotiations are carried out by people and the establishment of a good relationship and rapport can be very beneficial. A knowledge of the other party's cultural background and business norms can help to put the other side at ease, especially where social rituals are important to them. Past co-operative ventures should be mentioned, and a discussion of common acquaintances, alliances, and interests all help to break the ice and tend to put all parties at their ease. A quick overview of the common goals as well as the differences may enable the parties to focus on the important issues, which can then be categorized for the subsequent stages.

Phase 4: Opening Proposal

One of the parties must make an initial offer that sets out their case and requirements. The wording of this opening would give some indication of the flexibility as an inducement to reaching a mutually acceptable settlement. Often the requirements of the opening gambit are inflated to increase the negotiation margin, but the other party will probably adopt the same tactics. It is at this opening stage that the other party's body language such as hand gestures, posture, eye movements, and facial expressions can give clues as to the acceptance or non-acceptance of particular suggestions or offers. The common identification of important points will help to lead the discussion into the next phase.

Phase 5: Bargaining

The purpose of bargaining is to reach an agreement that lies somewhere between the initial extreme positions taken by the parties. Both parties may employ well-known tactics such as veiled threats, artificial explosions of anger or outrage, threats of walking out, or other devices, but this is all part of the process. Often a concession on one aspect can be balanced by an enhancement on another. For example, a supplier may reduce his price to the level required by the buyer, provided his production (and hence his delivery) period can be increased by a few weeks or months. The buyer has to decide which aspect takes priority: money or time.

Concessions should always be traded for a gain in another area, which may not be necessarily in the same units or terms. For example, a reduction in price can be balanced by an increase in the number of units ordered or a later delivery. There should always be a number of issues on the table for discussion, so that quid pro quo deals can be struck between them.

Phase 6: Agreement

Negotiations are only successful if they end with an agreement. If both parties walk away without an agreement, one or other (or possibly both) of the negotiators have not done their job and the case will probably end up in adjudication, arbitration, or litigation. Concessions, which are not just given away, should not be regarded as a sign of weakness, but a realization that the other party has a valid point of view that merits some consideration. Both parties should be satisfied enough to wish to continue working or trading together and both are probably aware that there is always the risk that the legal costs of an action can exceed the amount in dispute. This realization often concentrates the mind to agree on a settlement. It may even be prudent, if there is no great time pressure, to leave the door open for a further discussion at a later date or allow the future discussions to take place at a higher level of management.

Phase 7: Finalizing

When an agreement has been reached, the deal has to be formalized by a written statement setting out the terms of the agreement. This must be signed by both parties attending the negotiations. In some cases the agreement reached will be subject to ratification by senior management, but if the settlement is reasonable, such confirmation is usually given without

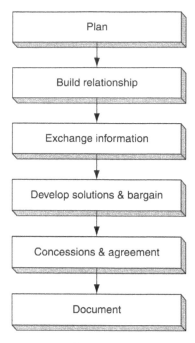

Figure 41.1
Negotiation stages

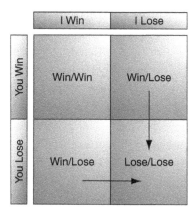

Figure 41.2
Negotiation outcomes

question. It is a fact that the further a person is removed from the 'coal face' of the dispute, the more likely he or she is to ratify a settlement.

It must be pointed out that negotiations involving labour disputes are best carried out by specialist negotiators with experience in industrial relations and national agreements, local working practices, and labour laws. The procedures for such negotiations, which often end up with applications to conciliation boards or tribunals, are outside the scope of this book.

However, differences of opinion can sometimes be reconciled by resorting to mediation involving the help of an independent third party.

When an agreement between the parties appears to be impossible, but neither party relishes the idea of potentially expensive and drawn-out arbitration or litigation, a practical next step would be for both parties to consider resorting to the relatively inexpensive and quick process of mediation. If this procedure fails, there is still the option of adjudication. Both these dispute resolution procedures are described more fully in Chapter 42.

Conflict Management and Dispute Resolution

Chapter Outline

Conflict management covers a wide range of areas of disagreement from smoothing out a simple difference of opinion to settling a major industrial dispute.

Projects, as life in general, tend to have conflicts. Wherever there are a wide variety of individuals with different aspirations, attitudes, views, and opinions there is a possibility that what may start out as a misunderstanding escalates into a conflict. It is one of the functions of a project manager to sense where such a conflict may occur and, once it has developed, to resolve it as early as possible to prevent a full-blown confrontation that may end in a strike, mass resignations, or a complete stoppage of operations.

Conflicts can be caused by differences in opinions, cultural background or customs, project objectives, political aspirations, or personal attitudes. Other factors that tend to cause conflicts are poor communications, weak management, competition for available resources, unclear objectives, and arguments over methods and procedures.

Conflict between organizations can often be traced back to loose contractual arrangements, sloppy or ambiguous documentation, and non-confirmation in writing of statements or instructions.

Thomas and Kilman published a study on conflict management and suggested five techniques that can be employed for resolving conflicts. These are:

- Forcing
- Confronting the problem
- Compromising
- Smoothing
- Withdrawing

Forcing involves one party using its authority acquired by virtue of position in the organization, rank, or technical knowledge to force through its point of view. While such a situation is not uncommon in the armed forces where it is backed up by strict discipline, it should only be used in a project environment where there is a health and safety issue or in an emergency posing a risk of serious physical damage. In most other situations where forcing has been used to solve a conflict, one party almost certainly feels aggrieved with a consequent adverse effect on morale and future co-operation.

Confronting the problem is, by contrast, a more positive method. In this situation both parties will try to examine what the actual issue is and will make a concerted effort to resolve it by reasoning and showing mutual respect for the other's point of view. The most likely situation where this method will succeed is where both parties realize that failure to agree will be disastrous for everybody and where success will enhance both their positions, especially when it is understood that future co-operation is vital for the success of the project. Often there are useful by-products such as innovative solutions or a better understanding of the wider picture.

Compromising is probably the most common method to resolve disputes, but generally both parties have to give up something or part with something, whether it is a point of principle, a financial claim, pension rights, or an improvement in conditions. This means that the settlement may only be temporary and the dispute may well flare up again when one of the parties believes itself to be in a better bargaining position. No one really wins, yet both lose something and it may well be the subject of regret later when the effects of the compromise become apparent. Often commercial or time (programme) pressures make it necessary to reach a quick compromise solution, which means that if these pressures had not existed a more rational discussion could have produced a more lasting result.

Smoothing is basically one party acceding to the other party's demands because a more robust stance would not be in their best interest. This could occur where one party has more authority or power (financial, political, or organizational) or where the arguments put forward are more cogent. Smoothing does not mean complete surrender as it may just not be opportune or politically wise at this particular time to be more assertive.

Withdrawing in effect means avoiding the issue or ignoring it. While this may appear to be a sign of weakness, there may be good reasons for taking this stance. One may be aware that the dispute will blow over when the other party's anger has cooled down or when a confrontation is likely to inflame the situation even more. One may also feel that the possibility of winning the argument is small, so that by making what may be considered a small concession, good relations are maintained. In practice this procedure is only suitable for minor issues since by ignoring important ones, the problem is only shelved and will have to be resolved at a later date. If the issue is a major one and unlikely to be resolved by the other four options, it may still be correct for one or both parties to withdraw and agree to take the dispute to adjudication, arbitration, or litigation as described later.

Whatever techniques are adopted in resolving disputes, the personality of the project manager or facilitator plays a major role. Patience, tact, politeness, and cool-headedness are essential irrespective of the strength or weakness of the technical case. Any agreement or decision made by a human being is to a large extent subjective, and human attributes (or even failings) such as honour, pride, status, or face-saving must be taken into account. It is good politics to allow the losing party to keep their self-respect and self-esteem. Team members may or may not like each other, but any such feelings must not be allowed to detract from the professionalism required to do their job.

In general, confrontation is preferable to withdrawal, but to follow such a course, project managers should practise the following:

- Be a role model and set an example to the team members in showing empathy with the conflicting parties.
- Keep an open door policy and encourage early discussion before it festers into a more serious issue.
- Hear people out and allow them to open up before making comments.
- Look for a hidden agenda and try to find out what is really going on as the conflict may have different (very often personal) roots.

Where a dispute involves organizations outside the project team as with suppliers, subcontractors, or labour unions, professional specialized assistance is essential in the form of commercial lawyers or industrial (labour) relations officers.

Where the conflict is between two organizations and no agreement can be reached by either discussions or negotiations between the parties, it may be necessary to resort to one of the following five established methods of dispute resolution available to all parties to a contract. These, roughly in order of cost and speed, are:

- Conciliation
- Mediation
- Adjudication
- Arbitration
- Litigation

Conciliation

The main purpose of conciliation, which is not used very often in commercial disputes, is to establish communications between the parties so that negotiations can be resumed. Conciliators should not try to apportion blame, but to focus on the common interests of the parties and the systemic reasons for the breakdown of relationships.

Mediation

In mediation, the parties in dispute contact and engage a third party either directly or via the mediation service of one of the established professional institutions. Although the parties retain control over the final outcome, which is not enforceable, the mediator, who is impartial and often experienced in such disputes, has control over the proceedings and pace of the mediation process. The mediator must on no account show him/herself to be judgemental or give advice or opinions, even if requested to do so. His or her main function is to clarify and explore all the common interests and issues as well as possible options, which may lead to a mutually beneficial and acceptable settlement. Once an agreement has been reached, it must be recorded in writing.

If mediation is started early enough before the differences become entrenched, the possibility of an amicable settlement is high. Provided legal advisers are not employed, the only costs are the fees of the mediator, which makes the procedure much cheaper and certainly quicker than any of the three more formal and legally binding dispute resolution procedures described below.

Adjudication

Although adjudication has always been an option in resolving disputes, before 1996 it required the agreement of both parties. This also meant that both parties had to agree about who would be the adjudicator. As this in itself was often a source of disagreement, it was not a common method of dispute resolution until the 1996 Construction Act, more accurately called *Housing Grants, Construction and Regeneration Act 1996*, was passed. This Act allowed one party to apply to one of a number of registered institutions called the *Adjudicator Nominating Body* (ANB) to appoint an independent adjudicator. The other party is then obliged by law to accept both the adjudication process and the nominated adjudicator.

Once appointed, the adjudicator invites the instigating party, called the *referring party,* to submit details of the dispute, called the *referral*, and the opposing side, called the *responding party,* to answer with a *response*. The adjudicator then reviews these submissions together with any other papers or evidence he may request, and unless extended by special circumstances, is obliged to give a ruling (called a *decision*) within 28 days. Although in the early days of adjudication, the adjudicator had to be requested to give *reasons* for his decision, in most cases giving reasons is now the norm.

One of the main advantages of adjudication is that the dispute can be resolved while the contract is still running and any financial awards by the adjudicator must be paid within the stipulated period as decided by the adjudicator. Both parties are responsible for their own costs and this, together with the time limit of the process, makes adjudication relatively inexpensive. However, the losing party can resort to arbitration (where there is an arbitration clause in the contract) or litigation to reverse the decision after the contract has been completed. In case of non-payment of the money awarded to it, the winning party can ask the courts for *enforcement*.

Arbitration

Many contracts contain an arbitration clause, which, in the case of a dispute, requires the parties to either agree to the appointment of an *arbitrator*, or ask one of the recognized chartered institutions to appoint an independent arbitrator. The arbitrator asks for submissions from both parties (preferably in writing), and has the power to open up all the books and documents relating to the dispute, call witnesses, and seek expert opinions. In most cases both sides will be assisted by legal and technical advisers, which could generate considerable costs. Unlike an adjudicator, the arbitrator has the right to award all or part of the costs of the case against one or both of the parties as he sees fit. Generally there is a three-month time limit, but this can be extended by the arbitrator if necessary.

In some cases, especially in overseas contracts, it may be necessary to appoint two arbitrators. If these cannot agree on an award, the matter has to be resolved by a third person called an *umpire*.

In a technical dispute, the arbitrator should ideally be an expert in that field, but if the dispute is of a non-technical nature, it may in the end be better to have the matter resolved by litigation, i.e., in the courts. This, however, means that the privacy afforded by arbitration will be lost.

Arbitration was designed to be speedier and cheaper than court proceedings, but nowadays both parties appoint a galaxy of legal advisers and expert witnesses, which may make arbitration as, if not more, costly than court action. As with adjudication, it is now necessary for the arbitrator to give reasons with his decision.

Litigation

Any dispute can be taken to court whether it is technical, contractual, financial, legal, environmental, personal, etc., provided that there is an applicable basis for legal action. A court procedure, which is more formal, and, being in open court, lacks the privacy of arbitration, involves the employment of solicitors and barristers who present the case to the judge. In addition there will be expert witnesses recruited by both parties whose evidence may be given under oath and be subject to cross-examination. If such proceedings are not settled before they go to trial, they tend to be very expensive both in terms of the award and subsequent damages, if the case is lost, but also in costs, i.e., the court fees and especially the legal fees incurred by both sets of legal teams, which can soon escalate if the case takes many weeks or months. Further the rules of what evidence can be presented and how it is presented are not as flexible in court as they are in arbitration. Finally, the court timetable and the need to comply with certain pre-trial procedures may mean it can take some months or even years for a litigation case to get to trial. For these reasons, every effort should be made to settle technical and contractual disputes by one of the other two methods of dispute resolution. In fact where an arbitration clause is part of the contract, a court will require the arbitration procedure to be followed before permitting it to be heard by a judge.

The benefits of litigation are that the services of the judge are free and the ruling could be of public interest thus acting as an important precedent for future cases. Furthermore, although there is certainty of enforcement of the award, there is a right of appeal to a higher court.

Needless to say, a well-managed project, benefiting from the use of tight but fair and equitable contract documents and change procedures, should never require the project manager to invoke any of these stages. Most arguments and disagreements should be resolved as early as possible by discussions and negotiations before the dispute festers and anger turns into hostility.

CHAPTER 43

Governance

Chapter Outline

Governance of Project Management (GoPM)

Organizations require their objectives set and the determination of the means of attaining these objectives and of monitoring performance. Governance provides the necessary structure for this. It is not the product of any one body but involves relationships between interested parties.

For corporations these interested parties include the board, management, and shareholders as well as the legislature and other stakeholders. For public sector organizations, social enterprises, and other non-corporate organizations the parties providing governance will differ.

Governance applies to all the ongoing and once-off activities of an organization. The delivery of governance requires such activities as annual general meetings, board meetings, benchmarking, standard procedures, assurance, marketing, press releases, and core documents control such as memorandum of understanding, adopted policies and delegated authority schedules, and management reports.

Once-off activities are best managed through the discipline of project management, hence the importance to all organizations of governance of their project management. Governance of project management could well be defined as the framework of the overall strategic requirements of the sponsoring organisation within which projects are managed. The governance of any one particular project is therefore a specific subset of the governance of all the project management in an organisation.

To assist directors and equivalents to ensure that good governance applies to project management throughout their enterprise 13 principles can be identified. These are published by the APM in 'Directing Change, A Guide to Governance of Project Management'. With the kind permission of APM, these principles are repeated below:

- The Board of Directors has the overall responsibility for the governance of project management.
- The organisation differentiates between projects and non-project-based activities.
- Roles and responsibilities for the governance of project management are defined clearly. (For a particular project this should be enshrined in the project management plan.)
- Disciplined governance arrangements, supported by appropriate cultures, methods, resources, and controls are applied throughout the project life cycle. Every project has a sponsor. (From conception to termination.)
- There is a demonstrably coherent and supporting relationship between the project portfolio and the business strategy and policies; for example, ethics and sustainability.
- All projects have an approval plan containing authorisation points at which the business case, inclusive of cost, benefit, and risk is reviewed. Decisions made at authorisation points are recorded and communicated. (These review points can be shown on the project schedule or bar chart.)
- Members of delegated authorization bodies have sufficient representation, competence, authority, and resources to enable them to make appropriate decisions. (These are normally covered in the terms of reference and responsibilities of the individuals and committees.)
- Project business cases are supported by relevant and realistic information that provides a reliable basis for making authorisation decisions. (This is the basis of producing and maintaining the business case.)
- The board or its delegated agents decide when independent scrutiny of projects or project management systems is required and implement such assurance accordingly.
- There are clearly defined criteria for reporting project status and for the escalation of risks and issues to the levels required by the organisation. (These are set out in the project management plan.)
- The organisation fosters a culture of improvement and of frank internal disclosure of project information.
- Project stakeholders are engaged at a level that is commensurate with their importance to the organisation and in a manner that fosters trust.
- Projects are closed when they are no longer justified as part of the organisation's portfolio.

- The main components of portfolio, programme, and project management that need to be examined to ensure compliance with the principles of good governance are:
  - Portfolio direction
  - Project sponsorship
  - Project management capability
  - Disclosure and reporting

Portfolio Direction

It must be confirmed that:

- The project portfolio and its constituent parts are aligned with the business objectives of the organization, which cover profitability, reputation, customer service, sustainability, security, and growth and the impact of the various projects is acceptable to the ongoing operations of the organization.
- The financial controls of the portfolios and individual projects are compatible with those of the organisation. In addition, the planning, expenditure, and review procedures are aligned to the organisation's procedures.
- The organisation discriminates effectively between activities requiring project management and other activities that should be managed as non-project operations. Priorities of projects in the portfolio are periodically reviewed to ensure that the mix meets the corporate strategy.
- Project risks are regularly assessed to determine their combined impact on the organisation as a whole.
- The organisation's capacity is consistent with the project portfolio.
- The sponsoring organisation and external stakeholders such as suppliers, customers, backers, finance providers, and regulators must be sufficiently engaged and involved to ensure a sustainable development of the organisation as well as being aligned with project successes.

Project Sponsorship

To ensure that governance is applied it is essential that:

- All projects have competent sponsors who represent the whole organisation and who can make clear and timely decisions.
- Sponsors devote sufficient time to their projects from concept to closeout and final handover.
- Sponsors keep up to date with the progress and general management of the project by convening regular meetings with their project managers and are prepared to seek independent advice to assist them in the appraisal process.

- Sponsors ensure that sufficient resources (financial, material, and human) are available and appropriate skilled personnel are supplied when required.
- Sponsors maintain the business case (which they own) and accept accountability for the realisation of the specified benefits.

Project Management Capability

The success of projects depends on the experience and leadership qualities of the managers, the skills of team members, the timely availability of the necessary resources, and the processes and procedures employed. It is necessary to ensure therefore that:

- The business case fully reflects the strategic objectives of the organization.
- Projects have clear objectives, scope definitions, envisaged business outcomes, and critical success criteria that will enable realistic decisions to be taken.
- The organisation's project management processes, procedures, and management tools are appropriate for the project in question.
- The project manager, team members, and operatives are competent, aware of their roles and responsibilities, and motivated to improve the performance and delivery of the project.
- The organisation develops and maintains a dynamic stakeholder management procedure with particular emphasis on communications to, from, and between stakeholders.
- The suppliers, contractors, and providers of services (internal and external) have competent staff and sufficient resources to meet the project's requirements in terms of time, cost, and performance.
- Establish procedures for change and risk management and ensure that they are adhered to and implemented.
- Allowances have been made for contingencies that are controlled by authorised persons.

Disclosure and Reporting

Project management information flows up, down, and across as well as to and from organisations. Regular, timely, and reliable disclosure and reporting is an essential part of good project management. Checks should be carried out to ensure that the reporting procedures comply with company policies. Top-level reports on any project should cover:

- Progress
- Financial forecasts
- Completion forecasts
- Major risk-related issues
- Major quality problems
- Overall performance problems

- Compliance or deviations with Key Performance Indicators
- Any independent assurance

Periodic appraisal is recommended by peer and/or external groups of the complexity and value of information flows including reports to ensure that the minimum effort is expended to produce only the necessary information.

The culture of the organisation should encourage honest and open reports. It should not deter whistle-blowers.

Ethics

Organisations should have a policy on ethics. As noted under Portfolio Direction above, governance of project management requires that project management complies with organisational policies including that on ethics. It is the duty of project management practitioners to be fully informed about their organisation's and their profession's policies on ethics.

A policy of openness and disclosure together with a basic trait of honesty will ensure that attempts of bribery and corrupt practices are exposed and firmly rejected. There are however contexts where such standards just do not exist. Indeed, it may be difficult if not impossible to carry out any business operations in certain countries unless 'on-costs', often euphemistically called mobilisation costs, introduction fees, facilitation payments, etc., are added to the contract sum. Such allowances are often added to the fees paid to the agent representing the organisation in that country and that is the end of the matter as far as the company is concerned. Care must be taken in all cases to comply, where applicable, with the requirements of the UK Bribery Act 2010 or where applicable the U.S. Foreign Corruption Practices Act 1977.

On a more personal level, there is always the dilemma of how far one can go when giving or receiving seasonal gifts, entertaining clients, or being entertained by suppliers. Perhaps the clearest answer to this question was given by a senior manager of a construction company to a project manager who asked for clarification as to what gifts are acceptable or permissible. The advice given was: 'You can accept anything, provided you can carry it home in your stomach'.

Evidence

With the increased focus on governance, second and third parties such as clients, investors, suppliers, and regulators are looking for evidence of good governance. Some have developed scoring systems for the maturity of governance, inclusive of the governance of change. Difficulties must be recognised in this external scoring.

Governance arrangements for an organisation should be tailored to its own needs. It is unlikely that any standard scheme of measurement will usefully apply across organizations at different stages of development or in differing sectors and cultures. Difficulties arising in applying standard governance models include allowing for the relatively common situation where one organization does not have sole control of one or more projects. Also, the particular governance issues arising in the public sector require different treatment to that in the private sector.

The evidence is mixed as to whether organizations that score higher on objective governance schemes perform better for stakeholders. Some sectoral studies such as in the Oil and Gas sector show a positive relationship, but some wider studies do not. Certainly there is anecdotal evidence from highly successful technology companies such as Apple and Google that the best judges of appropriate governance are their own leaders rather than third-party standard setters.

The quality of applied governance is also difficult to measure. The reality is that dynamic governance decisions can be made under stress and subject to group dynamics that are not recorded. Assurance schemes too often rely entirely on verifiable documented evidence. These capture only a part of the significant reality, such as in the case of Royal Bank of Scotland in UK in 2007 where excellence in reported governance compliance preceded a major crisis, which was subsequently attributed to 'a real failure in corporate governance'.

Conclusion

Too many projects fail not because of poor project management, but because the project's governance regime and application is poorly developed, not integrated into the wider governance of the organisation, and therefore damaging. Implementing the principles above through the four components described with the proper ethical approach will go a long way to ensure the reliable delivery of projects contributing to organisational success. A conditional questioning approach is preferred to any one standard. More specific guidance is available in the APM publications Directing Change, Co-Directing Change and Sponsoring Change.

CHAPTER 44

Project Close-Out and Handover

Chapter Outline

Close-Out

Most projects involving construction or installation work include a *commissioning* stage during which the specified performance tests and operating trials are carried out with the objective of proving to the client that the deliverables are as specified and conform to the required performance criteria. The *snagging process*, which should have taken place immediately prior to the start of commissioning, often overlaps the commissioning stage so that adjustments and even minor modifications may be necessary. Often commissioning is carried out with the assistance of the client's operatives and this has the advantage that in this way the persons who will run the plant or system will learn how to operate the controls and make necessary adjustments. This is as true for a computer installation as a power station.

On the more complex projects, it may be necessary to run special training and familiarization programmes for clients' staff and operatives, both in the workplace and in classrooms.

When the project is complete and all the deliverables have been tested and approved, the project must be officially closed out. This involves a number of checks that have to be made and documents that have to be completed to ensure that there is no 'drip' of manhours being booked against the project. Unless an official, dated close-out instruction is issued to all members of the project team, there is always a risk of time and money being expended on additional work not originally envisaged. Even where the work was envisaged, there is the possibility of work being dragged out because no firm cut-off date has been imposed.

All contracts (and subcontracts) must be properly closed out and (if possible) all claims and back charges (including liquidated damages) agreed and settled.

One of the most unpopular, but necessary, tasks prior to commissioning is the collation, indexing, and binding of all the operating and maintenance manuals, drawings, test certificates, lubrication schedules, guarantees, and priced spares lists that should have been collected and stored during the course of the project. Whether this documentation is in electronic format or hard copy, the process is the same. Indeed some client organizations require both and the cost of preparing this documentation is often underestimated.

Many of these documents obtained and collated during the various phases of the project have to be bound and handed over to the client to enable the plant or systems to be operated and maintained. It goes without saying that all these documents have to be checked and updated to reflect the latest version and as-built condition.

The following list gives some of the documents that fall into this category:

- Stage acceptance certificates
- Final handover certificate
- Operating instructions in electronic or hard copy format or both
- Maintenance instructions or manuals
- A list of operational and strategic spares with current price lists and anticipated delivery periods as obtained from the individual suppliers. These are divided into operating and strategic spares
- Lubrication schedules
- Quality control records and audit trails
- Material test certificates including confirmation of successful testing of operatives' (especially welders') test certificates and performance test results
- Radiography and other NDT (non-destructive testing) records
- A dossier of the various equipment, material, and system guarantees and warranties
- Equipment test and performance certificates

On completion, the site must be cleared, all temporary buildings, structures, and fences have to be removed, and access roads must be made good.

Arrangements will have to be made to dispose of unused equipment or surplus materials. These may be sold to the client at a discounted rate or stored for use on another project. However, certain materials, such as valves, instruments, and even certain piping and cables, cannot be used on other jobs unless the specified test certificates and certificates of origin are literally wired to the item being stored. Materials that do not fall into these categories will have to be sold for scrap and the proceeds credited to the project.

Those project managers who wish to show some appreciation to their team may decide to use this money for a closing-down party. The team will now have to be disbanded, a process that is the 'mourning' stage of the Tuckman team phases. On large projects that required the team to work together for many months or years, the close-out can be a terrible anticlimax and the human aspect must be handled diplomatically and sympathetically.

Handover

The formal handover involves an exchange of documents, which confirm that the project has been completed by the contractor or supplier and accepted by the client. These documents, which include the signed acceptance certificate, will enable the contractor to submit his final payment certificate, subject to agreed retentions. If a retention bond has been accepted by the client, payment has to be made in full.

CHAPTER 45

Project Close-Out Report and Review

Chapter Outline

Close-Out Report

Most organizations nowadays require the project manager to produce a close-out report at the end of the project. This is often regarded by some project managers as a time-consuming chore, as in many cases the project manager will already have been earmarked for a new project which he or she is keen to start as soon as possible.

Provided a reasonably detailed project diary has been kept by the project manager throughout the various stages of the project the task of producing a close-out report is not as onerous as it would appear. Certainly if the project included a site construction stage, the site manager's diary, which is in most companies an obligatory document, will yield a mass of useful data for incorporation in the close-out report. The information given in the report should cover not only what went wrong and why, but also the successes and achievements in overcoming any particularly interesting problem.

The following is a list of some of the topics that should be included in a close-out report:

- Degree to which the original objectives have been met
- Degree of compliance with the project brief (business case)
- Degree to which the original KPIs have been achieved
- Level of satisfaction expressed by client or sponsor
- Comparison between original (budgeted) cost and actual final cost
- Reasons for cost overruns (if any)
- Major changes incorporated due to:
  - (a) client's approved requirements
  - (b) internal modifications caused by errors or omissions
  - (c) other possible reasons (statutory, environmental, legal, health, and safety, etc.)
- Comparison between original project time and actual total time expended
- Reasons for time overruns or underruns
- Major delays and the causes of these delays

Project Management, Planning, and Control. http://dx.doi.org/10.1016/B978-0-08-098324-0.00045-7
409

- Special actions taken to reduce or mitigate particular delays
- Important or interesting or novel methods adopted to improve performance
- Performance and attitude of project team members in general and some in particular
- Performance of consultants and special advisers
- Performance of contractors, subcontractors, and suppliers
- Attitude and behaviour of client's project manager (if there was one)
- Attitude and behaviour of client's staff and employees
- Comments on the effectiveness of the contract documents
- Comments on the clarity or otherwise of specifications, data sheets, or other documents
- Recommendations for actions on future similar projects
- Recommendations for future documentation to close loopholes
- Comments on the preparation and application/operation of major project management tools such as CPA, EVA, and data gathering/processing

The report will be sent to the relevant stakeholders and discussed at a formal close-out meeting at which the stakeholders will be able to express their views on the success (or otherwise) of the project. At the end of this meeting the project can be considered to be formally closed.

Close-Out Review

Using the close-out report as a basis, the final task of the project manager is to carry out a post-project review (or a post-implementation review), which should cover a short history of the project and an analysis of the successes and failures together with a description of how these failures were handled.

The review will also discuss the performance of the project team and the contributions (positive and negative) of the other stakeholders. All this information can then be examined by future project managers employed on similar projects or working with the same client/stakeholders, so that they can be made aware of the difficulties and issues encountered and ensure (as far as is practicable) that the same problems do not arise. Learning from previous mistakes is a natural process developed from childhood. Even more beneficial and certainly wider reaching, is learning from other people's mistakes. For example, where a new project manager finds that he has to deal with people, either in the client's or contractor's camps, who were described as 'difficult' in a previous close-out report, he or she should contact the previous project manager and find out the best ways of 'handling' these people.

For this reason the close-out review, together with the more formal close-out report, has to be properly indexed and archived in hard copy or electronic format for easy retrieval.

The motto is: 'Forewarned is Forearmed'.

CHAPTER 46

Stages and Sequence

Chapter Outline

Summary of Project Stages and Sequence

The following pages show the stages and sequences in diagrammatic and tabular format:

1. Figure 46.1 shows the normal sequence of controls of a project from business case to close-out;
2. Figure 46.2 gives a diagrammatic version of the control techniques for the different project stages;
3. Figure 46.3 is a hierarchical version of the project sequence, which also shows the chapter numbers in the book where the relevant stage or technique is discussed;
4. Table 46.1 is a detailed tabular breakdown of the sequence for a project control system, again from business case to project close-out.

While the diagrams given will cover most types of projects, it must be understood that projects vary enormously in scope, size, and complexity. The sequences and techniques given may therefore have to be changed to suit any particular project. Indeed certain techniques may not be applicable in their entirety or may have to be modified to suit different requirements. The principles are, however, fundamentally the same.

Project Management, Planning, and Control. http://dx.doi.org/10.1016/B978-0-08-098324-0.00046-9
411

Project Stage Control Techniques

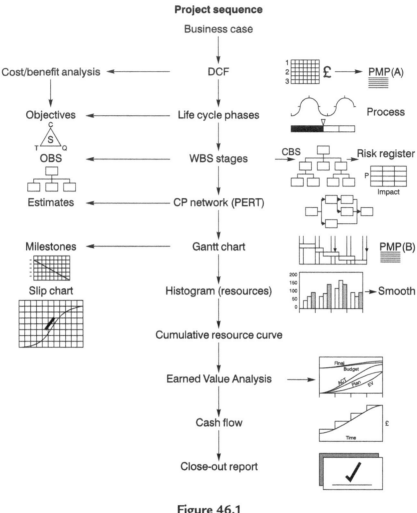

Figure 46.1
Project sequence

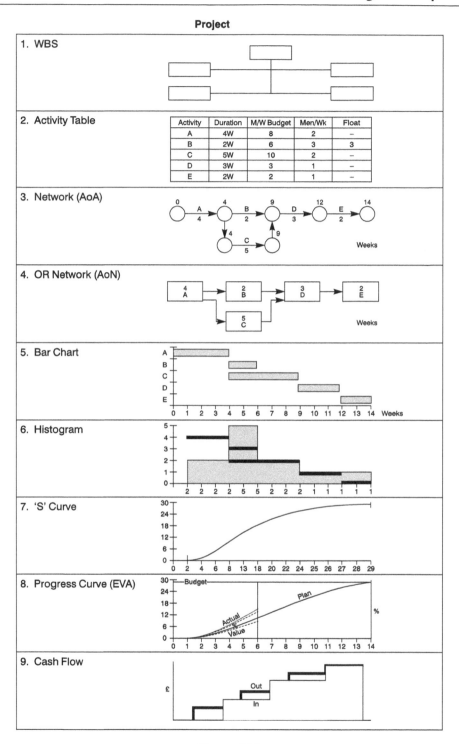

Figure 46.2
Control techniques

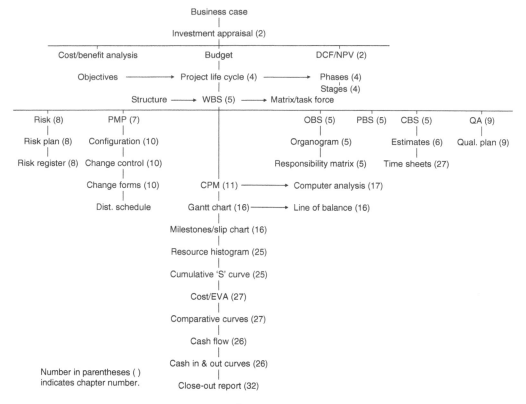

Figure 46.3
Detailed project sequence

Table 46.1: Sequence for project control system

Business case
Cost/benefit analysis
Set objectives
DCF calculations
Establish project life cycle
Establish project phases
Produce project management plan (PMP)
Produce budget (labour, plant, materials, overheads, etc.)
Draw work breakdown structure (WBS)
Draw product breakdown structure
Draw organization breakdown structure
Draw responsibility matrix
List all possible risks
Carry out risk analysis
Draw up risk management plan
Produce risk register
Draw up activity list

Table 46.1: Sequence for project control system—*cont'd*

Draw network logic (CPM) (freehand)
Add activity durations
Calculate forward pass
Revise logic (maximize parallel activities)
Calculate 2nd forward pass
Revise activity durations
Calculate 3rd forward pass
Calculate backward pass
Mark critical path (zero float)
Draw final network on grid system
Add activity numbers
Draw bar chart (Gantt chart)
Draw milestone slip chart
Produce resource table
Add resources to bar chart
Aggregate resources
Draw histogram
Smooth resources (utilize float)
Draw cumulative 'S' curve (to be used for EVA)
List activities in numerical order
Add budget values (person hours)
Record weekly actual hours (direct and indirect)
Record weekly % complete (in 5% steps)
Calculate value hours weekly
Calculate overall % complete weekly
Calculate overall efficiency weekly
Calculate anticipated final hours weekly
Draw time/person hour curves (budget, planned, actual, value, anticipated final)
Draw time/% curves (% planned, % complete, % efficiency)
Analyse curves
Take appropriate management action
Calculate cost per activity (labour, plant, materials)
Add costs to bar chart activities
Aggregate costs
Draw curve for plant and material costs (outflow)
Draw curve for total cash OUT (this includes labour costs)
Draw curve for total cash IN
Analyse curves
Calculate overdraft requirements
Set up information distribution system
Set up weekly monitoring and recording system
Set up system for recording and assessing changes and extra work
Set up reporting system
Manage risks
Set up regular progress meetings
Write close-out report
Close-out review

Worked Example 1: Bungalow

Chapter Outline

The previous chapters described the various methods and techniques developed to produce meaningful and practical network programmes. In this chapter most of these techniques are combined in two fully worked examples. One is mainly of a civil engineering and building nature and the other is concerned with mechanical erection – both are practical and could be applied to real situations.

The first example covers the planning, manhour control, and cost control of a construction project of a bungalow. Before any planning work is started, it is advantageous to write down the salient parameters of the design and construction, or what is grandly called the 'design and construction philosophy'. This ensures that everyone who participates in the project knows not only what has to be done but why it is being done in a particular way. Indeed, if the design and construction philosophy is circulated *before* the programme, time- and cost-saving suggestions may well be volunteered by some recipients which, if acceptable, can be incorporated into the final plan.

Design and Construction Philosophy

1. The bungalow is constructed on strip footings.
2. External walls are in two skins of brick with a cavity. Internal partitions are in plasterboard on timber studding.
3. The floor is suspended on brick piers over an oversite concrete slab. Floorboards are T & G pine.
4. The roof is tiled on timber-trussed rafters with external gutters.
5. Internal finish is plaster on brick finished with emulsion paint.
6. Construction is by direct labour specially hired for the purpose. This includes specialist trades such as electrics and plumbing.
7. The work is financed by a bank loan, which is paid four-weekly on the basis of a regular site measure.
8. Labour is paid weekly. Suppliers and plant hire are paid four weeks after delivery. Materials and plant must be ordered two weeks before site requirement.

Project Management, Planning, and Control. http://dx.doi.org/10.1016/B978-0-08-098324-0.00047-0
417

9. The *average* labour rate is £5 per hour or £250 per week for a 50-hour working week. This covers labourers and tradesmen.
10. The cross-section of the bungalow is shown in Figure 47.1 and the sequence of activities is set out in Table 47.1, which shows the dependencies of each activity. All durations are in weeks. The network in Figure 47.2 is in AoA format and the equivalent network in AoN format is shown in Figure 47.3.

The activity letters refer to the activities shown on the cross-section diagram of Figure 47.1, and on subsequent tables only these activity letters will be used. The total float column can, of course, only be completed when the network shown in Figure 47.2 has been analysed (see Table 47.1).

Table 47.2 shows the complete analysis of the network including TL_e (latest time end event), TE_e (earliest time beginning event), total float, and free float. It will be noted that none of the activities have free float. As mentioned in Chapter 21 free float is often confined to the dummy activities, which have been omitted from the table.

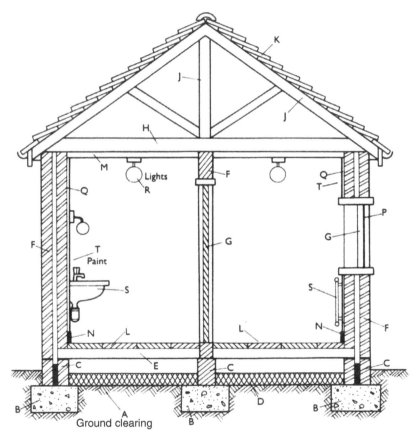

Figure 47.1
Bungalow (six rooms)

Table 47.1

Activity Letter	Activity – Description	Duration (weeks)	Dependency	Total Float
A	Clear ground	2	Start	0
B	Lay foundations	3	A	0
C	Build dwarf walls	2	B	0
D	Oversite concrete	1	B	1
E	Floor joists	2	C and D	0
F	Main walls	5	E	0
G	Door and window frames	3	E	2
H	Ceiling joists	2	F and G	4
J	Roof timbers	6	F and G	0
K	Tiles	2	H and J	1
L	Floorboards	3	H and J	0
M	Ceiling boards	2	K and L	0
N	Skirtings	1	K and L	1
P	Glazing	2	M and N	0
Q	Plastering	2	P	2
R	Electrics	3	P	1
S	Plumbing and heating	4	P	0
T	Painting	3	Q, R, and S	0

0 = Critical

To enable the resource loading bar chart in Figure 47.4 to be drawn it helps to prepare a table of resources for each activity (Table 47.3). The resources are divided into two categories:

A   Labourers
B   Tradesmen

This is because tradesmen are more likely to be in short supply and could affect the programme.

The total labour histogram can now be drawn, together with the total labour curve (Figure 47.5). It will be seen that the histogram has been hatched to differentiate between labourers and tradesmen, and shows that the maximum demand for tradesmen is eight men in weeks 27 and 28. Unfortunately, it is only possible to employ six tradesmen due to possible site congestion. What is to be done?

The advantage of network analysis with its float calculation is now apparent. Examination of the network shows that in weeks 27 and 28 the following operations (or activities) have to be carried out:

Activity Q	Plastering	3 men for 2 weeks
Activity R	Electrics	2 men for 3 weeks
Activity S	Plumbing and heating	3 men for 4 weeks

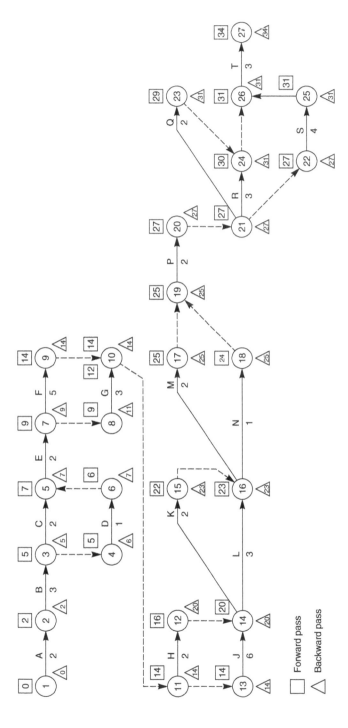

Figure 47.2
Network of bungalow (duration in weeks)

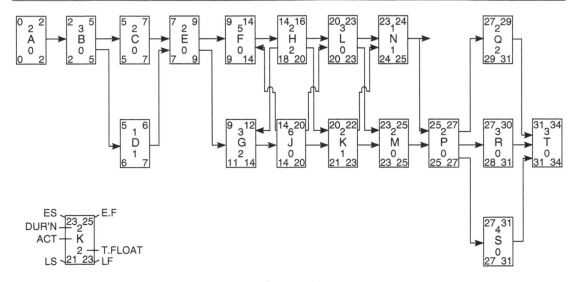

Figure 47.3
Network diagram of bungalow AoN format

Table 47.2

a	b	c	d	e	f	g	h
Activity Letter	Node no.	Duration	TL$_e$	TE$_e$	TE$_b$	d-f-c Total Float	e-f-c Free Float
A	1–2	2	2	2	0	0	0
B	2–3	3	5	5	2	0	0
C	3–5	2	7	7	5	0	0
D	4–6	1	7	6	5	1	0
E	5–7	2	9	9	7	0	0
F	7–9	5	14	14	9	0	0
G	8–10	3	14	12	9	2	0
H	11–12	2	20	16	14	4	0
J	13–14	6	20	20	14	0	0
K	14–15	2	23	22	20	1	0
L	14–16	3	23	23	20	0	0
M	16–17	2	25	25	23	0	0
N	16–18	1	25	24	23	1	0
P	19–20	2	27	27	25	0	0
Q	21–23	2	31	29	27	2	0
R	21–24	3	31	30	27	1	0
S	22–25	4	31	31	27	0	0
T	26–27	3	34	34	31	0	0

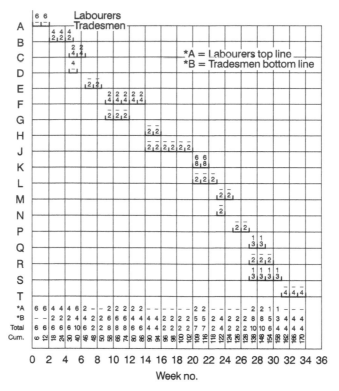

Figure 47.4

Resource loaded bar chat

Table 47.3: Labour Resources per Week

Activity Letter	Resource A Labourers	Resource B Tradesmen	Total
A	6		6
B	4	2	6
C	2	4	6
D	4	–	4
E	–	2	2
F	2	4	6
G	–	2	2
H	–	2	2
J	–	2	2
K	2	3	5
L	–	2	2
M	–	2	2
N	–	2	2
P	–	2	2
Q	1	3	4
R	–	2	2
S	1	3	4
T	–	4	4

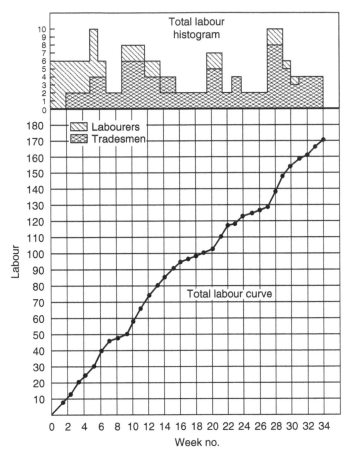

Figure 47.5
Histogram and 'S' curve

The first step is to check which activities have float. Consulting Table 47.2 reveals that Q (Plastering) has 2 weeks float and R (Electrics) has 1 week float. By delaying Q (Plastering) by 2 weeks and accelerating R (Electrics) to be carried out in 2 weeks by 3 men per week, the maximum total in any week is reduced to 6. Alternatively, it may be possible to extend Q (Plumbing) to 4 weeks using 2 men per week for the first two weeks and 1 man per week for the next two weeks. At the same time, R (Electrics) can be extended by one week by employing 1 man per week for the first two weeks and 2 men per week for the next two weeks. Again, the maximum total for weeks 27–31 is 6 tradesmen.

The new partial disposition of resources and revised histograms after the two alternative smoothing operations are shown in Figures 47.6 and 47.7. It will be noted that:

1. The overall programme duration has not been exceeded because the extra durations have been absorbed by the float.
2. The total number of man weeks of any trade has not changed – i.e., Q (Plastering) still has 6 man weeks and R (Electrics) still has 6 man weeks.

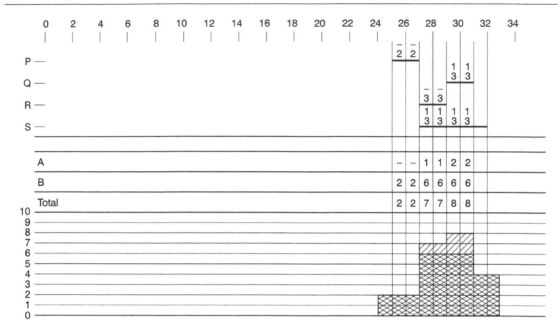

Figure 47.6
Resource smoothing 'A'

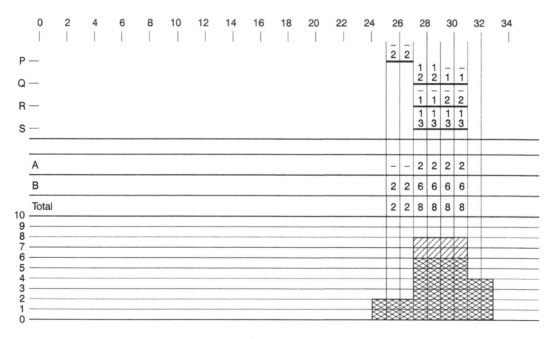

Figure 47.7
Resource smoothing 'B'

If it is not possible to obtain the necessary smoothing by utilizing and absorbing floats the network logic may be amended, but this requires a careful reconsideration of the whole construction process.

The next operation is to use the EVA system to control the work on site. Multiplying for each activity the number of weeks required to do the work by the number of men employed on that activity yields the number of man weeks. If this is multiplied by 50 (the average number of working hours in a week), the manhours per activity are obtained. A table can now be drawn up listing the activities, durations, number of men, and budget hours (Table 47.4).

As the bank will advance the money to pay for the construction in four-weekly tranches, the measurement and control system will have to be set up to monitor the work every 4 weeks. The anticipated completion date is week 34, so that a measure in weeks 4, 8, 12, 16, 20, 24, 28, 32, and 36 will be required. By recording the *actual* hours worked each week and assessing the percentage complete for each activity each week the value hours for each activity can be quickly calculated. As described in Chapter 32, the overall percentage complete, efficiency, and predicted final hours can then be calculated. Table 47.5 shows a manual EVA analysis for four sample weeks (8, 16, 24, and 32).

Table 47.4

a	b	c	d
Activity Letter	Duration (weeks)	No. of Men	b × c × 50 Budget Hours
A	2	6	600
B	3	6	900
C	2	6	600
D	1	4	200
E	2	2	200
F	5	6	1500
G	3	2	300
H	2	2	200
J	6	2	600
K	2	5	500
L	3	2	300
M	2	2	200
N	1	2	100
P	2	2	200
Q	2	4	400
R	3	2	300
S	4	4	800
T	3	4	600
Total			8500

In practice, this calculation will have to be carried out every week either manually as shown or by computer using a simple spreadsheet. It must be remembered that only the activities actually worked on during the week in question have to be computed. The remaining activities are entered as shown in the previous week's analysis.

For purposes of progress payments, the *value* hours for every 4-week period must be multiplied by the average labour rate (£5 per hour) and, when added to the material and plant costs, the total value for payment purposes is obtained. This is shown later in this chapter.

At this stage it is more important to control the job, and for this to be done effectively, a set of curves must be drawn on a time base to enable all the various parameters to be compared. The relationship between the actual hours and value hours gives a measure of the efficiency of the work, while that between the value hours and the planned hours gives a measure of progress. The actual and value hours are plotted straight from the EVA analysis, but the planned hours must be obtained from the labour expenditure curve (Figure 47.5) and multiplying the labour value (in men) by 50 (the number of working hours per week). For example, in week 16 the total labour used to date is 94 man weeks, giving 94 × 50 = 4700 manhours.

The complete set of curves (including the efficiency and percentage complete curves) is shown in Figure 47.8. In practice, it may be more convenient to draw the last two curves on a separate sheet, but provided the percentage scale is drawn on the opposite side to the manhour scale no confusion should arise. Again, a computer program can be written to plot these curves on a weekly basis as shown in Chapter 32.

Once the control system has been set up it is essential to draw up the cash flow curve to ascertain what additional funding arrangements are required over the life of the project. In most cases where project financing is required the cash flow curve will give an indication of how much will have to be obtained from the finance house or bank and when. In the case of this example, where the construction is financed by bank advances related to site progress, it is still necessary to check that the payments will, in fact, cover the outgoings. It can be seen from the curve in Figure 47.10 that virtually permanent overdraft arrangements will have to be made to enable the men and suppliers to be paid regularly.

When considering cash flow it is useful to produce a table showing the relationship between the usage of a resource, the payment date, and the receipt of cash from the bank to pay for it – even retrospectively. It can be seen in Table 47.6 that

1.   Materials have to be ordered 4 weeks before use.
2.   Materials have to be delivered 1 week before use.
3.   Materials are paid for 4 weeks after delivery.
4.   Labour is paid in week of use.

Table 47.5

Period	Budget	Week 8 Actual cum.	%	V	Week 16 Actual cum.	%	V	Week 24 Actual cum.	%	V	Week 32 Actual cum.	%	V
A	600	600	100	600	600	100	600	600	100	600	600	100	600
B	900	800	100	900	800	100	900	800	100	900	800	100	900
C	600	550	100	600	550	100	600	550	100	600	550	100	600
D	200	220	90	180	240	100	200	240	100	200	240	100	200
E	200	110	40	80	180	100	200	180	100	200	180	100	200
F	1500	–	–	–	1200	80	1200	1550	100	1500	1550	100	1500
G	300	–	–	–	300	100	300	300	100	300	300	100	300
H	200	–	–	–	180	60	120	240	100	200	240	100	200
J	600	–	–	–	400	50	300	750	100	600	750	100	600
K	500	–	–	–	–	–	–	500	100	500	550	100	500
L	300	–	–	–	–	–	–	250	80	240	310	100	300
M	200	–	–	–	–	–	–	100	60	120	180	100	200
N	100	–	–	–	–	–	–	50	40	40	110	100	100
P	200	–	–	–	–	–	–	–	–	–	220	100	200
Q	400	–	–	–	–	–	–	–	–	–	480	100	400
R	300	–	–	–	–	–	–	–	–	–	160	60	180
S	800	–	–	–	–	–	–	–	–	–	600	80	640
T	600	–	–	–	–	–	–	–	–	–	100	10	60
Total	8500	2280	27.8%	2360	4450	52%	4420	6110	70.6%	6000	7920	90.4%	7680
Efficiency			103%			99%			98%			96%	
Estimated final hours			8201			8557			8654			8761	

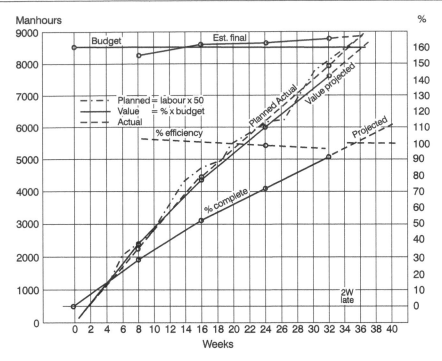

Figure 47.8
Control curves

Table 47.6

Week Intervals	1	2	3	4	5	6	7	8
Order date			X	X				
Material delivery				X				
Labour use				X				
Material use								
Labour payments								
Pay suppliers							O	
Measurement							M	
Receipt from bank								R
Every 4 weeks								
Starting week no. 5								
First week no.	–3	–2	–1	1	2	3	4	5

5. Measurements are made 3 weeks after use.
6. Payment is made 1 week after measurement.

The next step is to tabulate the labour costs and material and plant costs on a weekly basis (Table 47.7). The last column in the table shows the total material and plant cost for every

Table 47.7

Activity	No. of Weeks	Labour Cost per Week	Material and Plant per Week	Material Cost and Plant
A	2	1500	100	200
B	3	1500	1200	3600
C	2	1500	700	1400
D	1	1000	800	800
E	2	500	500	1000
F	5	1500	1400	7000
G	3	500	600	1800
H	2	500	600	1200
J	6	500	600	3600
K	2	1300	1200	2400
L	3	500	700	2100
M	2	500	300	600
N	1	500	200	200
P	2	500	400	800
Q	2	1000	300	600
R	3	500	600	1800
S	4	1000	900	3600
T	3	1000	300	900
Material total				33600

activity, because all the materials and plant for an activity are being delivered one week before use and have to be paid for in one payment. For simplicity, no retentions are withheld (i.e., 100% payment is made to all suppliers when due).

A bar chart (Figure 47.9) can now be produced, which is similar to that shown in Figure 47.4. The main difference is that instead of drawing bars, the length of the activity is represented by the weekly resource. As there are two types of resources – men and materials and plant – each activity is represented by two lines. The top line represents the labour cost in £100 units and the lower line the material and plant cost in £100 units. When the chart has been completed the resources are added vertically for each week to give a weekly total of labour out (i.e., men being paid, line 1) and material and plant out (line 2). The total cash out and the cumulative outflow values can now be added in lines 3 and 4, respectively.

The chart also shows the measurements every 4 weeks, starting in week 4 (line 5), and the payments one week later. The cumulative total cash is shown in line 6. To enable the outflow of materials and plant to be shown separately on the graph in Figure 47.10, it was necessary to enter the cumulative outflow for material and plant in row 7. This figure shows the cash flow curves (i.e., cash in and cash out). The need for a more-or-less permanent overdraft of approximately £10,000 is apparent.

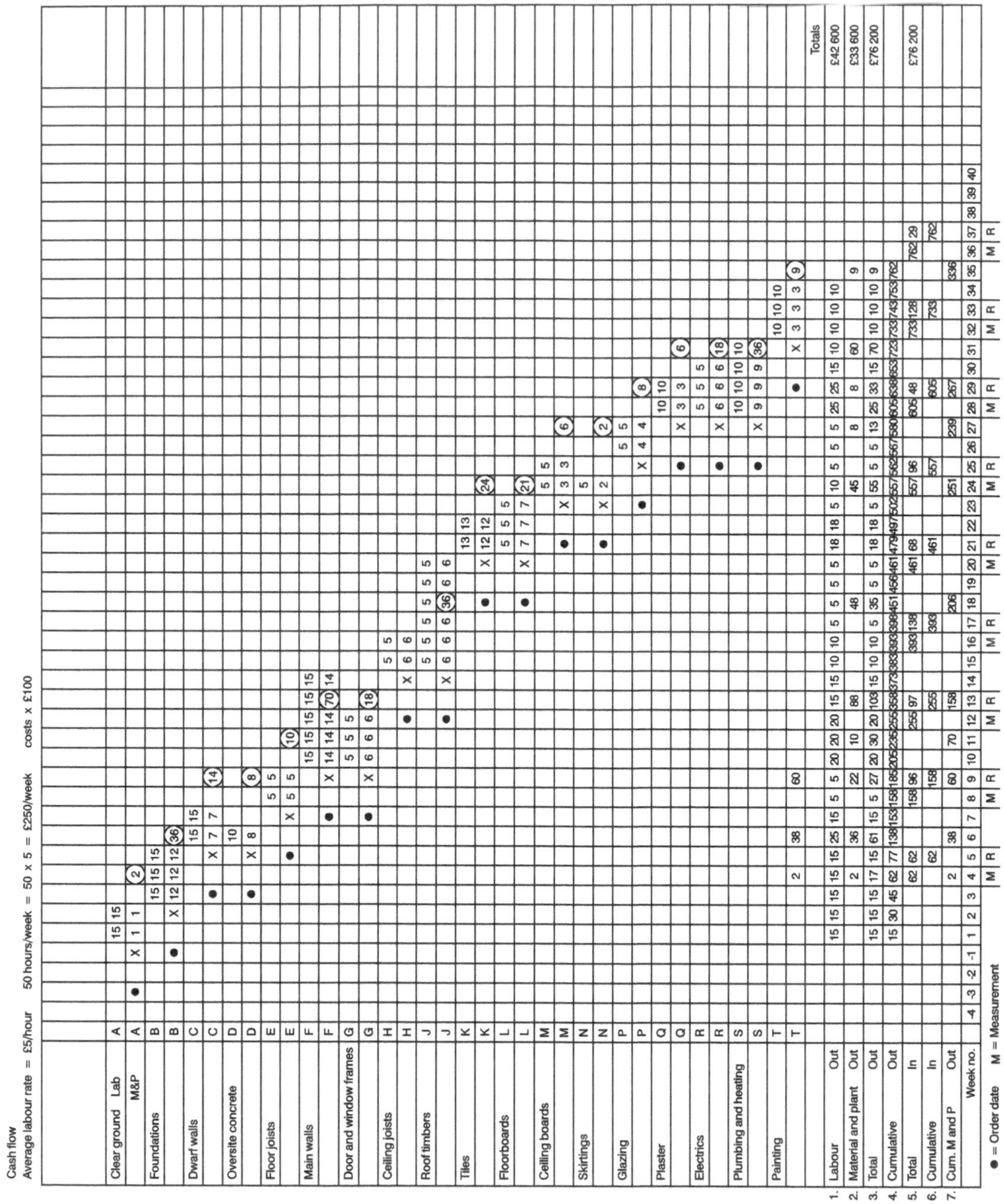

Cash flow
Average labour rate = £5/hour   50 hours/week = 50 × 5 = £250/week   costs × £100

Figure 47.9
Resource bar chart

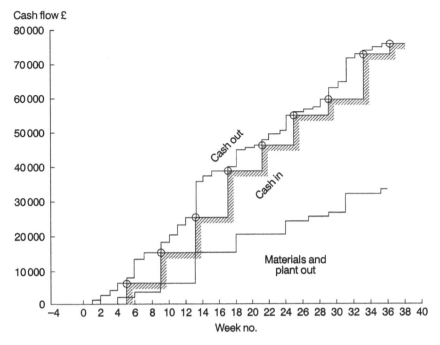

Figure 47.10
Cash flow curves

CHAPTER 48

Worked Example 2: Pumping Installation

Chapter Outline

Design and Construction Philosophy

1. 3-tonne vessel arrives on-site complete with nozzles and manhole doors in place.
2. Pipe gantry and vessel support steel arrive piece small.
3. Pumps, motors, and bedplates arrive as separate units.
4. Stairs arrive in sections with treads fitted to a pair of stringers.
5. Suction and discharge headers are partially fabricated with weldolet tees in place. Slip-on flanges to be welded on-site for valves, vessel connection, and blanked-off ends.
6. Suction and discharge lines from pumps to have slip-on flanges welded on-site after trimming to length.
7. Drive, couplings to be fitted before fitting of pipes to pumps, but not aligned.
8. Hydro test to be carried out in one stage. Hydro pump connection at discharge header end. Vent at top of vessel. Pumps have drain points.
9. Resource restraints require Sections A and B of suction and discharge headers to be erected in series.
10. Suction to pumps is prefabricated on-site from slip-on flange at valve to field weld at high-level bend.
11. Discharge from pumps is prefabricated on-site from slip-on flange at valve to field weld on high-level horizontal run.
12. Final motor coupling alignment to be carried out after hydro test in case pipes have to be re-welded and aligned after test.
13. Only pumps No. 1 and 2 will be installed.

In this example it is necessary to produce a material take-off from the layout drawings so that the erection manhours can be calculated. The manhours can then be translated into man days and, by assessing the number of men required per activity, into activity durations. The manhour assessment is, of course, made in the conventional manner by multiplying the operational units, such as numbers of welds or tonnes of steel, by the manhour norms used by the construction organization. In this exercise the norms used are those published by the

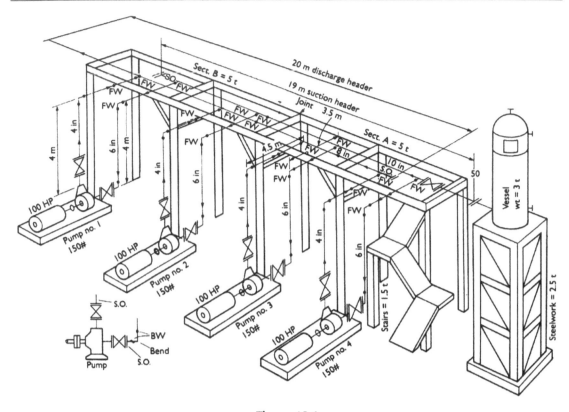

Figure 48.1
Isometric drawing. FW = field weld, BW = Butt weld, SO = Slip-on

OCPCA (Oil & Chemical Plant Contractors Association). These are base norms that may or may not be factorized to take account of market, environmental, geographical, or political conditions of the area in which the work is carried out. It is obvious that the rate for erecting a tonne of steel in the UK is different from erecting it in the wilds of Alaska.

The sequence of operations for producing a network programme and EVA analysis is as follows:

1. Study layout drawing or piping isometric drawings (Figure 48.1).
2. Draw a construction network. Note that at this stage it is only possible to draw the logic sequences (Figure 48.2) and allocate activity numbers.
3. From the layout drawing, prepare a take-off of all the erection elements such as number of welds, number of flanges, weight of steel, number of pumps, etc.
4. Tabulate these quantities on an estimate sheet (Figure 48.3) and multiply these by the OCPCA norms given in Table 48.1 to give the manhours per operation.
5. Decide which operations are required to make up an activity on a network and list these in a table. This enables the manhours per activity to be obtained.

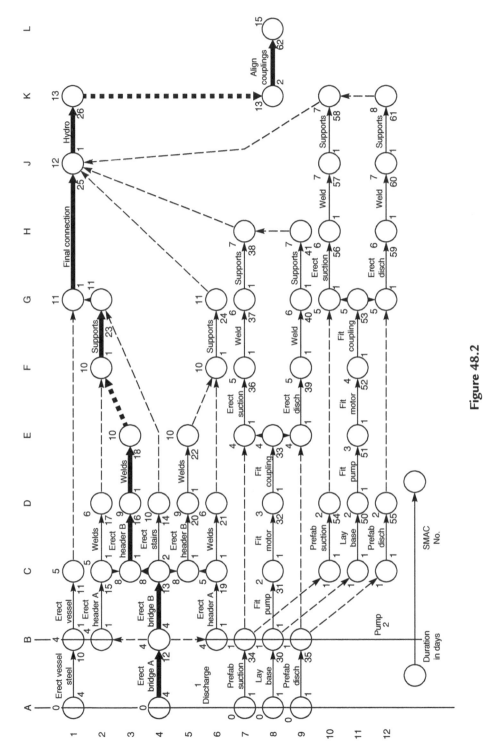

Figure 48.2
Network (using grid system)

ESTIMATE SHEET / SMAC ALLOCATION

A Item	B Unit	C Quant 1 set	D Hours rate	E =C × D man hours 1 set	F Pump man hours 2 sets	SMAC no. 1 set	SMAC man hours 1 set	SMAC no. pump no. 2	SMAC man hours pump no. 2	Duration days 1 set 2 men/act
Erect vessel steelwork	Tonne	2.5	24.7	61.75						4
Erect vessel 3 T	No. + Tonne	1	6.5 + 3.9	10.40						1
Erect bridge sect A	Tonne	5	12.3	61.50						4
Erect bridge sect B	Tonne	5	12.3	61.50						4
Erect stairs	Tonne	1.5	19.7	29.55						2
10" Suct. head erect sect. A	Metre	10	0.90	9.00		10	62			1
10" Suct. head erect sect B	Metre	9	0.90	8.10		11	11			1
10" Suct. head slip-on (valve)	No	2	2.92	5.84		12	62			1
10" Suct. head butt joint	No	1	3.25	5.25		13	62			–
10" Suct. head fit valve	No	1	3.41	2.41		14	30			–
10" Suct. head slip-on (vessel)	No	1	2.92	2.92		15	9			–
10" Suct. head slip-on (end)	No	1	0.90	0.90		16	8			1
10" Suct. head fit blank	No	1	1.44	5.76		17.1	15			1
10" Suct. head fit supports	No	4	0.90	0.90		17.2	–			–
10" Suct. head final conn.	No	1	0.80	6.40		17.3	–			–
8" Disch. head erect sect. A	Metre	8	0.80	9.60		17.4	4			1
8" Disch. head erect sect. B	Metre	12	2.77	2.77		18.1	–			1
8" Disch. head butt joint	No	1	2.49	2.49		18.2	6			1
8" Disch. head slip-on (end)	No	1	0.50	0.50		23	1			1
8" Disch. head fit blank	No	1	1.44	5.76		25	6			1
8" Disch. head fitt supports	No	4	4.00	4.00		19	10			1
Erect base plate	No	1	4.00	4.00	8.00	20	3	50	4	1
Fit pump 100 HP	No	1	14.00	14.00	28.00	22	–	51	14	1
Fit motor	No	1	14.00	14.00	28.00	21.1	3	52	14	1
Fit coupling	No	1	10.00	10.00	20.00	21.2	–	53	10	1
Fit 2 valves 6" & 4"	No	2	0.77	1.54	3.08	24	6	56.1	7	1
6" Suction erect	Metre	7.5	0.70	5.25	10.50	30	4	56.2	–	1
6" Suction make joint	No	1	0.44	0.44	0.88	31	14	56.3	7	–
6" Suction butt bend	No	2	2.30	4.60	9.20	32	14	57.1	–	1
6" Suction butt header	No	1	2.30	2.30	4.60	33	10	57.2	7	1
6" Suction fit supports	No	3	1.44	4.32	8.64	36.1	7	58	4	1
6" Suction 2 butts bend	* No	2	2.41	4.82	9.64	36.2	–	54.1	4	1
6" Suction slip-on	* No	1	1.44	1.44	2.88	36.3	7	54.2	6	1
4" Disch. erect	Metre	8.5	0.59	5.01	10.03	37.1	–	59	6	1
4" Disch. make joint	No	1	0.37	0.37	0.74	37.2	4	60.1	4	1
4" Disch. butt joint	No	1	1.82	1.82	3.64	38	4	60.2	–	–
4" Disch. butt header	No	3	1.82	1.82	3.64	34.1	6	60.3	–	–
4" Disch. fit supports	No	2	1.44	4.32	8.64	34.2	–	61	4	1
4" Disch. 2 butts bend	* No	1	1.89	3.78	7.56	39	6	55.1	5	1
4" Disch. slip-on	* No	1	1.14	1.14	2.28	40.1	4	55.2	–	–
Hydro-test 54 m	No	1	12.00	12.00		26	12			2
Align couplings	No	2	25.00	50.00		62	50		†	–
Total							445	+	85	41 12 = 530

* Pre-fabricate on site
† Item 62 is performed in 1 day due to overtime working

No. of man days = (41 + 12)2 = 53 × 2 = 106

Average hours/man day = 530/106 = 5

Figure 48.3

Table 48.1: Applicable rates from OCPCA norms

Steel erection		Hours
Pipe gantries		12.3/tonne
Stairs		19.7/tonne
Vessel support		24.7/tonne
Vessel (3 tonne)		6.5 + 1.3/tonne
Pump erection (100 hp)		14
Motor erection		14
Bedplate		4
Fit coupling		10
Align coupling		25
Prefab. piping (Sch. 40)		
6-inch suction prep.		0.81/end ⎫
4-inch discharge prep.		1.6/butt ⎬ 2.41
suction welds		1.44/flange ⎭
4-inch discharge prep.		0.62/end ⎫
discharge welds		1.27/butt ⎬ 1.89
discharge slip-on		1.14/flange ⎭
Pipe erection	(10-inch)	$0.79 \times 1.15 = 0.90$/m
Pipe erection	(8-inch)	$0.70 \times 1.15 = 0.80$/m
Pipe erection	(6-inch)	$0.61 \times 1.15 = 0.70$/m
Pipe erection	(4-inch)	$0.51 \times 1.15 = 0.59$/m
Site butt welds	(10-inch)	$2.83 \times 1.15 = 3.25$/butt
	(8-inch)	$2.41 \times 1.15 = 2.77$/butt
	(6-inch)	$2.0 \times 1.15 = 2.30$/butt
	(4-inch)	$1.59 \times 1.15 = 1.82$/butt
Slip-ons	(10-inch)	$3.25 \times 0.9 = 2.92$/butt
	(8-inch)	$2.77 \times 0.9 = 2.49$/butt
	(6-inch)	$2.30 \times 0.9 = 2.07$/butt
	(4-inch)	$1.82 \times 0.9 = 1.64$/butt
Fit valves	(10-inch)	$2.1 \times 1.15 = 1.045$/item
	(6-inch)	$0.9 \times 1.15 = 1.04$/item
	(4-inch)	$0.45 \times 1.15 = 0.51$/item
Flanged connection	(10-inch)	$0.78 \times 1.15 = 0.90$/connection
	(8-inch)	$0.43 \times 1.15 = 0.50$/connection
	(6-inch)	$0.38 \times 1.15 = 0.44$/connection
	(4-inch)	$0.32 \times 1.15 = 0.37$/connection
Supports		$1.25 \times 1.15 = 1.44$/support
Hydro test	Set up	$6 \times 1.15 = 6.9$
	Fill and drain	$2 \times 1.15 = 2.3$
	Joint check	$0.2 \times 1.15 = 0.23$/joint
	Blinds	$0.5 \times 1.15 = 0.58$/blind
	Hydrotest Total	$= 6.9 + 2.3 + (0.23 \times 12)$
		$= 9.2 + 2.76 = 11.96$ (say 12)

6. Assess the number of men required to perform any activity. By dividing the activity manhours by the number of men the actual working hours and consequently working days (durations) can be calculated.
7. Enter these durations in the network programme.
8. Carry out the network analysis, giving floats and the critical path (Table 48.2).
9. Draw up the EVA analysis sheet (Table 48.3) listing activities, activity (SMAC) numbers, and durations.
10. Carry out EVA analysis at weekly intervals. The basis calculations for value hours, efficiency, etc., are shown in Table 48.4.
11. Draw a bar chart using the network as a basis for start and finish of activities (Figure 48.4).
12. Place the number of men per week against the activities on the bar chart.
13. Add up vertically per week and draw the labour histogram and 'S' curve.
14. Carry out a resource-smoothing exercise to ensure that labour demand does not exceed supply for any particular trade. In any case, high peaks or troughs are signs of inefficient working and should be avoided here (Figure 48.5). (Note: This smoothing operation only takes place with activities which have float.)
15. Draw the project control curves using the weekly EVA analysis results to show graphically the relationship between:
    * Budget hours
    * Planned hours
    * Actual hours
    * Value hours
    * Predicted final hours (Figure 48.6)
16. Draw control curves showing:
    * Percentage complete (progress)
    * Efficiency (Figure 48.6)

The procedures outlined above will give a complete control system for time and cost for the project as far as site work is concerned.

Cash Flow

Cash flow charts show the difference between expenditure (cash outflow) and income (cash inflow). Since money is the common unit of measurement, all contract components such as manhours, materials, overheads, and consumables have to be stated in terms of money values.

It is convenient to set down the parameters that govern the cash flow calculations before calculating the actual amounts. For the example being considered:

1. There are 1748 productive hours in a year (39 hours/week × 52) – 280 days of annual holidays, statutory holidays, sickness, and travelling allowance and induction.
2. Each manhour costs, on average, £5 in actual wages.

Table 48.2: Total float

M SMAC no.	Duration (days)	Backward pass TL$_e$	Forward pass TE$_e$	TE$_e$	Total float	Welding activity
10	14	10	4	0	6	
11	1	11	5	4	6	
12	4	4	4	0	0	
13	4	8	8	4	0	
14	2	11	10	8	1	
15	1	8	5	4	3	
16	1	9	9	8	0	
17	1	10	6	5	4	X
18	1	10	10	9	0	X
19	1	9	5	4	4	
20	1	10	9	8	1	
21	1	11	6	5	5	X
22	1	11	10	9	1	X
23	1	11	11	10	0	
24	1	12	11	10	1	
25	1	12	12	11	0	X
26	1	13	13	12	0	
30	1	5	1	0	4	
31	1	7	2	1	5	
32	1	8	3	2	5	
33	1	9	4	3	5	
34	1	8	1	0	7	X
35	1	8	1	0	7	X
36	1	10	5	4	5	
37	1	11	6	5	5	X
38	1	12	7	6	5	
39	1	10	5	4	5	
40	1	11	6	5	5	X
41	1	12	7	6	5	
50	1	6	2	1	4	
51	1	7	3	2	4	
52	1	8	4	3	4	
53	1	9	5	4	4	
54	1	9	2	1	7	X
55	1	9	2	1	7	X
56	1	10	6	5	4	
57	1	11	7	6	4	X
58	1	12	8	7	4	
59	1	10	6	5	4	
60	1	11	7	6	4	X
61	1	12	8	7	4	
62	1	15	15	13	0	

Table 48.3: EVA analysis

EVA no.	EVA budget manhours	Day 5			Day 10			Day 15		
		A	%	V	A	%	V	A	%	V
Erect vessel steelwork	10	62	70	100	62	70	100	62	70	100	62
Erect vessel	11	11	12	100	11	12	100	11	12	100	11
Erect bridge sect. A	12	62	60	100	62	60	100	62	60	100	62
Erect bridge sect. B	13	62	40	50	31	65	100	62	65	100	62
Erect stairs	14	30	–	–	–	35	100	30	35	100	30
10-inch suct. head. erect A	15	9	10	100	9	10	100	9	10	100	9
10-inch suct. head. erect B	16	8	–	–	–	8	100	8	8	100	8
10-inch suct. head. welds A	17	15	–	–	–	18	100	15	18	100	15
10-inch suct. head. welds B	18	4	–	–	–	5	100	4	5	100	4
8-inch disch. head. erect A	19	6	6	80	5	6	100	6	6	100	6
8-inch disch. head. erect B	20	10	–	–	–	11	80	8	12	100	8
8-inch disch. head. welds A	21	3	–	–	–	3	100	3	3	100	3
8-inch disch. head. welds B	22	3	–	–	–	–	–	–	3	100	3
Suction header supports	23	6	–	–	–	7	60	4	8	100	6
Discharge header supports	24	6	–	–	–	–	–	–	6	100	6
Final connection	25	1	–	–	–	–	–	–	1	100	1
Hydro test	26	12	–	–	–	–	–	–	10	100	12
Base plate pump 1	30	4	3	100	14	3	100	4	3	100	4
Fit pump 1	31	14	14	100	14	14	100	14	14	100	14
Fit motor	32	14	12	100	14	12	100	14	12	100	14

33	Fit coupling 1	10	12	100	10	12	100	10	12	100	10
34	Prefab. suction pipe 1	6	10	100	6	10	100	6	10	100	6
35	Prefab. discharge pipe 1	5	4	80	4	5	100	5	5	100	5
36	Erect suction pipe 1	7	–	–	–	8	100	7	8	100	7
37	Weld suction pipe 1	7	–	–	–	5	100	7	5	100	7
38	Support suction pipe 1	4	–	–	–	4	80	3	5	100	5
39	Erect discharge pipe 1	6	5	70	4	7	100	6	7	100	6
40	Weld discharge pipe 1	4	–	–	–	4	100	4	4	100	4
41	Support discharge pipe 1	4	–	–	–	2	50	2	3	100	4
50	Basic plate pump 2	4	3	100	4	3	100	4	3	100	4
51	Fit pump 2	14	14	100	14	14	100	14	14	100	14
52	Fit motor 2	14	12	100	14	12	100	14	12	100	14
53	Fit coupling 2	10	10	100	10	10	100	10	10	100	10
54	Prefab. suction pipe 2	6	10	100	6	10	100	6	10	100	6
55	Prefab. discharge pipe 2	5	6	100	5	6	100	5	6	100	5
56	Erect suction pipe 2	7	5	60	4	8	100	7	8	100	7
57	Weld suction pipe 2	7	–	–	–	5	100	7	5	100	7
58	Support suction pipe 2	4	–	–	–	2	40	2	4	100	4
59	Erect discharge pipe 2	6	6	70	4	8	100	6	8	100	6
60	Weld discharge pipe 2	4	–	–	–	5	100	4	5	100	4
61	Support discharge pipe 2	4	–	–	–	3	70	3	4	100	4
62	Align couplings 1 & 2	50	–	–	–	–	–	–	16	10	20
	Totals	530	324	56%	297	482	84%	448	525	94%	500

Table 48.4: EVA calculations

	Day 5	Day 10	Day 15
Budget manhours	530	530	530
Actual manhours	324	482	525
Value manhours	297	448	500
Percentage complete	$\dfrac{297}{530}$	$\dfrac{448}{530}$	$\dfrac{500}{530}$
	= 56%	= 85%	= 94%
Est. final manhours	$\dfrac{324}{0.56}$	$\dfrac{482}{0.85}$	$\dfrac{525}{0.94}$
	= 597	= 567	= 559
Efficiency	$\dfrac{297}{324}$	$\dfrac{448}{482}$	$\dfrac{500}{525}$
	= 92%	= 93%	= 95%

A = Actual manhours

B = Budget manhours

V = Value manhours

V = Value manhours = Percentage complete × B of activity

$$\Sigma\% \text{ complete} = \frac{\Sigma V}{\Sigma B}$$

$$\text{efficiency} = \frac{V}{A}$$

$$\text{Est. final} = \frac{A}{\% \text{ complete}}$$

Activities shifted: 17, 12, 22, 35, 55, 19

3. After adding payments for productivity, holiday credits, statutory holidays, course attendance, radius, and travel allowance, the taxable rate becomes £8.40/hour.

4. The addition of other substantive items such as levies, insurance, protective clothing and non-taxable fares, and lodging increases the rate by £2.04 to £10.44/hour.

5. The ratio of other substantive items to taxable costs is $\dfrac{2.04}{8.40} = 0.243$

6. An on-cost allowance of 20% is made up of

Consumables	5%
Overheads	10%
Profit	5%
Total	20%

7. The total charge-out rate is, therefore, $10.44 \times 1.2 = £12.53$ per hour.

8. In this particular example:

   (a) the men are paid at the end of each day at a rate of £8.40/hour.

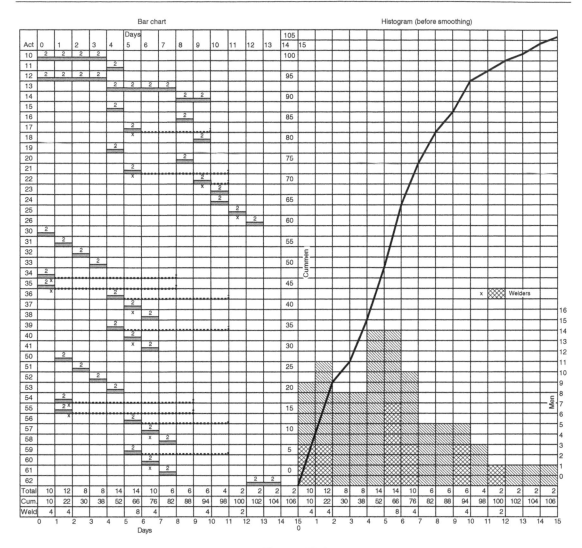

Figure 48.4
Bar chart

(b) the other substantive items of £2.04/hour are paid weekly.

(c) income is received weekly at the charge-out rate of £12.53/hour.

9. A week consists of 5 working days.

To enable the financing costs to be calculated at the estimate stage, cash flow charts are usually only drawn to show the difference between *planned* outgoings and *planned* income.

However, once the contract is under way a constant check must be made between *actual* costs (outgoings taken from time cards) and *valued* income derived from valuations of useful work done. The calculations for days 5 and 10 in Table 48.5 show how this is carried out. When these figures are plotted on a chart as in Figure 48.7 it can be seen that for

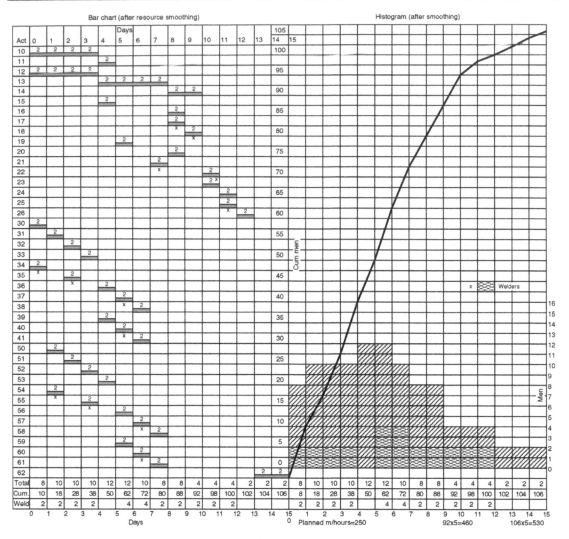

Figure 48.5

Bar chart after resource smoothing

days 0–5	the cash flow is negative (i.e., outgoings exceed income).
days 5–8	the cash flow is positive.
days 8–10	the cash flow is negative.
days 10–15	the cash flow is positive.
On day 15,	the total value is recovered assuming there are no retentions.

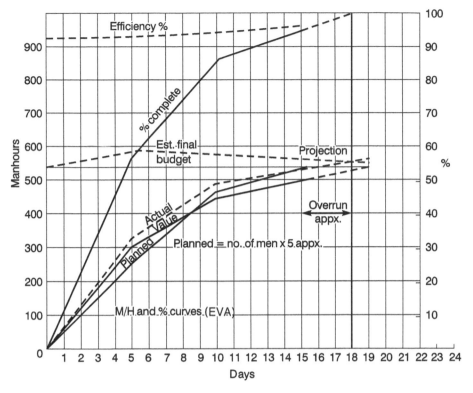

Figure 48.6
Control curves

The planned costs of the other substantives can be calculated for each period by multiplying the planned cumulative outgoings by the ratio of 0.243.

For day 5	the substantive costs are 2391 × 0.243 = £581.
For day 10	the substantive costs are 3826 × 0.243 = £930.
For day 15	the substantive costs are 4455 × 0.243 = £1083.

These costs are plotted on the chart and, when added to the planned labour costs, give total planned outgoings of:

£2391 + 581 = 2972 for day 5.

£3826 + 930 = 4756 for day 10.

£4455 + 1083 = 5538 for day 15.

To obtain the *actual* total outgoings it is necessary to multiply the *actual* labour costs by 1.243: e.g., for day 5, the *actual* outgoings will be

Table 48.5: Cash values

Activity	EVA no.	Duration (days)	EVA (budget) man-hours	Planned cost at £8.40 per hour	Planned price at £12.53 per hour	Day 5				Day 10			
						Actual man-hours	Actual cost at £8.40	Value hours	Value (price) at £12.53	Actual man-hours	Actual cost at £8.40	Value hours	Value (price) at £12.53
Erect vessel steelwork	10	4	62	521	777	70	588	62	777	70	588	62	777
Erect vessel	11	1	11	92	138	12	101	11	138	12	101	11	138
Erect bridge sect. A	12	4	62	521	777	60	504	62	777	60	504	62	777
Erect bridge sect. B	13	4	62	521	777	40	336	31	388	65	546	62	777
Erect stairs	14	2	30	252	376	–	–	–	–	35	294	30	376
10-inch suct. head. erect A	15	1	9	76	113	10	84	9	113	10	84	9	113
10-inch suct. head. erect B	16	1	8	67	100	–	–	–	–	8	67	8	100
10-inch suct. head. welds A	17	1	15	126	188	–	–	–	–	18	151	15	188
10-inch suct. head. welds B	18	1	4	34	50	–	–	–	–	5	42	4	50
8-inch disch. head. erect A	19	1	6	50	75	6	50	5	63	6	50	6	75
8-inch disch. head. erect B	20	1	10	84	125	–	–	–	–	11	92	8	100
8-inch disch. head. welds A	21	1	3	25	38	–	–	–	–	3	25	3	38
8-inch disch. head. welds B	22	1	3	25	38	–	–	–	–	–	–	–	–
Suction header supports	23	1	6	50	75	–	–	–	–	7	59	4	50
Discharge header supports	24	1	6	50	75	–	–	–	–	–	–	–	–

No.	Description												
25	Final connection	1	1	8	13	—	—	—	—	—	—	—	—
26	Hydro test	1	12	101	150	—	—	—	—	—	—	—	—
30	Base plate pump 1	1	4	34	50	3	25	4	50	3	25	4	50
31	Fit pump 1	1	14	118	175	14	118	14	175	14	118	14	175
32	Fit motor 1	1	14	118	175	12	101	14	175	12	101	14	175
33	Fit coupling 1	1	10	84	125	12	101	10	125	12	101	10	125
34	Prefab. suct. pipe 1	1	6	50	75	10	84	6	75	10	84	6	75
35	Prefab. discharge pipe 1	1	5	42	63	4	34	4	50	5	42	5	63
36	Erect suction pipe 1	1	7	59	88	—	—	—	—	8	67	7	88
37	Weld suction pipe 1	1	7	59	88	—	—	—	—	5	42	7	88
38	Support suction pipe 1	1	4	34	50	—	—	—	—	4	34	3	38
39	Erect discharge pipe 1	1	6	50	75	5	42	4	50	7	59	6	75
40	Weld discharge pipe 1	1	4	34	50	—	—	—	—	4	34	4	50
41	Support discharge pipe 1	1	4	34	50	—	—	—	—	2	17	2	25
50	Base plate pump 2	1	4	34	50	3	25	4	50	3	25	4	50
51	Fit pump 2	1	14	118	175	14	118	14	175	14	118	14	175
52	Fit motor 2	1	14	118	175	12	101	14	175	12	101	14	175
53	Fit coupling 2	1	10	84	125	10	84	10	125	10	84	10	125
54	Prefab. suction pipe 2	1	6	50	75	10	84	6	75	10	84	6	75
55	Prefab. discharge pipe 2	1	5	42	63	6	50	5	63	6	50	5	63

Continued

Table 48.5: Cash values—cont'd

Activity	EVA no.	Duration (days)	EVA (budget) man-hours	Planned cost at £8.40 per hour	Planned price at £12.53 per hour	Day 5				Day 10			
						Actual man-hours	Actual cost at £8.40	Value hours	Value (price) at £12.53	Actual man-hours	Actual cost at £8.40	Value hours	Value (price) at £12.53
Erect suction pipe 2	56	1	7	59	88	5	42	4	50	8	67	7	88
Weld suction pipe 2	57	1	7	59	88	–	–	–	–	5	42	7	88
Support suction pipe 2	58	1	4	34	50	–	–	–	–	2	17	2	25
Erect discharge pipe 2	59	1	6	50	75	6	50	4	50	8	67	6	75
Weld discharge pipe 2	60	1	4	34	50	–	–	–	–	5	42	4	50
Support discharge pipe 2	61	1	4	34	50	–	–	–	–	3	25	3	38
Align couplings 1 & 2	62	2	50	420	627	–	–	–	–	–	–	–	–
			530	4455	6640	324	2722	297	3719	482	4049	448	5613

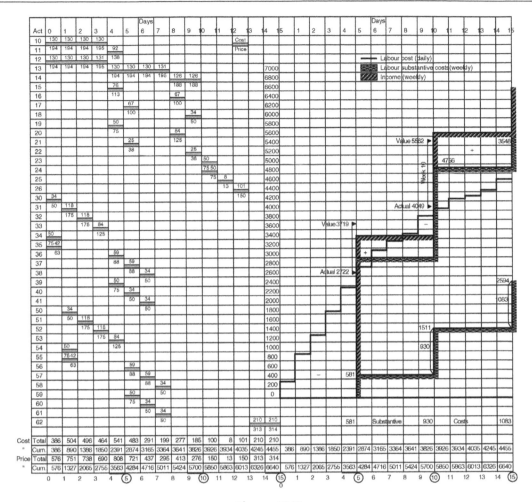

Figure 48.7
Bar chart and stepped 'S' curve

$2722 \times 1.243 = £3383$

and for day 10 they will be

$4049 \times 1.243 = £5033.$

The total planned and actuals can therefore be compared on a regular basis.

Worked Example 3: Motor Car

Chapter Outline

The example in this chapter shows how all the tools and techniques described so far can be integrated to give a comprehensive project management system. The project chosen is the design, manufacture, and distribution of a prototype motor car and while the operations and time scales are only indicative and do not purport to represent a real-life situation, the examples show how the techniques follow each other in a logical sequence.

The prototype motor car being produced is illustrated in Figure 49.1 and the main components of the engine are shown in Figure 49.2. It will be seen that the letters given to the engine components are the activity identity letters used in planning networks.

The following gives an oversight of the main techniques and their most important constituents.

As with all projects, the first document to be produced is the *Business Case*, which should also include the chosen option investigated for the *Investment Appraisal*. In this exercise, the questions to be asked (and answered) are shown in Table 49.1.

It is assumed that the project requires an initial investment of £60 million and that over a five-year period, 60,000 cars (units) will be produced at a cost of £5000 per unit. The assumptions are that the discount rate is 8%. There are two options for phasing the manufacture:

(a) That the factory performs well for the first two years but suffers some production problems in the next three years (option 1); and
(b) That the factory has teething problems in the first three years but goes into full production in the last two (option 2).

Project Management, Planning, and Control. http://dx.doi.org/10.1016/B978-0-08-098324-0.00049-4

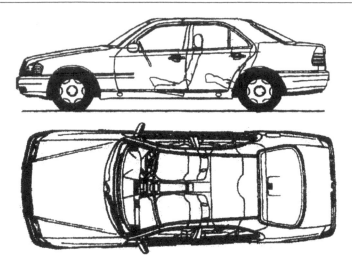

Figure 49.1
Motor car

The *Discounted Cash Flow* (DCF) calculations can be produced for both options as shown in Tables 49.2 and 49.3.

To obtain the *Internal Rate of Return* (IRR), an additional discount rate (in this case 20%) must be applied to both options. The resulting calculations are shown in Tables 49.4 and 49.5 and the graph showing both options is shown in Figure 49.3. This gives an IRR of 20.2% and 15.4%, respectively.

It is now necessary to carry out a cash flow calculation for the distribution phase of the cars. To line up with the DCF calculations, two options have to be examined. These are shown in Tables 49.6 and 49.7 and the graphs in Figures 49.4 and 49.5 for option 1 and option 2, respectively. An additional option 2a in which the income in years 2 and 3 is reduced from £65000K to £55000K is shown in the cash flow curves of Figure 49.6.

All projects carry an element of *Risk* and it is prudent to carry out a risk analysis at this stage. The types of risks that can be encountered, the possible actual risks, and the mitigation strategies are shown in Table 49.8. A risk log (or risk register) for five risks is given in Figure 49.7.

Once the decision has been made to proceed with the project, a *Project Life Cycle* diagram can be produced. This is shown on Figure 49.8 together with the constituents of the seven phases envisaged.

The next stage is the *Product Breakdown Structure* (Figure 49.9), followed by a combined *Cost Breakdown Structure* and *Organization Breakdown Structure* (Figure 49.10). By using these two, the *Responsibility Matrix* can be drawn up (Figure 49.11).

The parts of an overhead-camshaft engine

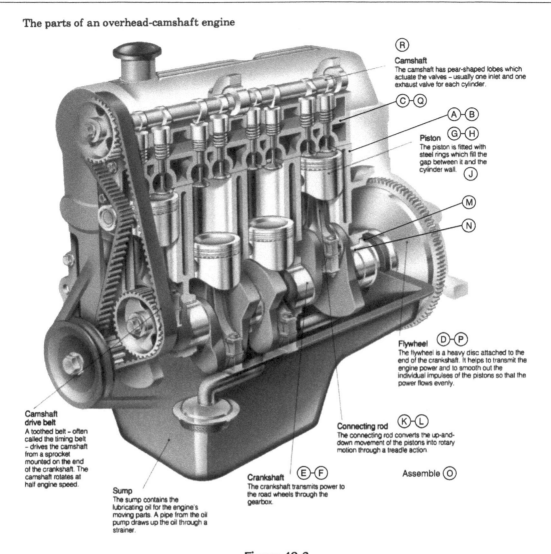

Figure 49.2
The parts of an overhead-camshaft engine

It is now necessary to produce a programme. The first step is to draw an *Activity List* showing the activities and their dependencies and durations. These are shown in the first four columns of Table 49.9. It is now possible to draw the *Critical Path Network* in either *AoN* format (Figure 49.12), *AoA* format (Figure 49.13), or as a *Lester diagram* (Figure 49.14).

After analysing the network diagram, the *Total Floats* and *Free Floats* of the activities can be listed (Table 49.10).

Apart from the start and finish, there are four milestones (days 8, 16, 24, and 30). These are described and plotted on the *Milestone Slip Chart* (Figure 49.15).

Table 49.1

Business Case
Why do we need a new model?
What model will it replace?
What is the market?
Will it appeal to the young, the middle aged, families, the elderly, women, trendies, yobos?
How many can we sell per year in the UK, the USA, the EEC and other countries?
What is the competition for this type of car and what is their price?
Will the car rental companies buy it?
What is the max. and min. selling price?
What must be the max. manufacturing cost and in what country will it be built?
What name shall we give it?
Do we have a marketing plan?
Who will handle the publicity and advertising?
Do we have to train the sales force and maintenance mechanics?
What should be the insurance category?
What warranties can be given and for how long?

What are the main specifications regarding
Safety and theftproofing?
Engine size (cc) or a number of sizes?
Fuel consumption?
Emissions (pollution control)?
Catalytic converter?
Max. speed?
Max. acceleration?
Size and weight?
Styling?
Turning circle and ground clearance?

What 'extras' must be fitted as standard?
ABS
Power steering
Air bags
Electric windows and roof
Cruise control
Air conditioning
What % can be recycled

Investment Appraisal (options)
Should it be a Saloon, Coupé, Estate, People Carrier, Convertible, 4 × 4, Mini?
Will it have existing or newly designed engine?
Will it have existing or new platform (chassis)?
Do we need a new manufacturing plant or can we build it in an existing one?
Should the engine be cast iron or aluminium?
Should the body be steel, aluminium or fibreglass?
Do we use an existing brand name or devise a new one?
Will it be fuelled on petrol, diesel, electricity or hybrid power unit?
DCF of investment returns, NPV, cash flow?

Table 49.2

DCF of Investment Returns (Net Present Value)							
Initial Investment £60000K 5 Year period							
Total car production 60000 Units @ £5000/Unit							
Option 1							
Year	Production Units	Income £K	Cost £K	Net Return £K	Discount Rate	Discount Factor	Present Value £K
1	15000	100000	75000	25000	8%	0.926	23150
2	15000	100000	75000	25000	8%	0.857	21425
3	10000	65000	50000	15000	8%	0.794	11910
4	10000	65000	50000	15000	8%	0.735	11025
5	10000	65000	50000	15000	8%	0.681	10215
Totals				95000			77725
Net Present Value (NPV) = 77725 – 60000 = £17725K							
Profit = £95000K – £60000K = £35000K							
Average Rate of Return (undiscounted) = £95000/5 = £19000K per annum							
Return on Investment = £19000/£60000 = 31.66%							

The network programme can now be converted into a bar chart (Figure 49.16) on which the resources (in men per day) as given in the fifth column of Table 49.9 can be added. After summating the resources for every day, it has been noticed that there is a peak requirement of 12 men in days 11 and 12. As this might be more than the available resources, the bar chart can be adjusted by utilizing the available floats to *smooth* the resources and eliminate the peak demand. This is shown in Figure 49.17 by delaying the start of activities D and F.

In Figure 49.18, the man days of the unsmoothed bar chart have been multiplied by 8 to convert them into manhours. This was necessary to carry out *Earned Value Analysis*. The daily manhour totals can be shown as a *histogram* and the cumulative totals are shown as an 'S' curve. In a similar way Figure 49.19 shows the respective histogram and 'S' curve for the smoothed bar chart.

It is now possible to draw up a table of *Actual Manhour* usage and *% complete* assessment for reporting day nos. 8, 16, 24, and 30. These, together with the *Earned Values* for these periods are shown in Table 49.11. Also shown are the efficiency (CPI), SPI, and predicted final completion costs and times as calculated at each reporting day.

Using the unsmoothed bar chart histogram and 'S' curve as a *Planned man hour* base, the Actual manhours and Earned Value manhours can be plotted on the graph in Figure 49.20. This graph also shows the *% complete* and *% efficiency* at each of the four reporting days.

Finally, Table 49.12 shows the actions required for the *Close-Out* procedure.

Table 49.3

DCF of Investment Returns (Net Present Value)

Initial Investment   £60000K   5 Year period

Total car production 60000 Units @ £5000/Unit

Option 2

Year	Production Units	Income £K	Cost £K	Net Return £K	Discount Rate	Discount Factor	Present Value £K
1	10000	65000	50000	15000	8%	0.926	13890
2	10000	65000	50000	15000	8%	0.857	12855
3	10000	65000	50000	15000	8%	0.794	11910
4	15000	100000	75000	25000	8%	0.735	18375
5	15000	100000	75000	25000	8%	0.681	17025
Totals				95000			74055

Net Present Value (NPV) = 74055 – 60000 = £14055K

Profit = £95000K – £60000K = £35000K

Average Rate of Return (undiscounted) = £95000/5 = £19000K per annum

Return on Investment = £19000/£60000 = 31.66%

Table 49.4

Internal Rate of Return (IRR)

Option 1

Year	Net Return £K	Disc. Rate	Disc. Factor	Present Value £K	Disc. Rate	Disc. Factor	Present Value £K
1	25000	15%	0.870	21750	20%	0.833	20825
2	25000	15%	0.756	18900	20%	0.694	17350
3	15000	15%	0.658	9870	20%	0.579	8685
4	15000	15%	0.572	8580	20%	0.482	7230
5	15000	15%	0.497	7455	20%	0.402	6030
Totals	60000			66555			60120
Less Investment				–60000			–60000
Net Present Value				£6555K			£120K

Internal Rate of Return (from graph) = 20.2%

Table 49.5

Internal Rate of Return (IRR)							
Option 2							
Year	*Net Return £K*	*Disc. Rate*	*Disc. Factor*	*Present Value £K*	*Disc. Rate*	*Disc. Factor*	*Present Value £K*
1	15000	15%	0.870	13050	20%	0.833	12495
2	15000	15%	0.756	11340	20%	0.694	10410
3	15000	15%	0.658	9870	20%	0.579	8685
4	25000	15%	0.572	14300	20%	0.482	12050
5	25000	15%	0.497	12425	20%	0.402	10050
Totals	60000			60985			53690
Less Investment				−60000			−60000
Net Present Value				£985K			−£6310K
Internal Rate of Return (from graph) = 15.4%							

Graph to obtain IRR

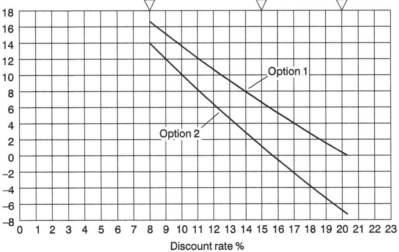

Figure 49.3
IRR curves

Table 49.6

Cash Flow							
Option 1							
Year		1	2	3	4	5	Cumulativi
Capital	£K	12000	12000	12000	12000	12000	
Costs	£K	75000	75000	50000	50000	50000	
Total	£K	87000	87000	62000	62000	62000	360000
Cumulative		87000	174000	236000	298000	360000	
Income	£K	100000	100000	65000	65000	65000	395000
Cumulative		100000	200000	265000	330000	395000	

Table 49.7

Cash Flow							
Option 2							
Year		1	2	3	4	5	Cumulative
Capital	£K	12000	12000	12000	12000	12000	
Costs	£K	50000	50000	50000	75000	75000	
Total	£K	62000	62000	62000	87000	87000	360000
Cumulative		62000	124000	186000	273000	360000	
Income	£K	65000	65000	65000	100000	100000	395000
Cumulative		65000	130000	195000	295000	395000	

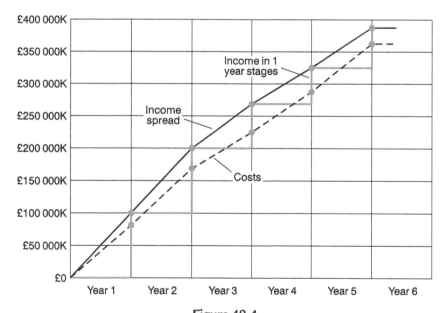

Figure 49.4
Cash flow chart, option 1

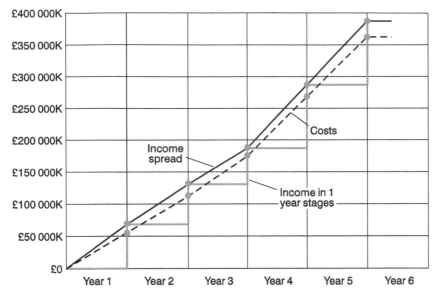

If income falls to £55 000K in years 2 and 3:

Income £K   =   65 000   55 000   55 000   100 000   100 000
Cumulative   =   65 000   120 000   175 000   275 000   375 000

Figure 49.5
Cash flow chart, option 2

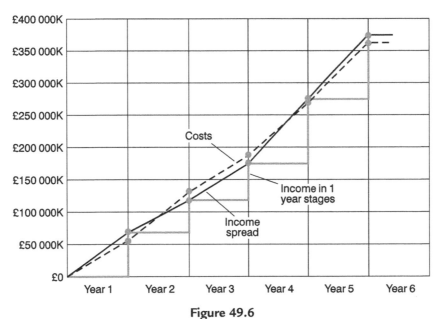

Figure 49.6
Cash flow chart, option 2a (with reduced income in years 2 and 3)

Table 49.8

Risk Analysis	Training problems
Types of risks	Suppliers unreliable
Manufacturing (machinery and facilities) costs	Rustproofing problems
Sales and marketing, exchange rates	Performance problems
Reliability	Industrial disputes
Mechanical components performance	Electrical and electronic problems
Electrical components performance	Competition too great
Maintenance	Not ready for launch date (exhibition)
Legislation (emissions, safety, recycling, labour, tax)	Safety requirements
Quality	Currency fluctuations
Possible risks	*Mitigation strategy*
Won't sell in predicted numbers	Overtime
Quality in design, manufacture, finish	More tests
Maintenance costs	More research
Manufacturing costs	More advertising/marketing
New factory costs	Insurance
Tooling costs	Re-engineering
New factory not finished on time	Contingency

RISK LOG

Project: Key: H – High; M – Medium; L – Low		Prepared by: *A.L.*						Reference: Date: *12.12.2000*		
Type of Risk	Description of Risk	Probability			Impact			Risk Reduction Strategy	Contingency Plans	Risk Owner
		H	M	L	Perf.	Cost	Time			
R1 *Manufact.*	*Factory not* *finished on time*			*10%*		*1M*	*3 months* *delay*	*Work more* *overtime*	*Cancel* *launch of car*	*PM*
R2 *Quality*	*Window mechanism* *faulty*		*50%*		*Not* *serious*	*5K*	*1 week to* *rectify*	*Test motor*	*Use manual* *winder*	*Chief* *Designer*
R3 *Safety*	*Air bags may* *explode*			*1%*	*Serious*	*10K*	*1 week to* *rectify*	*Run more* *tests*	*Remove Air* *bags*	*Chief* *Designer*
R4 *Legislation*	*Emission levels* *will be reduced*		*50%*		*Serious*	*3M*	*1 year to* *modify CC*	*Increase* *research*	*Buy another* *proven conv.*	*Chief* *Engineer*
R5 *Sales*	*Sales forecasts* *will not be met*		*30%*		*Very* *serious*	*10M*		*Increase* *advertising*	*Reduce* *price*	*PM*

Figure 49.7

Product and Project Life Cycle

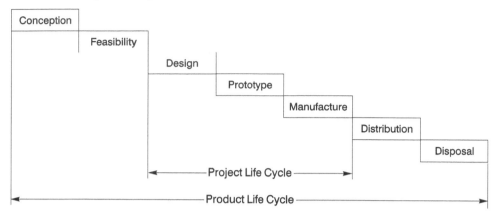

Phases

Conception: Original idea, high level discussions, preliminary market research

Feasibility: Consumer survey, market survey, type and size of car, production run and costs

Design: Vehicle design, tool design, development, component tests

Prototype: Tooling, production line, limited production, arctic and desert testing

Manufacture: Mass production, operator training, spares build up, customizing

Distribution: Deliveries, staff training, sales conferences, marketing, advertising, exhibitions

Disposal: Dismantling production line, selling tools, negotiating licences for spares

The phases could overlap.

The end of each phase could be a decision point to stop or proceed.

Figure 49.8

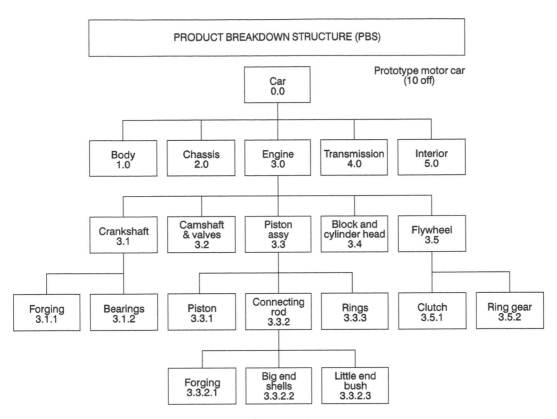

Figure 49.9

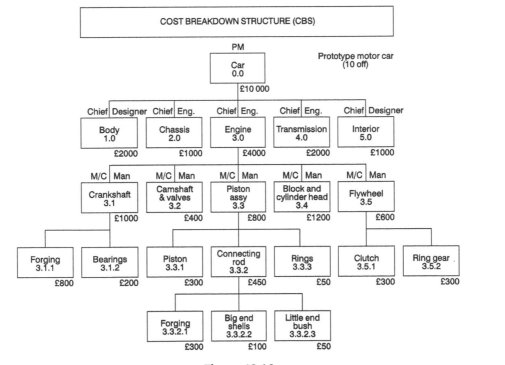

Figure 49.10

Responsibility Matrix

	Sponsor	Project manager	Chief designer	Chief engineer	M/C shop manager	Chassis manager	Styling manager
Body	B	B	A	D	D	D	C
Chassis		B		A	C	C	
Engine	B	B		A	C	D	
Transmission		B		A	C	C	
Interior	B	B	A	D	D	D	C

A Main responsibility
B Must be advised
C Must be consulted
D Requires updates

Figure 49.11

Table 49.9: Activity List of Motor Car Engine Manufacture and Assembly (10 off), 8 hours/day

Activ. Letter	Description	Dependency	Duration Days	Men per Day	Manhours per Day	Total Manhours
A	Cast block and cylinder head	Start	10	3	24	240
B	Machine block	A	6	2	16	96
C	Machine cylinder head	B	4	2	16	64
D	Forge and mc. flywheel	E	4	2	16	64
E	Forge crankshaft	Start	8	3	24	192
F	Machine crankshaft	E	5	2	16	80
G	Cast pistons	A	2	3	24	48
H	Machine pistons	G	4	2	16	64
J	Fit piston rings	H	1	2	16	16
K	Forge connecting rod	E	2	3	24	48
L	Machine conn. rod	K	2	2	16	32
M	Fit big end shells	L	1	1	8	8
N	Fit little end bush	M	1	1	8	8
O	Assemble engine	B, F, J, N	5	4	32	160
P	Fit flywheel	D, O	2	4	32	64
Q	Fit cylinder head	C, P	2	2	16	32
R	Fit camshaft and valves	Q	4	3	24	96
Total						1312

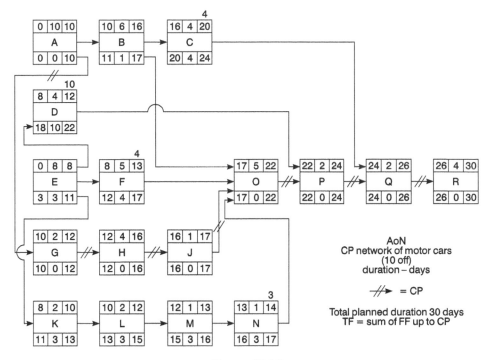

Figure 49.12

AoN
CP network of motor cars
(10 off)
duration – days

—//→ = CP

Total planned duration 30 days
TF = sum of FF up to CP

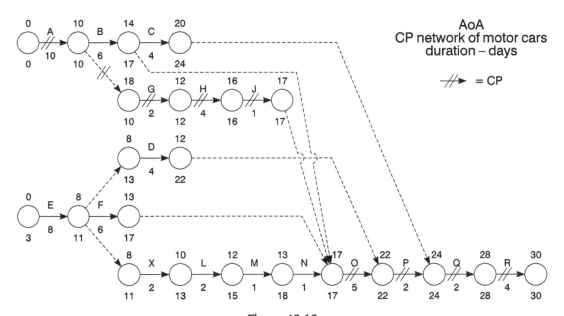

Figure 49.13

AoA
CP network of motor cars
duration – days

—//→ = CP

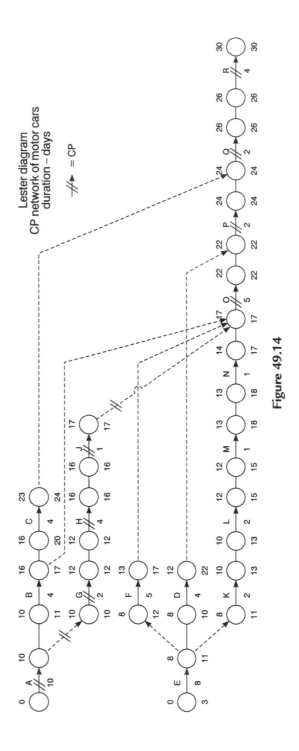

Lester diagram
CP network of motor cars
duration – days

—#—▶ = CP

Figure 49.14

Table 49.10: Activity Floats from CP Network

Activ. Letter	Description	Duration	Total Float	Free Float
A	Cast block and cylinder head	10	0	0
B	Machine block	6	1	0
C	Machine cylinder head	4	4	4
D	Forge and mc. flywheel	4	10	10
E	Forge crankshaft	8	3	0
F	Machine crankshaft	5	4	4
G	Cast pistons	2	0	0
H	Machine pistons	4	0	0
J	Fit piston rings	1	0	0
K	Forge connecting rod	2	3	0
L	Machine conn. rod	2	3	0
M	Fit big end shells	1	3	0
N	Fit little end bush	1	3	3
O	Assemble engine	5	0	0
P	Fit flywheel	2	0	0
Q	Fit cylinder head	2	0	0
R	Fit camshaft and valves	4	0	0

Milestones

Milestone 1 Forge crankshaft (E)   Day 8
Milestone 2 Machine pistons (H)   Day 16
Milestone 3 Fit flywheel (P)   Day 24
Milestone 4 Completion   Day 30

Milestone slip chart   **Programme**

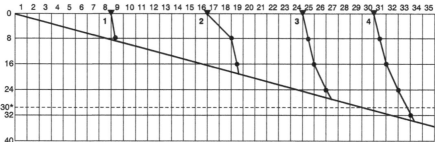

Assume:

* Reporting periods (8, 16, 24 and 30)
Milestone 1 slips ½ day
     "   2 "   2 days, then ½ day
     "   3 "   ½ day, then ½ day, then 1 day
     "   4 "   ½ day, then ½ day, then 1 day, then 1 day

Figure 49.15

Bar chart of prototype motor cars (10 off)

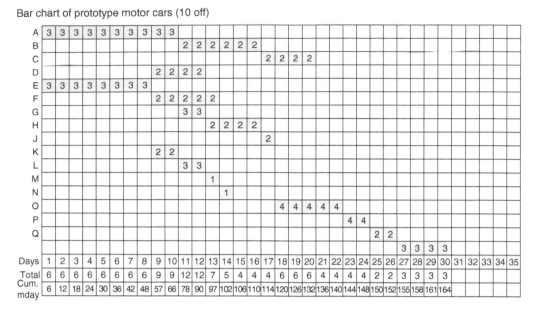

Figure 49.16
Unsmoothed

Bar chart of prototype motor cars (10 off)   After moving D to start at day 18
and moving F to start at day 12

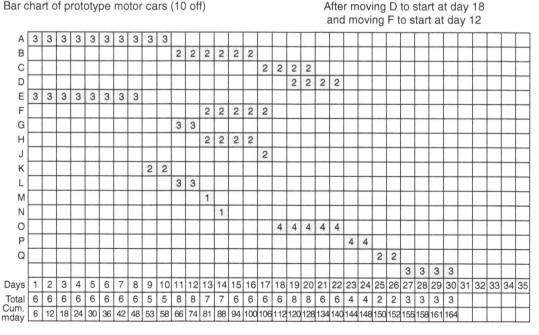

Figure 49.17
Smoothed

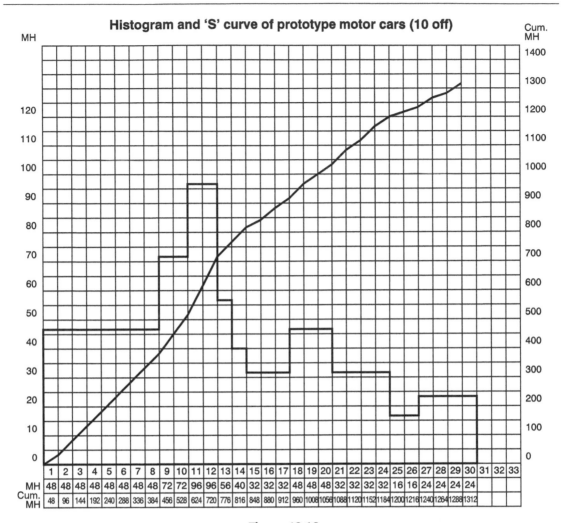

Figure 49.18

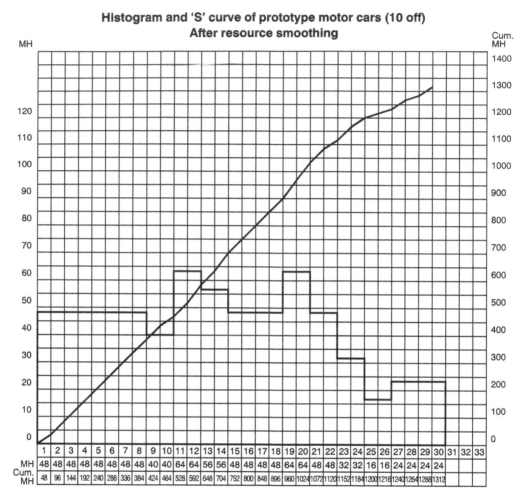

Figure 49.19

Table 49.11: Manhour Usage of Motor Car Engine Manufacture and Assembly (10 off) (unsmoothed)

Period Act.	Budget M/H	Day 8 Actual cum.	Day 8 % comp.	Day 8 EV	Day 16 Actual cum.	Day 16 % comp.	Day 16 EV	Day 24 Actual cum.	Day 24 % comp.	Day 24 EV	Day 30 Actual cum.	Day 30 % comp.	Day 30 EV
A	240	210	80	192	260	100	240	260	100	240	260	100	240
B	96				30	20	19	110	100	96	110	100	96
C	64							70	100	64	70	100	64
D	64				60	50	32	80	100	64	80	100	64
E	192	170	80	154	200	100	192	200	100	192	200	100	192
F	80				70	80	64	90	100	80	90	100	80
G	48				54	100	48	60	100	48	60	100	48
H	64				60	80	51	68	100	64	68	100	64
J	16							16	100	16	16	100	16
K	48				52	100	48	52	100	48	52	100	48
L	32				40	100	32	40	100	32	40	100	32
M	8				6	80	6	8	100	8	8	100	8
N	8				6	80	6	8	100	8	8	100	8
O	160							158	90	144	166	100	160
P	64										80	100	64
Q	32										24	60	19
R	96										52	40	38
Total	1312	380		346	838		738	1220		1104	1384		1241
% complete			26.3			56.2			84.1			94.6	
Planned manhours			384			880			1184			1312	
Efficiency (CPI) %			91			88			90			90	
Est. final manhours			1442			1491			1458			1458	
SPI (cost)			0.90			0.84			0.93			0.96	
SPI (time)			0.90			0.86			0.92			0.89	
Est. completion day			33			36			32			31	

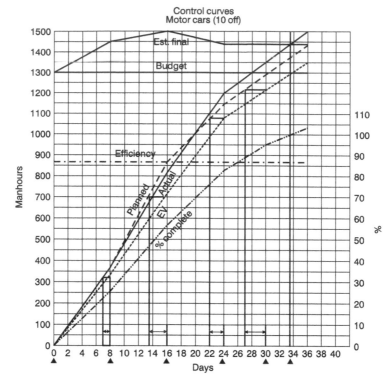

Figure 49.20
Unsmoothed resources

Table 49.12

Close-out
Close-out meeting Store standard tools
Sell special tools and drawings to Ruritania
Clear machinery from factory
Sign lease with supermarket that bought the site
Sell spares to dealers
Sell scrap materials
Write report and highlight problems
Press release and photo opportunity for last car
Give away 600000th production car to special lottery winner

Summary

Business Case

Need for new model. What type of car? Min./max. price. Manufacturing cost. Units per year. Marketing strategy. What market sector is it aimed at? Main specification. What extras should be standard? Name of new model. Country of manufacture.

Investment Appraisal

Options: Saloon, Coupé, Estate, Convertible, People carrier, 4 × 4. Existing or new engine. Existing or new platform. Materials of construction for engine, body. Type of fuel. New or existing plant. DCF of returns, NPV, Cash flow.

Project and Product Life Cycle

Conception:	Original idea, submission to top management
Feasibility:	Feasibility study, preliminary costs, market survey
Design:	Vehicle and tool design, component tests
Prototype:	Tooling, production line, environmental tests
Manufacture:	Mass production, training
Distribution:	Deliveries, staff training, marketing
Disposal:	Dismantling of plant, selling tools

Work and Product Breakdown Structure

Design, Prototype, Manufacture, Testing, Marketing, Distribution, Training.

Body, Chassis, Engine, Transmission, Interior, Electronics.

Cost Breakdown Structure, Organization Breakdown Structure, Responsibility Matrix.

AoN Network

Network diagram, forward and backward pass, floats, critical path, examination for overall time reduction, conversion to bar chart with resource loading, histogram, reduction of resource peaks, cumulative 'S' curve. Milestone slip chart.

Risk Register

Types of risks: manufacturing, sales, marketing, reliability, components failure, maintenance, suppliers, legislation, quality. Qualitative and quantitative analysis. Probability and impact matrix. Risk owner. Mitigation strategy, contingency.

Earned Value Analysis

EVA of manufacture and assembly of engine, calculate Earned Value, CPI, SPI, cost at completion, final project time, draw curves of budget hours, planned hours, actual hours earned value, % complete, efficiency over four reporting periods.

Close-Out

Close-out meeting
Close-out report
Instruction manuals
Test certificates
Spares lists
Dispose of surplus materials

CHAPTER 50

Worked Example 4: Battle Tank

Chapter Outline

Business Case for Battle Tank Top Secret

Memo: From: General Johnson

To: The Department of Defence

1 September 2006

Subject: new battle tank

It is imperative that we urgently draw up plans to design, evaluate, test, build, and commission a new battle tank.

The 'What'

A new battle tank which:

(a) Has a 90 mm cannon
(b) Has a top speed of at least 70 mph
(c) Weighs less than 60 tonnes fully loaded and fuelled
(d) Has spaced and active armour
(e) Has at least 2 machine gun positions including the external turret machine gun
(f) Has a crew of not more than 4 men (or women)
(g) Has a gas turbine engine and a fuel tank to give a range of 150 miles (240 km)
(h) In addition: The cost must not exceed $5500000 each
(j) 500 units must be ready for operations by February 2008

The 'Why'

(a) The existing battle tanks will be phased out (and worn out) in 2008
(b) Ruritania is developing a tank which is superior to our existing tanks in every way
(c) The existing tank at 80 tonnes is too heavy for 50% of our road bridges
(d) The diesel engine is too heavy and unreliable in cold weather
(e) The armour plate on our tanks can be penetrated by the latest anti-tank weapon
(f) A new tank has great export potential and could become the standard tank for NATO

Project Management, Planning, and Control. http://dx.doi.org/10.1016/B978-0-08-098324-0.00050-0

Major risks

(a) The cost may escalate due to poor project management
(b) The delivery period may be later than required due to incompetence of the contractors
(c) The fuel consumption of the gas turbine may not give the required range
(d) Ruritania will have an even better tank by 2008
(e) No matter how good our tank is, NATO will probably buy the new German Leopard Tank
(f) Heavy tanks may eventually be replaced by lighter airborne armoured vehicles

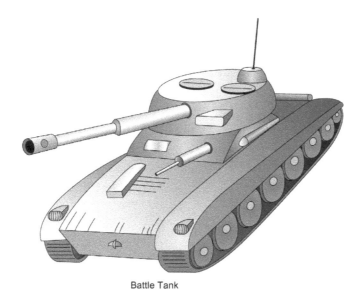

Battle Tank

Figure 50.1
Battle tank

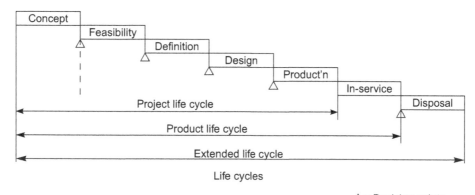

Life cycles

△ = Decision points

Figure 50.2
Life cycles and phases

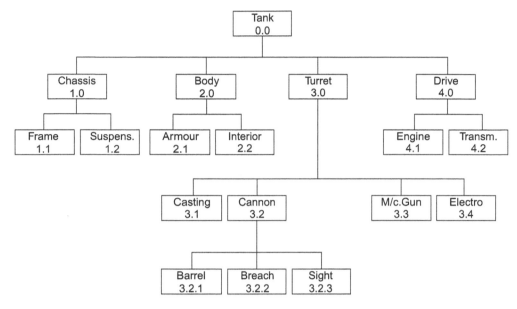

Figure 50.3
Product breakdown structure (PBS)

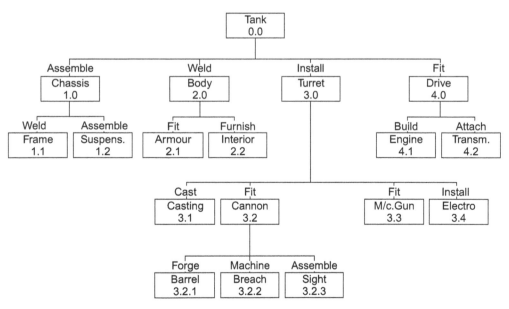

Figure 50.4
Work breakdown structure (WBS)

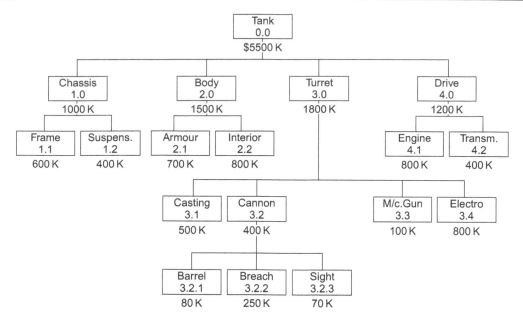

Figure 50.5
Cost breakdown structure (CBS)

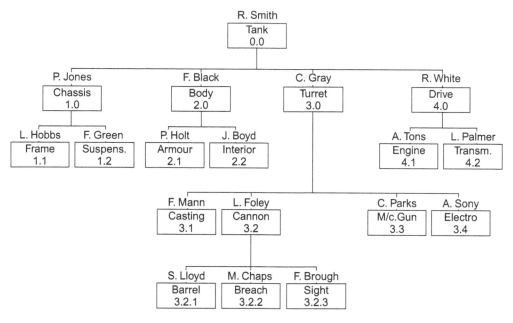

Figure 50.6
Organization breakdown structure (OBS)

	C. Gray	F. Mann	L. Foley	C. Parks	A. Sony	S. Lloyd	M. Chaps	F. Brough	
Turret	R	A	A	A	A	A	A	A	
Casting	C	R	C	C	C	A	A	C	
Cannon	C	C	R	A	C	C	C	C	
M/c Gun	A	C	C	R	C	A	A	–	
Electro.	C	C	C	–	R	–	–	C	
Barrel	A	C	C	–	–	R	C	–	
Breach	A	C	C	–	–	C	R	–	
Sight	A	C	C	–	C	A	–	R	

R = Responsible
C = Must be consulted
A = Must be advised
– = Not affected

Figure 50.7
Responsibility matrix

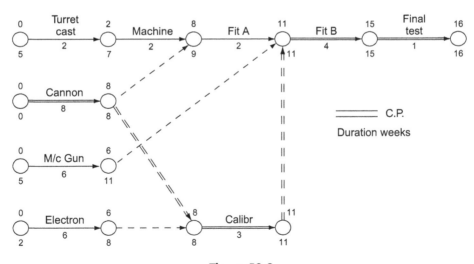

Figure 50.8
Activity on arrow network (AoA)

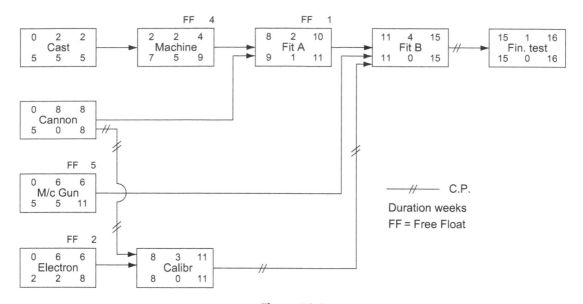

Figure 50.9

Activity on node network (AoN)

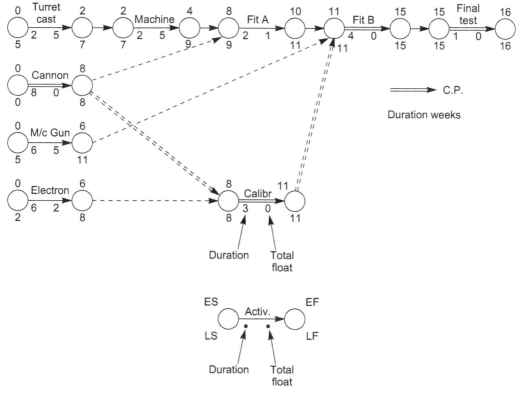

Figure 50.10

'Lester' diagram

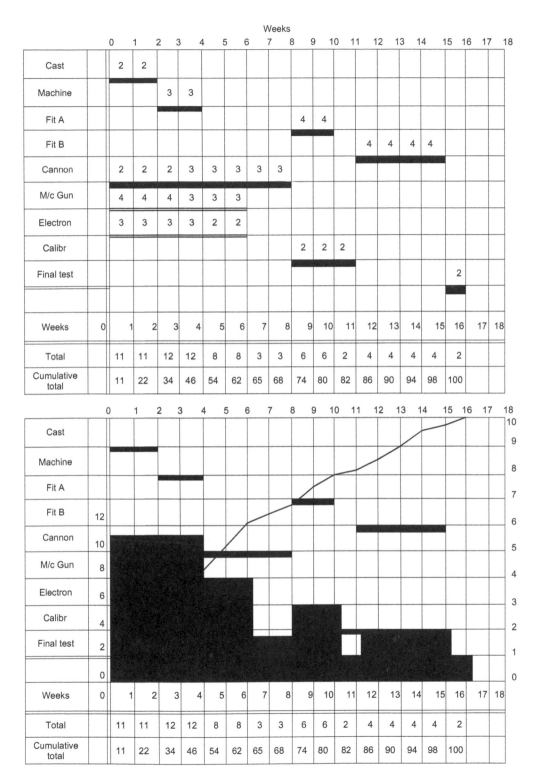

Figure 50.11
Histogram and 'S' curve

Activity	M/H Budg	Week 4 Plan	Week 4 Act	Week 4 %	Week 4 EV	Week 8 Plan	Week 8 Act	Week 8 %	Week 8 EV	Week 12 Plan	Week 12 Act	Week 12 %	Week 12 EV	Week 16 Plan	Week 16 Act	Week 16 %	Week 16 EV
Casting	160		180	100	160		180	100	160		180	100	160		180	100	160
Machine	240		180	80	192		200	100	240		200	100	240		200	100	240
Fit A	320		—	—	—		—	—	—		300	80	256		340	100	320
Fit B	640		—	—	—		—	—	—		80	10	64		600	50	320
Cannon	840		600	50	420		780	90	756		850	100	840		850	100	840
M/C Gun	840		500	60	504		700	90	756		820	100	840		820	100	840
Electronic	640		300	40	256		620	80	512		700	100	640		700	100	640
Calibrate	240		—	—	—		—	—	—		140	60	144		250	100	240
Test	80		—	—	—		—	—	—		—	—	—		110	20	32
Total	4000	1840	1760		1532	2720	2480		2424	3440	3270		3184	4000	4050		3632
% Complete		$\frac{1532}{4000}$	=	38.3%		$\frac{2424}{4000}$	=	60.6%		$\frac{3684}{4000}$	=	79.6%		$\frac{3632}{4000}$	=	90.8%	
CPI (Efficiency)		$\frac{1532}{1760}$	.870	=	87%	$\frac{2424}{2480}$	.977	=	98%	$\frac{3184}{3270}$	.973	=	97%	$\frac{3632}{4050}$	.896	=	90%
SPI (Cost)		$\frac{1532}{1840}$	=	.832		$\frac{2424}{2720}$	=	.891		$\frac{3184}{3440}$	=	.925		$\frac{3632}{4000}$	=	.908	
SPI (Time)		$\frac{3.2}{4}$	=	.80		$\frac{6.7}{8}$	=	.837		$\frac{10.3}{12}$	=	.858		$\frac{13.7}{16}$	=	.856	
Final cost		$\frac{4000}{.870}$	=	4598		$\frac{4000}{.977}$	=	4094		$\frac{4000}{.973}$	=	4110		$\frac{4000}{.896}$	=	4462	
Final time	(Cost)	$\frac{16}{.832}$	=	19.2		$\frac{16}{.891}$	=	17.9		$\frac{16}{.925}$	=	17.3		$\frac{16}{.908}$	=	17.6	

Budget man hours = duration × no. of men × 40 hrs/week

Duration in weeks

Figure 50.12
Earned value table

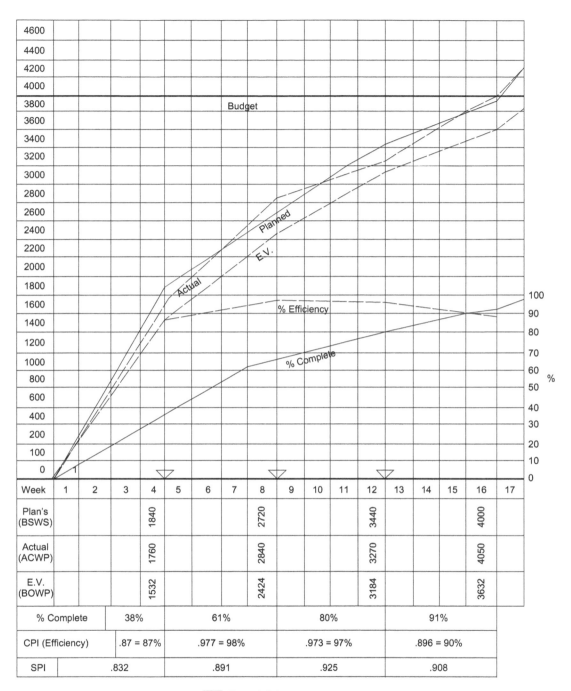

Week	1	2	3	4	5	6	7	8	9	10	11	12	13	14	15	16	17
Plan's (BSWS)				1840				2720				3440				4000	
Actual (ACWP)				1760				2840				3270				4050	
E.V. (BOWP)				1532				2424				3184				3632	

% Complete	38%	61%	80%	91%
CPI (Efficiency)	.87 = 87%	.977 = 98%	.973 = 97%	.896 = 90%
SPI	.832	.891	.925	.908

▽ Report dates

Figure 50.13
Control curves

CHAPTER 51

Primavera P6

Chapter Outline

Evolution of Project Management Software

Early Project Management Software

Early project management software packages were mostly scheduling engines. Users of the packages were specialists, and the software did automate some low-level tasks, such as calculating early and late dates, but overall the users had to understand how these worked in order to manipulate the software and make sense of the results.

As more refinements were introduced, they were typically geared towards refining the detailed understanding of the work, resource, and cost plans. As the models were getting more precise, the user-base also had to become more skilled to use these extra functions.

Also, simple IT limitations, such as price and availability of computer memory, usually meant that each project had its own files. This made it difficult to spread best practices through the organisations, led to duplication of effort to redefine data structures for each project, and made it difficult and time consuming to aggregate reporting to all levels of the organisation.

Project Management, Planning, and Control. http://dx.doi.org/10.1016/B978-0-08-098324-0.00051-2

Although those tools achieved a good modelling of projects, in accordance with principles described in this book, they had limitations for a wider use, in particular:

The production of reports to all levels of the organisation was labour intensive and used different tools and data sets for different reporting levels, and the full process was time consuming, meaning that decisions frequently had to be based on information that was already out of date.

Different specialists worked with different data sets, leading to disconnects between various plans. For instance, many companies had planning engineers and cost engineers working in different packages. This sometimes led to having multiple versions of the truth.

The need of a specialised workforce to use them restricted them mainly to large projects. Organisations working on smaller projects frequently were scared by the perceived complexity of the packages and avoided them.

The Enterprise-Level Database

To address the first point above, software companies started to look at ways to get all levels of information into a central database. As networks developed, it became feasible to have people sitting in different locations either publish all their information to a central data repository, or even work straight into a central database.

This helped reporting at various levels of the organisation, as well as preventing wasting effort by recreating data structures for each project. Although this improves timeliness of high-level reports while reducing their labour intensiveness, the improvement is not significant for individual projects.

Systems Integration

Working with centralised systems allowed companies to link them with automated interfaces. This makes it possible for each participant in the management of the project to work in the tool that best matches their needs, while providing views and reports that combine information from those different systems. This helps with the second point mentioned above, by making sure information has a single point of entry and that all reports are using the same original data set.

Getting systems that were designed and configured separately to share information usually requires making some compromises. Each system will lose some flexibility, and decisions made when initially setting up the systems may have to be reversed to ensure data compatibility. Additionally, upgrading any of the systems involved can only be done after making sure the integration still works, or is upgraded. As separate systems have separate upgrade cycles, this maintenance can be costly.

The Scalable Integrated System

Some project management software vendors now provide systems that, although they are developed individually to provide good functionality in their respective domains, include in their design connexion points with the other packages contained in the solution. This makes integration between the systems much easier to build and maintain. Each package can be implemented and run individually, but when used with the other packages, the link is natural enough to feel like the solution was developed as a single system.

These solutions are scalable both in terms of size of the content – from a single project, to a department or even all projects in a company – and in terms of functions to be covered. This also makes their implementation more flexible as benefits come quickly from the first functions implemented, and other modules can be added later on.

Oracle Primavera P6

Oracle Primavera P6 (referred to as P6 for the rest of this chapter) is such an integrated system. As of version 8, Oracle has embedded in P6 a number of other systems. Each of those can be implemented separately and will work as a standalone solution, but when used together they will behave as a single integrated system.

Project Planning

The core of P6 is its scheduling engine. This is based on the planning tools developed by Primavera Systems since 1983. The scheduling engine uses the critical path method of scheduling to calculate dates and total float. While incorporating advanced scheduling functions, Primavera made sure each of these functions remained simple enough so that the main skills required to properly analyse project information relate to project management rather than the software.

Work Breakdown Structure and Other Analysis Views

The default reporting in P6 is based on the WBS. Summarisation at all levels of the WBS is automatic, and managers can easily access this information to the level that is relevant to them. However, cost, resource usage, and day-to-day organisation of the work on the project sometimes require to view the project from different angles. For this, P6 lets users define as many coding structures as they like, at project, resource, or activity level. These codes can be hierarchical or flat.

With simple ways to summarise, group, and filter activities, this allows building views of the project plan relevant to any actor of the project. This goes from a client view relating to the contract all the way down to a resource-specific view by system, location, or any other relevant way to organise the information.

These views can be saved to access them more easily in the future, and they can be shared with any other user of the system. The same principles apply at all levels of the system, so that a user fluent with detailed activity level views can also build portfolio or resource assignment level ones.

Resource Usage

P6 offers a resource pool shared between the projects to make it possible to analyse resource requirements at any level from a single WBS element to a complete portfolio of projects, or even for the company as a whole.

P6 allows to model resources as labour, equipment, facilities, and material. Once described in the dictionary, resources can be assigned to activities in the project. Once the activities have been scheduled, this provides profiles of resource requirements.

Resources can be assigned unit prices, should the company decide to model costs on the schedule based on usage of the resources modelled. Costs can also be modelled as expense items, which are direct cost assignments that do not require resources to be created in the dictionary. Expense items can relate to vendors, cost category, cost account, or any other analysis angle deemed desirable for analysis or reporting.

Should there be a need to identify resources more precisely than they are known at the time the baseline is taken (e.g., analysis of named individual time is required, but only the skills needed are known at the time the baseline is taken), P6 also allows resourcing of activities by role. This allows the user to book a budget on the schedule without knowing exactly which resource is going to do the job.

Baseline and Other Reference Plans

P6 can save any number of versions of a schedule to be used as references and compared to the current schedule when required. This can include the baseline, any re-baseline following a change order, any periodic copy of the schedule, or what-if analysis versions of the schedule that someone may want to compare with the current schedule.

Each reference plan contains a full copy of all the information contained in the schedule. It is therefore possible to compare schedule information – such as dates, duration, and float – resource information – such as requirements or actual usage – and cost information – such as budget, earned value, or actual cost.

Baselines can also be updated by copying a selected subset of the current project information into an existing baseline. This makes it much easier to adjust the content of the baseline to the current scope without affecting in the baseline the part of the project that was not affected by change management. As any number of reference plans can be maintained, this enables earned value analysis against the original baseline, or against a current baseline reflecting the current scope of the project.

Progress Tracking

P6 has very flexible rules for tracking progress, to allow each organisation to track only information deemed to be relevant. Progress-related quantities such as remaining duration, actual and remaining units, and cost can be linked or entered separately. This allows each company to decide on what information should be tracked, while making reporting available for all based on the level of detail that can be gathered.

Depending on the requirements, progress can be displayed based on the current schedule or mapped on the baseline in the form of a progress line.

As different organisations find different information to be relevant, it is possible to choose from many percentages to report progress, such as (but not limited to) physical percent complete, labour unit percent complete, material cost percent complete, or cost percent of budget.

P6 also calculates the variance between the baseline and the current schedule in terms of duration, dates, units of resources by type, or costs.

Earned Value Analysis

As P6 tracks dates, resource usage, and costs at detailed level, and as reference plans, including the baseline, are maintained with the same level of detail, earned value analysis can be performed at any level required. As for any information in the system, earned value information can be summarised according to any angle considered useful for analysis, be it through the use of the WBS or any coding.

P6 calculates and aggregates earned value information automatically. This can be displayed as easily as any current plan information. If the decision is made to track the relevant information, earned value fields available in P6 include, for both labour units and total cost: actual, planned value, earned value, estimate to complete, estimate at completion, cost performance index, schedule performance index, as well as cost, schedule, and at completion variances.

Based on the level of confidence in progress tracking, it is possible to define different rules for the calculation of earned value. It is also possible to consider the cost performance index and schedule performance index to date in the calculation of the estimate to complete.

Risk Management

As standard, P6 includes a risk register. As with the rest of P6, this risk register can be configured to contain the relevant information for a company. Probability, impact type, and impact ranges can be defined to reflect the important factors for a specific company. The system allows the definition of as many impacts categories as needed, to help reflect quantifiable impacts such as schedule and costs, as well as other impacts such as image, health and safety, or environment.

Once qualified with levels or probability and impacts, P6 will rate the risk based on a configurable risk rating matrix. The overall rating of a risk can combine the impact ratings in different ways, by selecting the highest one, the average of the impacts, or the average of the impact ratings.

Running a Monte Carlo analysis on a schedule requires Primavera Risk Analysis. It is possible to store three-point estimates in the P6 schedule though so the uncertainty can be maintained within the main schedule.

Multi-Project System

Even though schedules are split in projects in the database, P6 handles multiple projects as if they were just subsets of activities, part of the same total group. This means that reporting makes no difference between single and multi-project content, but also that scheduling can be done across all the projects.

Even when opening only one of a group of inter-dependent projects, the user has a choice between taking into account inter-dependencies or not during the scheduling calculations. Similarly, one can analyse resource utilisation based on only the schedule he is working with, or include requirements from other projects. If needed, projects can be prioritised so as to only consider projects of a high enough priority in the resource analysis.

Role-Based Access

There are two main ways to access P6: through the Web, or by using the optional Windows client. The later requires a high bandwidth between the client and the database server, or the use of virtualisation technology such as Citrix or Terminal Services. It is a powerful tool, but the high number of functions available makes it feel complex for users who do not have much time to learn it. This makes it a specialist tool, perfect for planners or Central Project Office people, but less appealing to people who only have limited interactions with planning.

The Web access of P6 lets users connect to the P6 database through a Web browser-based application. It is both simpler and more compete. It is simpler in that, when looking at a specific function, the interface isn't quite as busy as the Windows client. Yet it is more complete in that it provides views and functions that are geared toward the different roles participants of the project may play in the project.

From the resource assigned to a few tasks on a project to a company director interested only in traffic light type reporting on cost and schedule for each project, P6 Web can display relevant information to each person involved based on their involvement. Resources can see the tasks they are assigned to, including detailed information about those tasks, and documents that may be attached to the tasks to help complete them. Planners can see the schedule, with similar functionality to what the optional Windows client provides. Project managers can see high-level reporting and analysis on the schedule, costs, and resources, with drill-down capability to find where the problem is. Resource managers can view how busy their team is across all the projects in the company, as well as details of what each individual is working on. Executives and directors can have portfolio or programme level traffic light type reporting, with high-level schedule, cost, resource, or earned value figure summarised at any level that makes sense to them.

For each role identified, the administrator can provide a dashboard that will contain easy access to each report and function needed for that role. The content of those reports and function portlets will be based on the individual, the activities he is assigned to, the projects he is in charge of, or the portfolios that are relevant to that user. As users could have several roles, they can subscribe to several dashboards.

Reporting

Most of the reporting out of Oracle Primavera P6 is based on viewing or printing on-screen layouts. In agreement with the role-based access described above, P6 offers many ways to present and aggregate the project information based on the person accessing this information. These ways of viewing project information are usually enough for most people. However, should people prefer to get reports delivered in other formats, it is possible to use Oracle BI Publisher, which comes bundled with P6 licenses.

With BI Publisher, users can schedule reports to be run at specific times, and either made available on a website, or sent by e-mail. These reports can be created in a number of formats, including MS Office tools and Adobe PDF.

Using P6 through a Project Life Cycle

The following pages describe the planning, execution, and control of a project using P6.

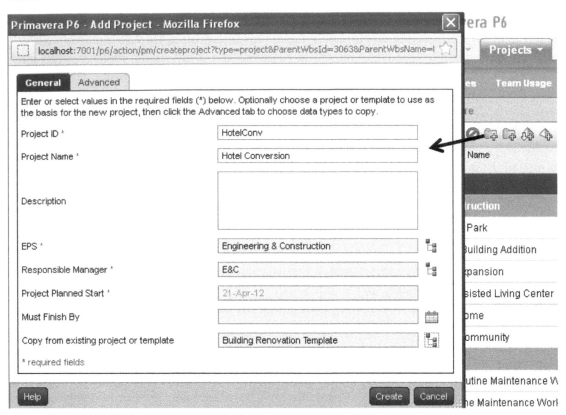

From the EPS tab of the Projects section, select Add a Project. If available, select a template, or copy an existing project to ensure high level preferences are common to all projects.

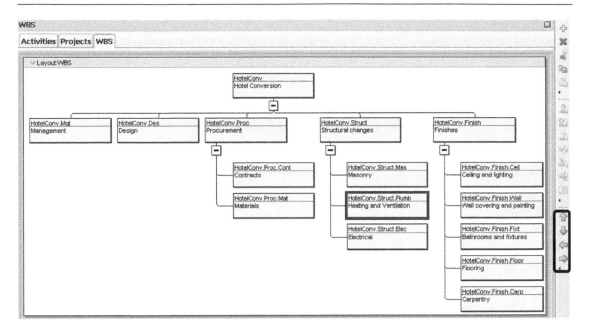

Build the Work Breakdown Structure in the WBS window. If necessary, adjust the structure by using the navigation arrows.

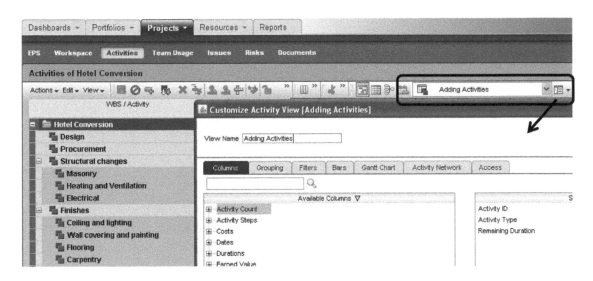

In the Activities tab of the Projects section, select a view or customise your own to simplify adding activities. Grouping by WBS and making sure the relevant columns are displayed help save time.

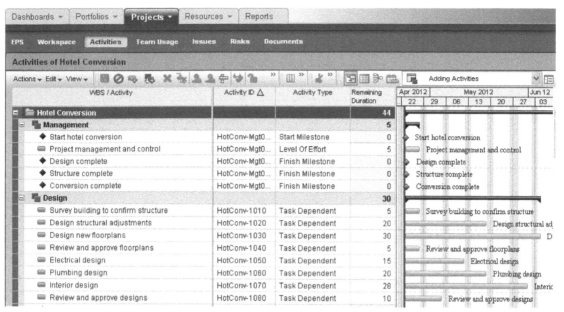

Add the activities necessary to complete each element of the WBS, as well as any milestone that can be useful for tracking the project progress. As the activities have not been scheduled yet, the bars only represent the duration of those activities.

Logic can be built by linking graphically the bars,

Adding successors in the Successors details tab at the bottom of the view,

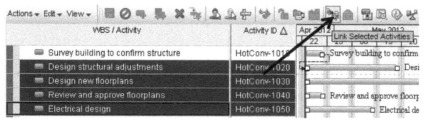

Selecting multiple activities, and then clicking the Link Selected Activities button at the top of the view,

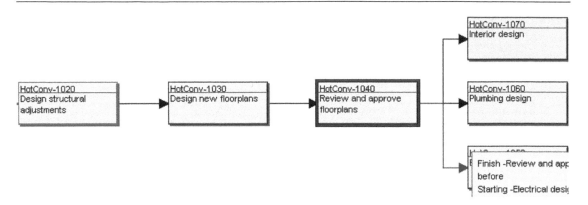

Or drawing the relationships in the activity Network view.

Double click the – sign at the top left corner of a group to collapse the content of that group.

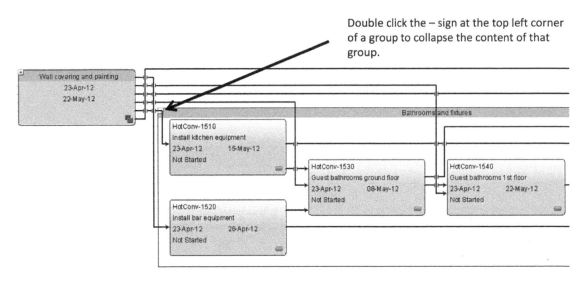

You can review the logic by displaying a PERT view of the project.

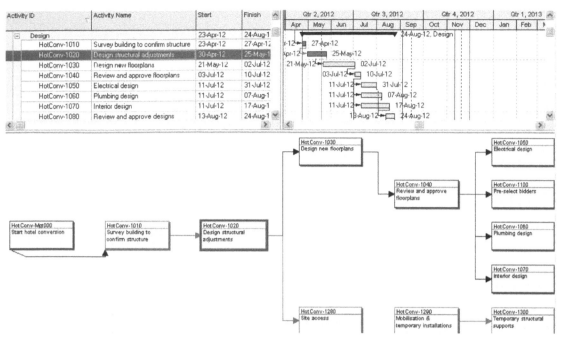

It is also possible to trace logic with a combined Gantt Chart and PERT view.

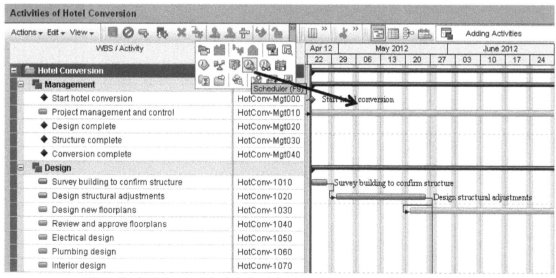

Once the logic is correct, click the Scheduler icon or press F9 to schedule the project.

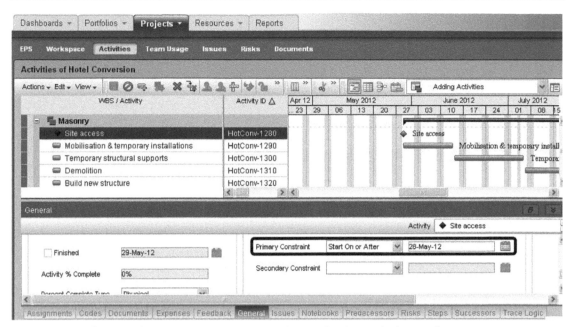

If required, assign constraints on activities in the General tab, to reflect external constraints on the project.

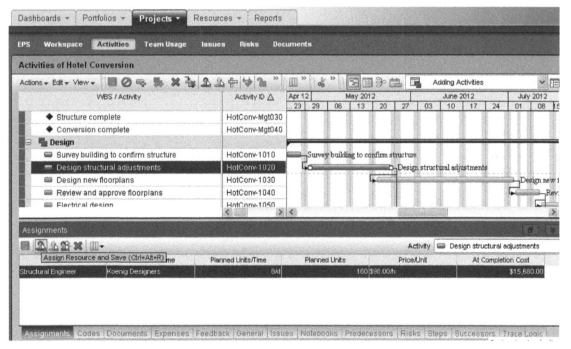

In the Assignments tab, assign the resources needed to complete each activity.

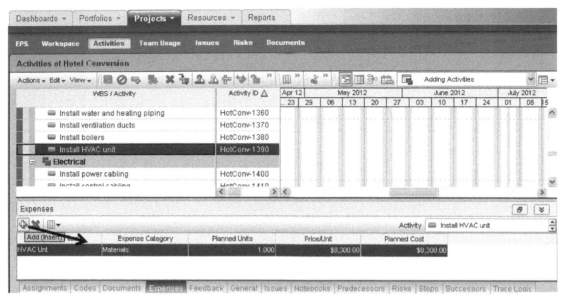

For resources that do not need to be track in detail or for subcontracts, it is possible to just assign a cost in the Expenses tab.

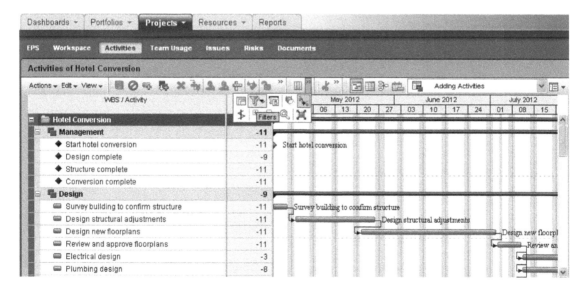

From the filters menu, run the critical activities filter and make required adjustments to the schedule for the project to finish on time.

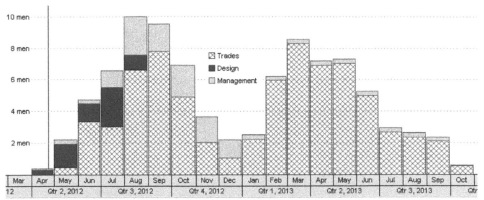

Analyse resource usage. If necessary, adjust the schedule.

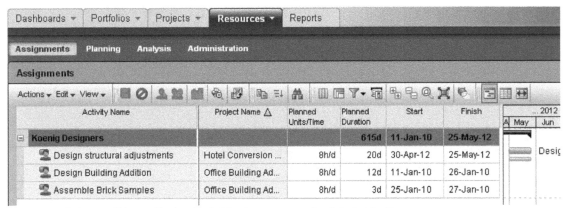

It is possible to display individual resource schedule in the Resources view. If resources are working on multiple projects, it is possible to show usage on any or all projects.

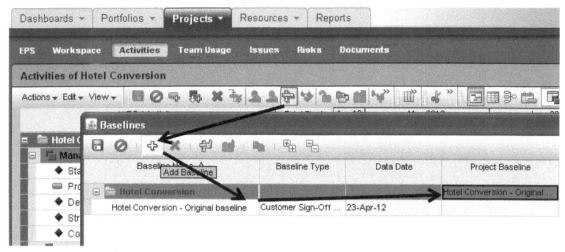

Once the project plan has been approved, save a copy as a baseline to keep as a reference and compare with the current schedule.

Track activity progress by entering actual dates for completed activities, actual start and percent complete for activities in progress,

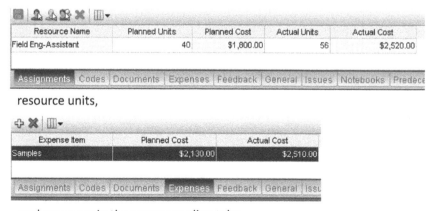

resource units,

Expense Item	Planned Cost	Actual Cost
Samples	$2,130.00	$2,510.00

and expenses in the corresponding tabs.

WBS		Qtr 2, 2012			Qtr 3, 2012			Qtr 4, 2012			Qtr 1, 2013			Qtr 2, 2013			Qtr 3, 2013			Q
		Apr	May	Jun	Jul	Aug	Sep	Oct	Nov	Dec	Jan	Feb	Mar	Apr	May	Jun	Jul	Aug	Sep	Oct
Hotel Conversion	Baseline	$3k	$31k	$67k	$126k	$208k	$285k	$359k	$413k	$426k	$453k	$527k	$613k	$683k	$754k	$803k	$832k	$855k	$866k	$866k
	Earned Value	$2k	$16k	$35k	$66k	$101k	$120k	$128k	$128k	$128k	$128k	$128k	$128k	$128k	$128k	$128k	$128k	$128k	$128k	$128k
	Actual Cost	$2k	$18k	$37k	$70k	$144k	$226k	$229k	$229k	$229k	$229k	$229k	$229k	$229k	$229k	$229k	$229k	$229k	$229k	$229k
	Estimate At Completion	$2k	$18k	$37k	$70k	$144k	$226k	$401k	$510k	$546k	$607k	$628k	$681k	$795k	$904k	$980k	$1,078k	$1,126k	$1,155k	$1,186k

Cost and earned value data can be analysed from many angles, including time-distributed tabular presentation,

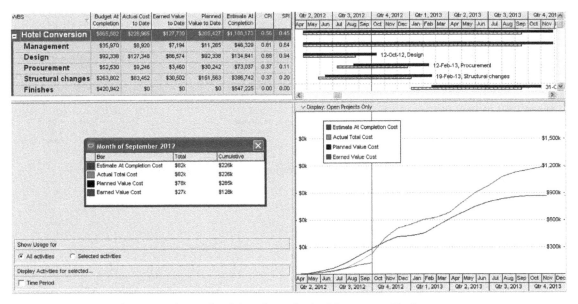

totals at any relevant level, bar chart of schedule progress Vs the baseline, and as S-curves.

WBS	Budget At Completion	Actual Cost to Date	Earned Value to Date	Planned Value to Date	Estimate At Completion	CPI	SPI		Apr-12	May-12	Jun-12	Jul-12	Aug-12	Sep-12	Oct-12
Hotel Conversion	$ 865,582	$ 228,965	$ 127,730	$ 285,427	$ 1,188,173	0.56	0.45	Baseline	$ 3,301	$ 31,466	$ 66,980	$ 126,019	$ 207,786	$ 285,427	$ 359,439
								Earned Value	$ 1,928	$ 16,150	$ 35,074	$ 66,397	$ 101,221	$ 127,730	
								Actual Cost	$ 2,021	$ 17,806	$ 36,586	$ 69,719	$ 144,062	$ 228,965	
								Estimate At Completion	$ 2,021	$ 17,806	$ 36,586	$ 69,719	$ 144,062	$ 226,455	$ 400,951
Management	$ 35,970	$ 8,920	$ 7,194	$ 11,285	$ 46,329	0.81	0.64	Baseline	$ 605	$ 2,821	$ 4,937	$ 7,053	$ 9,370	$ 11,285	$ 13,602
								Earned Value	$ 385	$ 1,799	$ 3,147	$ 4,496	$ 5,974	$ 7,194	
								Actual Cost	$ 478	$ 2,230	$ 3,902	$ 5,575	$ 7,407	$ 8,920	
								Estimate At Completion	$ 478	$ 2,230	$ 3,902	$ 5,575	$ 7,407	$ 8,920	$ 11,817
Design	$ 92,338	$ 127,348	$ 86,574	$ 92,338	$ 134,841	0.68	0.94	Baseline	$ 2,697	$ 25,831	$ 41,847	$ 76,012	$ 92,338	$ 92,338	$ 92,338
								Earned Value	$ 1,543	$ 14,351	$ 28,374	$ 39,051	$ 61,286	$ 86,574	
								Actual Cost	$ 1,543	$ 15,576	$ 30,804	$ 42,321	$ 82,555	$ 127,348	
								Estimate At Completion	$ 1,543	$ 15,576	$ 30,804	$ 42,321	$ 82,555	$ 127,348	$ 134,841
Procurement	$ 52,530	$ 9,246	$ 3,460	$ 30,242	$ 73,037	0.37	0.11	Baseline				$ 3,460	$ 17,688	$ 30,242	$ 40,651
								Earned Value					$ 3,460	$ 3,460	
								Actual Cost					$ 6,736	$ 9,246	
								Estimate At Completion					$ 6,736	$ 6,736	$ 38,506
Contracts	$ 28,034	$ 6,736	$ 3,460	$ 19,113	$ 38,682	0.51	0.18	Baseline				$ 3,460	$ 13,848	$ 19,113	$ 20,335
								Earned Value					$ 3,460	$ 3,460	
								Actual Cost					$ 6,736	$ 6,736	
								Estimate At Completion					$ 6,736	$ 6,736	$ 26,393

Reports can be send by e-mail using formats familiar to the recipients

BIM

Chapter Outline

Introduction

Technology to access and control construction information is constantly changing and the relevant information is required to be available on various devices and platforms. For example even preparing this chapter has involved storing information on the 'Cloud', interfaced from a smartphone, various laptops, and a PC, all on different platforms and this is just for controlling text. The construction industry is moving to digital integrated design team project delivery, with all of the necessary information being available at any stage of the contract and beyond. Each project design team can be formed from many different players depending on the actual work requirements and content. For example, a design team could include the client; architect; engineer; MEP (mechanical, electrical, and piping); contractors; etc. One of the processes that support these developments is Building Information Modelling (BIM). BIM is not a single 3D application; it is a process that streamlines the product model content and delivery.

Project Management, Planning, and Control. http://dx.doi.org/10.1016/B978-0-08-098324-0.00052-4

It is worth concentrating on the 'Information' part of BIM before addressing the BIM process. Information always needs to be collected and contained in a physical object or system. In history this could have been on a stone tablet, or nowadays a book or 'project library' would be made available to the reader or viewer. This really is the same with BIM, however, the whole project information is not just available as a great volume, but broken down to object-level nuggets of information for ease of reference.

A product model can be built or defined with a 3D or 4D application, where the latter is the 3D information which also encapsulates the time information. This can also include manufacturing information, start and completion dates, maintenance information, and even the required special demolition procedures, as the model can support the full life cycle of the building or project. However, the BIM model is not limited to just 4D information, as costing can be included, which is sometimes referred to as 5D information, or really anything that the user wants to track (nD information) and control.

Sometimes BIM is thought to be just another service that provides users with instant online access to an ever-increasing stream of constantly evolving, instantly updating digital data. However, if the processes are in place then real project control is possible. Many times it is said that: 'Nowadays to change a pump physically on a building site is relatively simple. However, changing all of the 3D models, drawings, sketches, specifications etc. is the hard and time-consuming part'. Adopting the BIM process will revolutionise this, as the information has only to be changed once, with the authoring application, after which the rest of the design team members can simply reinsert the new 'reference model', with all the latest information, ensuring all the models are up-to-date.

History of BIM

2D drawing systems have been used since the early 1990s and were really used only as 'electronic drawing boards', copying and pasting details or 'blocks' to reproduce drawings quicker than the older manual processes. However, as far as interoperability was concerned there were no advantages. Various applications allowed different drawings or blocks to be imported into the working drawing using some form of 'X-Ref' options, which really was the start of the reference model concept.

General Computer-Aided Design (CAD) tools were used for a while, leading to bespoke solutions which were developed to suit each industry or design team member's requirement. For example, tools used by architects would require different functionality from the toolkit required by engineers. Also a structural engineering application would have different drawing requirements from a Mechanical Electrical and Piping (MEP) solution. It is interesting to note that in many companies the technicians will produce the drawings, with engineers just resolving typical project sections or sketches.

In the building and construction sector, 'Information Modelling', is normally defined as the computer representation of a building or structure, including all the relevant information required for the manufacture and construction of the modelled elements. The elements or objects are required to be intelligent and should therefore know what they are; how they should behave in different circumstances; and their own properties and valid relationships. A simple example of this would be the reinforcement in a pad foundation. If the foundation size is modified then the embedded reinforcement should update and reposition itself accordingly. This is very different from a computer-generated model, which has been constructed for purely visualization purposes, where every objected is a non-related element.

The structural steelwork industry is always recognised as the lead sector in 3D modelling solutions and their developments can be traced back over the last 20 years with applications such as the forerunner of Tekla Structures. This type of application allowed the 3D steel frame to be modelled and the connection applied by the use of user-defined macros, which allowed the automatic production of general arrangement drawings, fabrication details, and then, after a few more years, the development of links to CNC (computer numerically controlled) machines, for the cutting and drilling of steel sections and fittings.

Once the 3D modelling technology was extended to include Parametric Modelling (true solid objects) then clash detections systems were able to be developed. Hard or soft clash detection allows applications to identify any possible material overlapping conditions or objects existing in the same space.

Nowadays Globally Unique IDs or Globally Unique Identifiers (GUIDs) are available in most applications, which are unique strings usually stored as a 128 bit integer, which is associated with the objects. The GUID is used to track the objects between applications for change management. Some systems allow the support for internal and external GUIDs. The term GUID usually, but not always, refers to Microsoft's Universally Unique Identifier (UUID) standard.

From the start of BIM, only drawings and reports were made available, as the actual BIM model was always confined to the office where the more powerful computers were situated. However, as computers have advanced, the models are now accessible on laptops and, over the last few years, even on tablets.

What Is BIM?

So what is BIM? It could simply be defined as rapidly evolving collaboration tools that facilitate integrated design and construction management. The importance of 'I' in BIM should never be underestimated, as this becomes a project or support for the company's enterprise framework and not just a means for 'building models'. This information means that more work is done earlier in the project to support green issue concepts, as less waste saves both materials and energy.

BIM enables multidimensional models including space constraints, time, costs, materials, design and manufacturing information, finishes, etc., to be created and even allows the support for information-based real-time collaboration. This information can be used to drive other recent technologies including city-sized models, augmented reality equipment used on site, radio-frequency identification (RFID) tags to track components from manufacture to site, and even the use of 3D printers.

It may be useful to consider the players who would want to have access to the BIM models. Not limiting the list they could include the clients, local authorities, architects, engineers (structural, civil, and MEP), main contractors, steelwork and concrete subcontractors, formwork contractors, and all site personnel. Until recent years BIM was only available as a solution for architects, engineers, and steelwork contractors, leaving everyone else just to work with 2D drawings that may be industry-specific but not totally readable without knowledge of that environment.

Various references have been made to the architects' BIM model or the structural BIM. However, they really are the same, as the boundaries between their models and their content are lessening all of the time. Architects' BIM models will include structural member sizes, but the models that they produce do not normally need to include the material grades, reactions, and finishes. Where the model is produced by the steelwork contractor, it will include at least the manufacturing details and all the information necessary to order, fabricate, deliver, and erect the members. The MEP contractor could also define the site fixings on his versions of the model, as the contractor will want to know when the member will be on site, where it will be fixed or poured, and how much the item costs. The client's view of the same member would be for control and for possible site maintenance. For this reason various models are created in the 'best-of-breed' authoring applications and shared with other design-team members as reference models, which are normally in the form of Industry Foundation Class (IFC) files for all structures except the plant and offshore markets, where CIMsteel Integration Standards (cis/2) and dgn format files are the dominant interoperability formats.

It is so much easier to work with a BIM model and to explore the building in 3D with rich information, than looking at hundreds of drawings and having to understand the industry drawing conversions. Now users can simply click on an object and obtain all the information that they require either through the native object, if in the authoring application, or through the reference model or even from a viewer or collaboration tool.

UK Government Recommendations

The UK and other governments have expressed support for BIM over recent years. However, in March 2011 the UK government published a BIM Strategy Paper. In fact the UK government has made it clear that BIM needs to be adopted on all of their projects within the lifetime of the present parliament to save the model objects from being rebuilt many times within the same project by different design team members.

The aims and objectives of the working group were:

- Identify how measured benefits could be brought to the construction industry through the increased use of BIM methodologies
- Identify what the UK government as a client needs to do to encourage the widespread adoption of BIM
- Review the international adoption of BIM including the U.S. federal government
- Look into government BIM policy to assist the UK consultant and contractor base to maintain and develop their standing in the international markets

The general recommendations were:

- Leave complexity and competition in the supply chain
- Be very specific with supply-chain partners
- Measure and make active use of outputs
- Provide appropriate support infrastructure
- Take progressive steps
- Have a clear target for the 'trailing edge' of the industry

The report also defines the project BIM maturity levels, from Level 0 to Level 3. As a quick summary Level 0 is where just CAD tools have been adopted; Level 1 is where 2D and 3D information is used to defined standards; Level 2 is where BIM applications are used with fully integrated model collaboration; and Level 3 is where BIM models are used for project/building lifecycle management. See Urn 11/948, which is a report commissioned by the Department of Business, Innovations and Skills, first published in July 2011, for further information.

What the UK government wants is:

- 20% reduction in costs
- Level 2 BIM by 2016
- COBie information available for decision making at critical points of the design and construction process

How BIM Is Applied in Practice

If a BIM model is being enhanced on the authoring application, it is normally referred to as a physical or native model, which can be enhanced using normal authoring tools. If the model is to be fixed by one design team member, then an IFC file or other reference models are normally adopted where objects can be commented upon but not changed by other members of the design team.

Tekla Structures

Tekla Structures is a multi-material BIM software tool that streamlines the construction design and delivery process from the planning stage through to design and manufacturing, providing a collaborative solution for the cast-in-place (in-situ)/precast concrete, steelwork, engineering, and construction segments.

The structural BIM is the part of the BIM process where the majority of multi-material structural information is created and refined. These are normally created by the structural engineer as architects work with space, mass, texture, and shapes and do not work with building objects in the same way as defined in the structural BIM. However, the connection between the architects' models and structural BIM is a very obvious way to help in the future development of intelligent integration, which should be always available in the form of reference models in the same way that the XREF function is used in a 2D drawing. These reference models could also be 2D information for collaboration with non-BIM applications.

The model starts to evolve during the engineering stage, where conceptual decisions of the structural forms are made. It is sometimes thought that the design portion of Analysis & Design (A&D) is just the pure physical sizing of the structural elements. It is in fact more than that, as it should also include the engineering and the 'value engineering' of the project, including all materials, their relationships, and their reference to the architectural and service objects, together with possible links to other design systems using .NET technology to form an Application Programming Interface, or API as it is more commonly known.

In its basic form, .NET is a flexible programming platform for connecting information, people, systems, and devices together using a modern programming environment and tools based upon the Microsoft® Visual Studio.NET developments. There are over 30 programming languages that are .NET enabled which allow true object information to be seamlessly transferred between systems. So for example element information and geometry can be passed from modelling applications to any other .NET-enabled system. These could be A&D systems, management information systems, cost control, or just systems used for internal bespoke company development.

One of the principal advantages of the structural BIM is that the project is built for the first time in the memory of the computer, before any physical materials are involved. This allows the scheme to be refined to a greater extent, allowing full clash checking facilities, automatic drawing, bar bending schedules, and report preparation. Modern applications also allow the drawings and reports to be automatically updated should the model be amended, thus allowing change management to be controlled and different design solution scenarios to be considered at any time together with having links back to the A&D systems if required.

For further information on the Tekla Corporation or Tekla Structures visit www.tekla.com. For Web tutorials on Tekla Structures or for general BIM lessons information see www.tekla.com/international/solutions/building-construction/videos/FirstSteps/index.html.

Linking Systems through Open .NET Interfaces

Sometimes using industry standard files is not appropriate when tight linking of applications is required. For example, a user may want to share design information between a modelling system and say A&D applications, management information systems (MIS), enterprise solutions, connection and fire design systems, project management systems, or planning systems.

In such a case the objects, a simplified form of the object, or just the object attributes, can be passed between systems using the Open .Net interface as defined above. This same interface could also be used for model transfer or for repetitive processes such as drawing and report creation.

Tekla BIMsight

Tekla BIMsight has been written by the Tekla Corporation, which is a Trimble company, and is a free BIM collaboration tool which is available for anyone to download and install. It is also an easy-to-access application that presents the complete project including all necessary building information from different construction disciplines. It is also much more than a viewer as one can communicate using the model with anyone, not just Tekla Structures or other BIM software users.

With Tekla BIMsight one can:

- Combine multiple models and file formats from a variety of BIM applications into one project
- Share building information for coordination between different trades and deliverables
- Identify and communicate problem areas, check clashes, manage changes, approve comments, and assign work in 3D by storing a history of different view locations and descriptions in the model
- Measure distances directly in the model to verify design requirements and construction tolerances
- Control the visualization and transparency of different types of parts in the model to make it easier to understand complex and congested areas of the project
- Query properties such as profile, material grade, length, and weight from parts
- Tekla BIMsight can be downloaded from www.TeklaBIMsight.com, which also includes video tutorials and a user forum

For interoperability the file formats supported by Tekla BIMsight currently are IFC (IFC, IFC XML, and IFC ZIP), dwg, dgn, and xml.

Savings with BIM

It is always hard to establish Return on Investment (ROI) and project savings with regards to any software systems. However, as the BIM project is created within the memory of the computer before any materials and site personnel's time are involved change is cheap. Various report and white papers regarding the cost of change and constant remodelling was one of the reasons for the UK government's BIM initiative.

Sample BIM Project – Alta Bates Summit Medical Centre - by DPR Construction Inc.

The addition to the Alta Bates Summit Medical Campus (ABSMC) is a $289 million, 13-storey patient care pavilion and future home of over 200 licensed beds and the facility is due to open in January 2014. DPR's highly collaborative Integrated Project Delivery (IPD) team was faced with many challenges. The use of Tekla Structures and Tekla BIMsight has grown significantly since the start of construction in 2010. Modelling scopes within Tekla Structures includes Structural Steel, Cast-in-Place (CIP) Concrete, Reinforcing Bar, Miscellaneous Steel and Light Gauge Drywall Framing. See Figures 52.1 to 52.7.

Seismic design requirements set forth by the Office of Statewide Healthcare Planning and Development (OSHPD) in addition to a hybrid steel and concrete shear wall structure required in-depth coordination between rebar and steel fabrication models. This coordination effort between multiple trade partners would not have been achieved without the use of this highly collaborative and detailed BIM platform. Since subcontractors Herrick Steel and

Figure 52.1
Actual drywall framing around the MEP equipment

Harris Salinas Rebar were both detailing with Tekla, it was very useful to provide the rebar detailer with the steel model during detailing to identify constructability issues with anchor bolts, stiffener plates, and other connection details that were not shown in the engineering design model.

More recently, the ABSMC Project has implemented the concept of Dynamic Detailing within Tekla Structures on both the reinforcing steel and drywall framing. By referencing in the structural steel Tekla model along with IFC models of the MEP&FP systems, Harris Salinas rebar and DPR Self-Perform Drywall were able to identify conflicts during the

Figure 52.2
Model of drywall framing around the MEP equipment

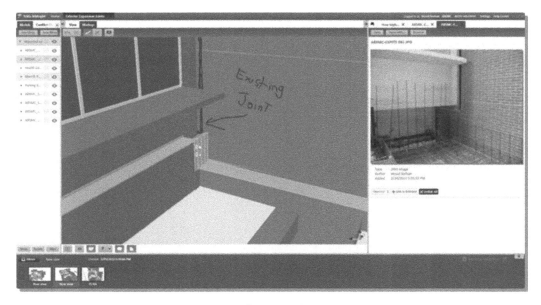

Figure 52.3
Tekla BIMsight showing model to site comparisons

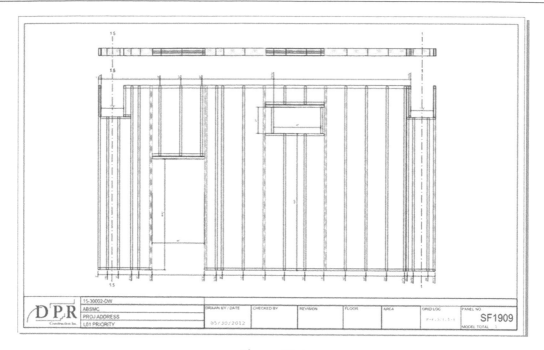

Figure 52.4
Drywall assembly 'spool sheet' from Tekla Structures

Figure 52.5
PCP tower adjacent to the existing hospital

Figure 52.6
Architect's rendering of the Alta Bates Summit Medical Centre

Figure 52.7
Rendering of the system through the tower

process of modelling, as opposed to the traditional method of modelling in a silo and having to rework the modelling after clash detection. This workflow results in a more efficient, streamlined workflow with fewer chances for modelling errors (see Figures 52.1 and 52.2).

The DPR Self-Perform Drywall detailer, Robert Cook, has developed an efficient workflow to detail the rebar using multiple custom components. Robert Cook is also creating drywall framing assembly spool sheets that can be printed to 11" × 17" or viewed on an iPad to help the site team to increase efficiency and quality while reducing rework. As all the ductwork is fabricated and assembled directly from the model, DPR drywall is now able to install framing around the duct and pipe openings prior to MEP installation and be confident that the framing is in the right location to align with the prefabricated ductwork and piping.

In addition to Tekla Structures, Tekla BIMsight has been used to convey differences between site conditions (see Figure 52.3), and the design model related to the exterior skin and expansion joint design. The Alta Bates project requires seismic expansion joints where surrounded on three sides by the existing hospital campus and around a five-storey pedestrian bridge. The comment tool, dimensioning tool, and photo attachments were used to convey coordination and constructability issues to the designer, fabricator, and installer.

Sample BIM Project – The National Museum of Qatar (Arup)

The National Museum of Qatar in Doha (see Figure 52.8) is the flagship project for an important series of cultural and educational projects which have being commissioned by the Qatari government. The project, which drew inspiration from the desert rose, has now started on site and has been in the planning phase since 2008. The desert rose is a crystalline formation found below-ground in saline regions of the desert. When imagined as a building, the result is a four-storey, 300 metre by 200 metre sculpture of intersecting discs that are up to 80 metres in diameter.

The evolved structural solution consists of radially and orthogonally framed steel trusses, supporting fibre-reinforced concrete cladding panels to create the required aesthetic and performance characteristics of the building envelope.

Figure 52.8
National Museum of Qatar

Figure 52.9
Showing the highly complex geometry

Challenges

The key challenge for the design was the highly complex geometry of the disc interactions. No two discs are the same and no two discs intersect each other in precisely the same way (see Figure 52.9). The galleries and other key spaces in the building are created by the interstices between the discs; any alteration to the architecture involves moving discs and thereby moving the structure within the discs. This has led to an evolution of systems and processes which were required to handle, manipulate, and develop geometric ideas from the architects, so that engineering solutions could be established before communicating these in their most useful form to the wider project community.

For this reason, the structural modelling (analysis, design, and manufacturing and construction) needed to address the following requirements:

- Position elements in the correct place in the 3D space within the cladding envelope
- Generate and model elements as efficiently and automatically as possible in order to keep up with iterations of the architectural arrangement
- Facilitate cross-discipline coordination, both with the Arup MEP design and 3D modelling teams in London, plus the architectural team based in Geneva and Paris together with the client in Qatar

Leeds Arena, UK. Fisher Engineering Limited

Leeds Arena is the United Kingdom's first purpose-built fan-shaped arena (see Figures 52.10 to 52.15). Using this form of geometry allows every spectator to have a perfect view of the centre stage. The main facades are rounded and have a domed effect, which terminates with a flat roof. Formed with two columns, one sloping outwards and the second spliced to the top

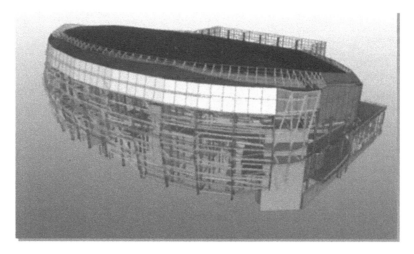

Figure 52.10
Leeds Arena project

Figure 52.11
Section through the Leeds Arena project

and cranked inwards, these curving elevations are clad with a honeycomb design of glazed panels that contain lights of changing colours.

The steel-framed structure of the roof is supported by a series of 13 trusses spanning up to 70 meters across the auditorium with the five central trusses being supported over the stage area by a 170-tonne trussed girder and plated columns, which form the 54m long × 10.5m deep proscenium arch. The proscenium arch truss was delivered to site in 32 separate sections; a total of nine trailer loads. Assembling the truss took three weeks, using two large mobile cranes.

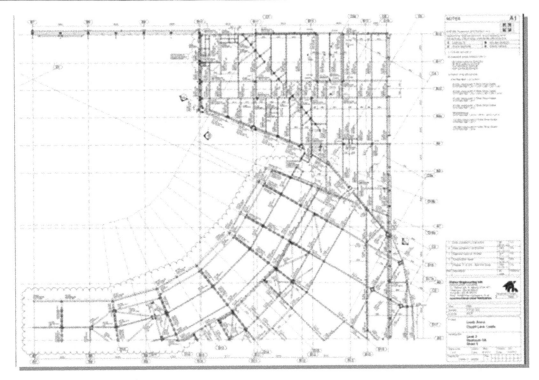

Figure 52.12
Typical general arrangement drawing

The bowl terraced seating is formed from precast concrete units supported on a radial steelwork structure, braced and tied into two main concrete stair cores that provide the required stability (see Figure 52.14). Acoustic resistance is a major design factor for a venue of this size, which is situated within a city centre, and required the structure to be shrouded in a skin of precast concrete wall panels and topped with a concrete roof topping on a metal deck.

A BIM strategy was essential for efficiency

With so many subcontractors providing major structural elements, all of which required prefabricated connections to interface with the complex geometry, it was clear from the start of the project that a BIM strategy was essential to obtain the necessary project efficiencies through the evolving design and collaboration process. The BIM model proved invaluable to all parties involved as it was passed between the design team and contractors for clash detection and for the resolution of incomplete design issues. This greatly assisted the project programming, sequencing, and general constructability.

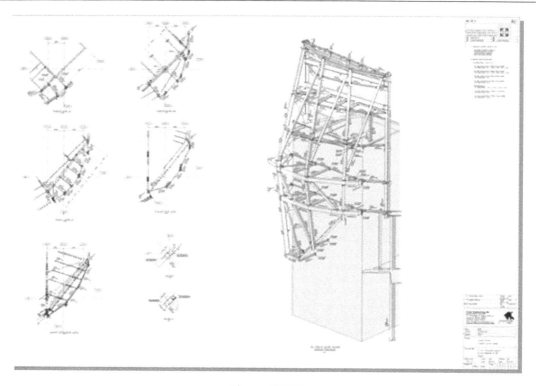

Figure 52.13
Steelwork arrangement detail

Figure 52.14
Model of the bowl terraced seating

Technical Detail

For readers who are interested in the more technical details, the following sections may be of interest and the various abbreviations and acronyms are explained below.

Figure 52.15
Arena during construction

Interoperability and Principle Industry Transfer Standards

3D interoperability between various building and construction applications are generally achieved through industry standard formats such as dwg, DXF, SDNF, cis/2, and IFC with the older systems being listed first. Other bespoke links have been adopted in the past based upon XML (extended mark-up language), which is basically is an extension of HTML and which is used for creating websites) or special file formats. Excel sheets have also been used in the form of reports or to enhance the various applications. It is generally accepted to adopt the full BIM process and then only IFC files are advanced enough to support all of the objects that are in building models.

DXF, DWG, DWF, and DGN Formats

DXF (Drawing eXchange Format) was developed by Autodesk® for enabling data interoperability between AutoCAD® and other programs. As the file format does not contain any form of part ID, it is not possible to track changes between different physical objects contained within different versions of a file.

DWG is used for 2D and 3D CAD data and is the standard file format for Autodesk® products.
DWF (Design Web Format) is a secure file format developed by Autodesk® for efficient distribution and communication of rich design data, normally created with DWG drawings. However, it is rarely seen within the BIM environment.

DGN has been the standard reference file transfer between plant design programs. Originally developed by Microstation, which is now part of Bentley Systems Inc., it is similar to DWG in that it is only a graphical data format, but does contain part IDs unique for that given model.

IGES and STEP

The Initial Graphics Exchange Specification (IGES) defines a neutral data format that allows the digital exchange of information among CAD systems. It was defined by the U.S. National Bureau of Standards and has largely been replaced by the Standard for the Exchange of Product Model Data (STEP) over recent years.

The International Standardisation Organisation (ISO) is concerned with creating standards for the computer interpretable representation and exchange of product manufacturing information, so STEP files are available across many manufacturing industries. In the construction market it is normal to only see files relating to ISO 103003 AP230, and these are generally treated as reference files.

SDNF Format

The Steel Detailing Neutral File (SDNF) format was originally defined for electronic data exchange between structural engineers, A&D, and design systems to steelwork modelling systems. Version 3.0 is the latest format supported by the software industry, and this format has been used for many years for transferring even complex plant structures between system like Tekla Structures and plant design systems such as Intergraph's PDS or Aveva's PDMS applications.

As a quick overview the SDNF files are split into packets and records. The main packets are defined as follows and generally not all items are supported by all applications:

- Packet 00 - Title Packet
- Packet 10 - Linear members
- Packet 20 - Plate elements
- Packet 22 - Hole elements
- Packet 30 - Member loads
- Packet 40 - Connection details
- Packet 50 - Grids
- Packet 60 - Curved members

CIS/2 Format

The CIS (CIMsteel Integration Standards) is one of the results of the Eureka CIMsteel project, which dates back to the start of this millennium. The current version 'cis/2' is an extended and enhanced second-generation release of the format, which was developed to facilitate a more integrated method of working through the sharing and management of information within, and between, companies involved in the planning, design, analysis, and construction of steel-framed buildings and structures. There are a number of different format versions: analysis, physical, and manufacturing formats for steel structures. The physical format has been widely used in the steel sector in the past.

The only downside of this format is that multi-material objects can't be defined as the standard just really concentrates on steel objects. For more information regarding this standard see www.cis2.org.

IFC Format

The latest and most complete transfer standard used within the BIM environment is the Industry Foundation Class (IFC) as defined by the buildingSMART organisation (www.buildingsmart.com), which was formally called the International Alliance of Interoperability. The buildingSMART organisation defines itself as 'buildingSMART is all about the sharing of information between project team members and across the software applications that they commonly use for design, construction, procurement, maintenance and operations. Data interoperability is a key enabler to achieving the goal of a buildingSMART process. buildingSMART has developed a common data schema that makes it possible to hold and exchange relevant data between different software applications. The data schema comprises interdisciplinary building information as used throughout its lifecycle. The name of this format is IFC, it is registered by ISO as ISO/PAS 16739 and is currently in the process of becoming an official International Standard ISO/IS 16739.' (http://www.buildingsmart.com)

The current version of the standard is 2x3, whilst the next version, IFC 4.0, is currently being defined.

With IFC true building objects as defined by architectural, engineering, MEP, and other systems can be shared. This allows the users to use the systems that they know, or that are best for creating the objects normally referred to as 'best-of-breed' systems. Adopting the IFC standard allows true object information to be shared between the major modelling applications. Different IFC formats and flavours are available including the ifcXML, ifcZIP, and standard IFC formats.

Agile Project Management

Graham Collins
Faculty of Engineering Sciences, University College London

Chapter Outline

Having an adaptable process to development that can respond to rapidly changing economic conditions is one way organisations can compete effectively. Agile project management enhances the ability of teams and organisations to react to these changes. Traditional approaches to project management often entail following a set plan and if there is a divergence from this plan it is often considered the role of the project manager to correct this and ensure no deviations. Agile approaches, however, recognise that goals will inevitably change and that achieving value for the client should be the most important consideration.

It is often the case that during a development process the requirements will have substantially changed. The longer the time interval from requirements gathering to delivery the more likely the client will indicate that what has been developed doesn't meet their needs. Also it is unlikely that many of the requirements will still be considered important after a few months or even at the start of development. During workshops with clients it often occurs that every participant volunteers at least one requirement, but if after outlining this list there is a voting opportunity for the priority of these requirements it is likely many are not voted for at all. It is not only that the client team needs to see a prototype, often to understand what they truly want, it is also likely that any software developed may change the business processes to such an extent as to make the original requirements even more obsolete.

One way to reduce the risk of development being detached from what is actually required by the client is to provide a more frequent feedback to the client including what is being developed. Agile project management accelerates the feedback cycle and actively involves the customer in the prioritisation of the requirements and design of the product. Delivering a product after 12 or 18 months runs the risk that the business needs have changed or that the client team realises on viewing the software it is not what they actually want. It is better to have a set regular feedback cycle continually prioritising the most valuable functionality and delivering some thread of working software for the client to comment on. Producing a tangible software or product at regular intervals and having a continual communication cycle, involving the client, is at the heart of agile methods.

Traditional planning is typically based on the concept of delivering a project within a set budget within a set schedule. The agile philosophy encompasses delivering as rapidly as possible high value products or software. Ensuring that this delivery benefits from regular feedback enhances this value. It also ensures that within a fixed time the greatest value in terms of the functionality as prioritised by the client is delivered. The shift for both the organisation and the project manager is one from delivery of a project to schedule and budget to one of delivery of the highest value within the time and other constraints set.

It is through a cycle of iterations and release, with continual working software of product developments, that trust is built and the client can see that every release is providing increased functionality and business value.

The Paradox

Managers typically wish to know how much and how long a project will take and yet they still want to have the flexibility to respond to the business environment and embrace innovation. How can we achieve flexibility and respond to change and at the same time follow a plan? The answer is partly in granularity, considering the capabilities at a high level of abstraction for planning and allowing development teams to define specific tasks according to the needs of the project. Agile is inherently measurable because of the clear regular cycles and internal and external measures. At a high level of abstraction these are the regular releases defined by the needs of the project, domain, and agile method, often every 90 to 120 days. Within these there are the iterations of one month or shorter cycles. And furthermore within a day there is the daily cycle identifying what has been delivered, what are the problems, and what will be done next. Within this especially in software development there are cycles that are achieved within even shorter periods by development teams using test-driven development and automated testing techniques.

Definitions of Success Are Part of the Problem

If we measure success in terms of achieving the original specifications, then measuring agile projects that are designed to incorporate changes in goals to achieve maximum business value to the client will bound to be problematic. What is needed is to base measures on what the client considers of value and for this to be updated continually. In agile methods and particularly with the Scrum method this is achieved via the Product Owner who is the client representative on the team. Typically the Product Owner is part of the development team and represents and works closely with the client or client team to determine the most valuable capabilities and constituent features. They will at the start of each iteration be involved in prioritising features from capabilities they have identified into a finer detail of functionality expected from the system. Dependent on the method, the requirements may be in the form of user stories, to gain some idea of how users will use the system. It is through this process that the Product Owner with the development team prioritises the most valuable stories for the coming iteration. This cycle continues so that at any given point the project has delivered the highest value functionality as defined by the client.

One advantage of reducing cycle time is that the team soon learns what is working and what is not and can correct the development as necessary. Another advantage is that valuable working software or products are brought into use earlier, starting to contribute economic benefits, so that the project reaches the breakeven point and provides a return on investment (ROI) sooner.

What Is Agile?

Jim Highsmith (2010) outlines that being agile is not a silver bullet to solving your development or project management problems. He characterizes agile in two statements: 'Agility is the ability to both create and respond to change in order to profit in a turbulent business environment. Agility is the ability to balance flexibility and stability' (Highsmith, 2002).

The concept of iterative and incremental development is not new, and developers in the 80s and 90s were designing light methods aimed at involving the development teams and customers in closer collaboration. An alliance of these developers met in Snowbird, Utah, in February 2001 to see if there was anything in common between the various methodologies being used at the time. They agreed on an agile manifesto and values (below) supporting the manifesto. The latest updates to this can be found at www.AgileAlliance.org:

- Our highest priority is to satisfy the customer through early and continuous delivery of valuable software
- Welcome changing requirements, even late in development. Agile processes harness change for the customer competitive advantage
- Deliver working software frequently, from a couple of weeks to a couple of months, with a preference to the shorter time scale

- Business people and developers must work together daily throughout the project
- Build projects around motivated individuals. Give them the environment and support they need, and trust them to get the job done
- The most efficient and effective method of conveying information to and within a development team is face-to-face conversation
- Working software is the primary measure of progress
- Agile processes promote sustainable development
- The sponsors, developer, and users should be able to maintain a constant pace indefinitely
- Continuous attention to technical excellence and good design enhances agility
- Simplicity – the art of maximizing the amount of work done – is essential
- The best architectures, requirements, and designs emerge from self-organizing teams
- At regular intervals the team reflects on how to become more effective, then tunes and adjusts its behaviour accordingly

One of the most important aspects of agile processes is the continual reflection and adaption. Many of these values have also been adopted by agile project management approaches, especially the concepts embedded within one of the methods, Scrum, including set iteration lengths termed sprints, daily stand-up meetings, and reviews, as part of the process. At the end of each sprint, a tangible product is delivered and at the end of a series of sprints, a release, the current working version of the product is released to the customer for detailed review.

Lean

Lean processes are typified by reduced inventories and cycle times. There are many concepts that agile and lean have in common particularly in processes to remove waste and rework. The background to the lean movement can be seen in the Japanese manufacturing sector and also in the Six Sigma quality improvement initiatives. As with many agile methods and approaches some of the ideas have their roots in previous practices.

> *Grant Rule (2011), during his guest lecture at UCL, 2010, outlined the similar concepts used in the production line techniques of building warships at the height of the Venetian empire, several hundred years prior to the concept of automotive production lines. During this period of expansion, the wooden warships were to protect the trading interest of the Venetians and due to the limited space in the Arsenale the inventory and waste was kept to a minimum and the pressure of conflicts meant that they had to reduce cycle times and release ships to protect their routes and territories as frequently as possible. This would be the equivalent of producing working software frequently to customers. There is evidence of continual improvement in the process although whether this allowed teams time for reflection and self-organisation is highly unlikely.*

The importance of frequent releases of software is central in agile project management. This enables feedback from the client to build trust and an effective product that provides the highest value within the given time. As requirements often fit into a profile familiar in Pareto analysis with only 20% of the functions used 80% of the time, then delivering the highest value and functionality first will often be sufficient as many of the additional requirements may be obsolete, seldom used, or need to be changed.

To take full advantage of agile methods lean practices need to be adopted across the organisation (Salloway, Beaver, and Trott, 2010). It is certainly the case that the major challenges in an organisation are often the cultural changes necessary, which include the empowerment of teams and creating an environment of trust to allow the teams to determine their own approach to development.

Agile is not a panacea. There is no set recipe to follow, but there are some patterns that perhaps all projects should follow. One pattern is the idea of time-boxing every aspect of the project cycle. This is not just a defined time but as short as possible time to speed up the feedback and associated learning cycle. This includes the discussions with clients as well as the review meetings.

Two Levels of Planning

Agile techniques encourage planning at two levels of abstraction. The customer or client, represented by the Product Owner, usually has an initial idea for the capabilities of the product being developed. This allows for a high-level capability plan to be developed in which the capabilities can be valued. Dean Leffingwell (2007) suggests a value feature approach in which the features are assigned priority and value, giving immediately some indication of value of the project as well as an approach to track progress in terms of value at a high level.

At an iteration level the capabilities need to be decomposed further into features and stories and prioritised. It is this that gives the second level of tracking and can be achieved in terms of stories and the size estimated in terms of story points. The team then decides how these tasks are broken down and commit at a daily level to delivery of this work.

Terminology

As with any professional practice there is often terminology that may be difficult for the outsider to interpret. In law, Latin phases are often used, although increasingly in other areas such as medicine it appears that the language is converging to more readily understandable terms. A glossary of agile terms can be found in the Guide to Agile Practices http://guide. agilealliance.org/. Perhaps it is that sometimes those not involved in agile project management and development are somewhat perplexed by terms such as Scrum and Planning Poker. Is this some kind of game we are playing? Well, part of the answer is yes, as the highly interactive and collaborative planning process has been termed a 'game' by some leading agile proponents (Cockburn, 2002). A game involving the customer at its core and delivering the highest business value within a given period.

How Does a Generic Agile Development and Project Run?

A workshop may be needed to determine the overall process and project management approach. Different agile techniques favour different approaches and use different terms. All outline the need for prioritisation of requirements. In some methods such as Feature Driven Development (FDD) these are known as features and are similar to other approaches that incorporate an understanding of how the user makes use of the system such as by user stories.

User stories tend to have more clarity if developed in conjunction with the client and time is spent on considering how they will be tested. The user stories are stored in a product backlog and the Product Owner or representative from the client organisation determines the value of each of these and prioritises them. The development team then determines which are done within the next iteration. This is achieved by an initial planning meeting at the start of the iteration often lasting only 2 hours in which the stories are estimated by the development team as to how long these will take. The team also takes into account the priority within the product backlog as to which stories are to be developed for the next iteration, and reviews this when the iteration is completed.

Stand-up Meeting

At the start of each day there is a stand-up meeting. The idea is to keep the meeting short to a maximum of 15 minutes and to encourage comments to be kept succinct. Here team members outline briefly what they did yesterday, what were the impediments to getting certain tasks done, and what they are going to achieve today. Tasks and who is doing them are summarised on the whiteboard and because these commitments are made to the group there tends to be motivation to achieve what individuals have outlined as their tasks. After the short meeting a technical problem discussed will often immediately be resolved by others in the group dependent on whether it is an architecture, programming, or resource issue. There is regular feedback. Typically there is a retrospective or review at the end of each iteration to outline what went well and what didn't go as well, but there are also daily reviews and the stand-up meeting helps highlight problems early.

The key is a built-in system of reviews at every level.

> *With my research groups after their projects, I asked them, if they were allowed to repeat their projects how long would they take to achieve the same results? Most had been working nearly 6 months and agreed that to repeat their work and achieve the same results would take less than half the time. Much of this is due to the learning curve, working out how to solve a problem, deciding the valid metrics, developing effective testing procedures, designing the tests and introducing automated testing procedures wherever possible. This is the importance of speeding up the cycle time to get results quickly and learn from them. To develop code and deliver this to the client after six months without a review with the client is a recipe for disaster. They will invariably say that this is not quite what they had envisaged. The feedbacks are necessary both for team learning and also to ensure that the teams are delivering the right product for the customer and ensuring that what they deliver is of the highest value and quality Collins (2013).*

It is important for research groups to consider the value of the project management approaches and consider these as assisting rather than being an overhead. Techniques in testing clarify the goals and if testing is automated should accelerate the process. Both in research and development a more facilitative approach to project management is required. Within the Scrum method this is supported by role of the ScrumMaster whose function is to ensure that the processes determined by the group are followed.

Estimation

Within traditional approaches managers estimate and try to establish predictable plans and any deviation from these is seen as a problem with the development team. This approach often combined with a waterfall sequence of requirements for gathering, design, development, and testing is often a failure. The long list of requirements that may have taken months to gather is often out of date before the design is started. This is why it is so important to build something for the client to see at regular intervals to check that what is being built meets requirements.

Mike Cohn (2004) humorously points out in his book on user stories, using an analogy based on shopping, that seemingly trivial tasks often take several hours to complete. This is a way of introducing several important points about overoptimism in estimation, and not allowing enough time for technical development or thorough testing.

Of course, there are times when project managers may deliberately tell senior management what they want to hear, and agreeing to an unrealistically short schedule that the development team has no hope of achieving. With agile project management these aspects are substantially reduced. Firstly, the 'death march' scenario of no time and unrealistic planning from the project manager is avoided because the teams estimate their own work. The priorities determined by the client in effect decide what should be drawn from the product backlog and incorporated into the next iteration. This backlog can then be used as a clear measure to protect the team from unreasonable additional demands. If further tasks are added to the product backlog it is the priority of each that determines the urgency. As only a certain number of stories in terms of relative difficulty and duration in story points can be achieved in any iteration, there is an immediate indication of whether it may be feasible or even if it is a priority to add the new stories into the current iteration. The work is then delivered at a sustainable pace in a predictable way and the project schedule is fine-tuned as more process and project metrics become available.

One aspect of agile is using the 'wisdom of crowds', tapping into the fact that groups can estimate with better predicted outcomes and faster than individual team members. Earlier approaches have used Delphi techniques based on averages and giving higher ranking to central values. In recent years a popular technique that has been adopted by the agile community is 'Planning Poker'. This allows the rapid estimation of user stories and will allow, based on these estimates, the team to plan a realistic set of user stories and constituent tasks for the current iteration. It is based on the concept that those doing the work are best placed to do the estimation. Also, as many of the benefits of estimation are quite quickly achieved, putting in more effort has decreasingly diminishing returns on accuracy.

Each of the development teams has a set of cards marked with a series of Fibonacci numbers. These are made up of the previous two numbers added, 1, 2, 3, 5, 8, 13, with the larger card not necessarily fitting in the sequence, say 100. There are other variants of this series that sum with 0 and a half, others based on multiples of 10, and others even with a question mark for the decisions that are difficult to initially make. The idea is that it is relatively easy to determine how long a task will take relative to another task, but this becomes more difficult as the size increases. As it becomes more difficult to differentiate between consecutive numbers as the task becomes bigger, so the widening gap makes it easier to estimate (e.g., it is easier to determine whether a task will take closer to 8 or 13 units than say 11 and 13 time units).

For each user story each member of the team estimates in story points the relative effort it is going take to develop and places their selected card with the value face down. As soon as everyone has done this they turn their cards over, to reveal their estimate at the same time. There may be consensus on the numbers say 3, 3, 3, and 5 in which case the card representing the generally agreed value is taken. However, if there is variation, a low and a high number say, those who selected these values outline their perspective and it may be that in light of this there are assumptions and technical issues that the others haven't considered, and there is another round of voting. If there is consensus then the team can move on to the next story. Rounds without discussion as outlined by Amr Elssamadisy (2009) allow reflection and consensus, but may have the consequence of individuals being too influenced by the perceived leaders in the group.

If the team is working with user stories, then they need to estimate the relative size of these stories at the beginning of the iteration as a team. Mike Cohn (2004) points out this can be done in story points, which can be relative to the team's own working practices and experience. However, the true measure of working becomes apparent from the team's own measures of velocity, how many of their story points they are completing within a week or iteration. This measure can then be used as yesterday's weather, to provide improved estimates of the rate of work, and better baseline the measures for completion of user stories and their constituent tasks to accomplish these (Cohn, 2006) (Figure A-1).

The team uses its own measures such as velocity and 'burndown' charts that show the remaining stories to be developed. As capabilities and features become increasingly clear and stable from management and the team acquires a sustainable development rate, the test data will support the figures to show when the project will be complete. The test data including percentage passing acceptance tests, the amount of code changes (churn), and the defect rates are

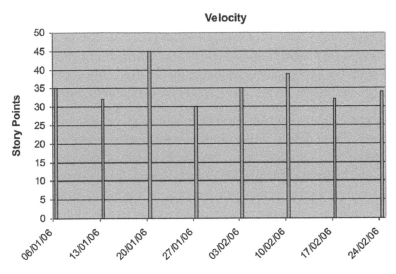

Figure A-1

Chart showing velocity or work achieved in story points during each iteration. This can help the team to determine the amount of story points they can realistically achieve in future iterations, and when the backlog is likely to be complete.

going to be of particular importance in software development and will be used in conjunction to establishing realistic estimates for the overall completion of the project.

Technical Debt

Within the planning process of the iteration it is essential to consider the quality of the product and ensure attributes such as scalability and security. These may not always be outlined by the customer or the Product Owner, but to avoid technical debt and the code being difficult to maintain and extend these issues need to be addressed. It is important within the planning process that architectural issues are dealt with. One way in which to help achieve this is to consider not just the user functionality or user stories but also tasks that have to be achieved in order to keep the architecture aspects required. This can be achieved through a dependency mapping as outlined by Brown, et al (2010).

Defining the Architecture of the System

The architecture of a system describes its overall structure, components, interfaces, and behaviour. Definitions vary but often include the perspectives or views of the structure (Bass, Clements, and Kazman, 2003). One area that is emphasised by one instance of the generic Unified Process (UP) and Rational Unified Process (RUP) is the concept of requirements. In this case it is often user functionality via user cases and subsets of scenarios that shows what functionality a user or actor requires of the system.

Some agile methods such as Extreme Programming (XP) favour the use of understanding of the system in an exploratory way, via development of code, improving the design without affecting the behavior, i.e., refactoring. So for instance, developing a security check in a banking system would necessitate understanding the structure and coding this feature would verify and quickly establish an architecture, if there wasn't a model already available.

Unless the architectural issues are addressed there will be inconsistencies in performance as the system is scaled. These defects will require ever more refactoring in order to avoid design problems. Therefore some consideration of the design is necessary to avoid later problems. This process of consideration of the planning of the architecture is termed architectural runway allowing for a smooth transition and rework and to avoid technical debt. Planning ahead, the architecture can be allocated on a planned process as advocated in agile architecture provisioning (Brown et al, 2010). Here architectural elements that are necessary for quality attributes (non-functional) such as security are allocated to the iteration backlog and provisioned within this period to carefully outline both functional and non-functional requirements, which are so important in determining scaling and performance factors. An alternative but similar approach would be to include design spikes at the Last Responsible Moment (LRM) to ensure the flexibility of the architecture. Where there is a choice of architecture this can then allow for a different approach to estimation of value. Instead of using a static cost-benefit analysis, which is normally based on static estimates and architecture, an investigation of alternative options and their relative future changes can be investigated and interpreted via real-options analysis (Bahsoon and Emmerich, 2004). This approach, which was originally developed in the financial markets, is increasingly used to determine the dynamically changing nature of viable options in areas such as provisioning of resources as in cloud provisioning within the IT sector (Collins, 2011). The nature of engineering is changing and increasingly 'composing systems from open source, commercial, and proprietary components' (Bosch, 2011), and in agile environments where the focus is on early exploration the 'selection and trade-off decisions' should be captured including the rationale that will help to understand why the product is better and why it is being built (CMMI, 2010). Pareto optimality and how this can be applied to balancing requirements as well as trade-off decisions for goals within project and programme management is a current area of research within the Faculty of Engineering Sciences at UCL.

Sharing knowledge and reflection

Research teams are well known for their synergy. Jeff Sutherland (2005) outlines this enthusiasm in agile projects and Dean Leffingwell (2007) outlines these concepts including how teams can create new points of view and resolve problems through dialogue. Within workshops, particularly in scientific research and innovation, it is useful to allow dialogue, not having to defend your idea, and allow further time to explore possibilities. This is refreshing, as often the stress is on discussions to resolve issues within a specified time period, i.e., time-boxing. Leffingwell also points out that in the spirit of Scrum, amongst its many attributes including commitment and autonomy, 'leaders provide creative chaos'. This is precisely the concept that Sir Paul Nurse, during the BBC David Dimbleby lecture in 2012, was conveying when he was outlining how to create a collaborative environment for researchers to excel at the future Crick Institute.

For each project the degree of understanding of goals, emergent design needs to be considered and appropriate patterns need to be in place. Agile development and research both need process and governance frameworks.

Earned Value

With traditional project management progress is based on tracking the intermediate tasks, such as production of the requirements document and the design artefacts, which can be achieved without a demonstrable or working product. Allowing measures to be unrelated to working products can give a false sense of progress.

With agile the focus is on whether the software works, whether it is what the client intended, and whether it is of value. This is done via continual feedback via releases to demonstrate and allow feedback to improve the product and value.

To track progress towards goals in technology and IT projects where there is emergent design, this needs to be achieved at a higher level of abstraction. In research projects a clear idea of initial goals or exploration of concepts still needs careful planning; otherwise it will be unlikely that funding will be granted. In projects in technology and IT when there is emergent design this needs to be done at a higher level of abstraction. In large research projects a clear idea of initial goals or exploration of concepts still needs careful planning, otherwise it will be unlikely that funding will be granted.

There may be an exploratory phase equivalent to a feasibility study in which one or two workshops are planned to outline strategy or develop architecture to develop the technology. This initial phase will have clear goals and should have clear rationale for those invited. As this is bounded by time and consultants and facilitator costs the budget can be easily ascertained. This is likely if parties agree to an exploratory design phase, then a development phase.

During the initial phase in order to define the scope or boundaries of the research or development it will cover the goals that will in many instances then be broken down further to capabilities according to the type of innovation or development project.

Capabilities, a term often used in military projects, determine what is required without determining how this will be achieved. If this is a software development project these may be subdivided into features and stories. Sometimes the term epic in agile development is often used for a larger aggregation of features.

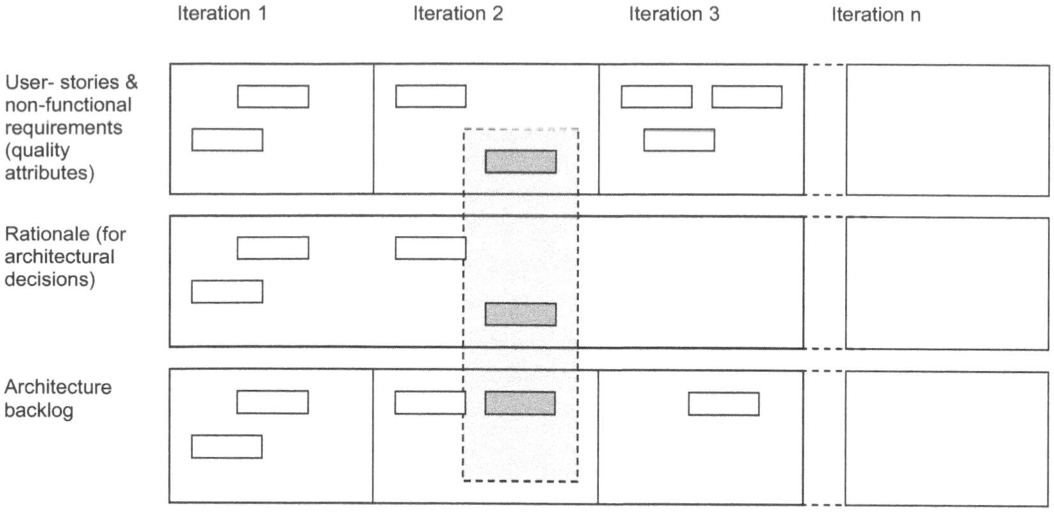

	Iteration 1	Iteration 2	Iteration 3	Iteration n

User- stories &
non-functional
requirements
(quality
attributes)

Rationale (for
architectural
decisions)

Architecture
backlog

Figure A-2

Architectural elements in iteration planning Collins [2011] based on Brown et al [2010].

User stories represent what the client or Product Owner requires. These requirements are often written as a short outline on a card and then decomposed further to tasks. The Product Owner discusses with the team the prioritisation and order of development with regard to development issues including software architecture.

The use of Earned Value Management (EVM) is well developed in certain sectors of project management including the construction industry and is being increasingly mandated in defence projects in the UK. This can be applied to agile and software development. The essence of agile development is this shared collaborative communication between client and development team, ensuring value for the client and a motivated development team. Although management may be used to using earned value measures this needs to be developed for the agile process so that the development team doesn't see this as an overhead and both groups can work collaboratively together. One criticism, however, of EVM is that this does not give an indication of the quality. One benefit of adapting this approach in conjunction with agile project management is that processes that improve quality, such as refactoring, are often incorporated into agile development practices.

If EVM is required by senior management or through the governance process of the project, one approach is to create a reporting on two levels; one for the capability tracking and one based on stories as shown in Figure A-3, which can be generated with minimal overhead as part of the planning process.

In agile development there is continual prioritisation at the iteration level, which is designed to be the same duration. Earned value can be applied to these user stories, but there is a subtle

Business value of story	Story points	Points earned	Planned developer hours	Actual logged hours	Earned value
3	10	10	100	120	100
2	8	8	60	60	60
2	4	4	60	80	60
1	2	0	20	0	0
	24	22	240	260	220

Figure A-3
Earned value derived from stories completed at the end of an iteration. Story points are an estimate of the relative time that the development team thinks the work will take to complete.

difference in application. The reconciliation seems difficult as the user stories are continually under prioritisation according to the client as to what they consider the most valuable user story. These stories are stored on a backlog for selection during each iteration and the team estimates how large they are in terms of story points.

The stories according to prioritisation are selected for the iteration by the client on their perceived value. If the table is prefaced with the functional value or business value, the order of priority can be much clearer, although the highest priority should be at the top of the backlog and the development process then pulls the next set of stories and set of tasks.

As can be seen from the figure the earned value can be derived from the amount of completed work. For the first story, with the highest priority on the list and business value, the story was completed and therefore earned all the points.

Although the story took slightly longer than expected (i.e., 120 hours) the story was complete and therefore earned the full earned value of 100. Where the story was not complete it was allocated zero. From these figures schedule performance index (SPI) and cost performance index (CPI) can be ascertained to give an indication of progress (Collins, 2006).

Some have argued that velocity is more important. This is not the same as earned value. The velocity is the rate at which a team works and is a useful internal measure as are burndown charts, which show the remaining work and can act as a motivator for the team.

However, while it can be seen that earned value is valid within an iteration there are different interpretations of how this could work at higher levels of abstraction and the value to project management at a project level. Craig Larman (2004) suggests the use of estimates in terms of budgets. This is most easily achieved in terms of hours of work. He also outlines the concept of re-calculating the planned work for each iteration and as more information arises then the baseline is updated. For tasks to earn value it is prudent to only consider these when fully completed. However assessing this needs careful consideration as the development of each story is usually considered finished when all tests including integration and acceptance tests are completed.

It can be seen that using a simple spreadsheet as a by-product of the team's estimates an EV system can be used as an indicator of progress in terms of business value.

Summary

In agile development it is often about trends emerging rather than making guesses about the future. Walker Royce has written about the indicators for converging on the solution and the indicators for value and progress (Royce, 2011). The way to gain credibility is through working with the client on a joint understanding of a problem, how to measure progress, and when to converge on the solution. This creates the real value, not only to the business, but to the self-worth of the team.

Agile project management is about achieving value collaboratively for the project and client team and the organisations concerned. This is not just about the bottom line but achieving something at work, feeling valued, and sharing knowledge with colleagues to achieve that next breakthrough. This is the true value of agile for individuals, the team, and the organisation.

> *At UCL at the front of projects are different types of workshops, examples of agile patterns that bring the right mix of researchers and project managers and support staff to solve problems. These vary from the Town Hall meetings where the challenge and opportunity is outlined to more detailed workshops allowing discussions and dialogue. The agile project management and development approaches favour the timeboxed approach to discussions, and estimation and often adding more time does not necessarily give a better outcome. It is these workshops which are often the driver for new innovations and approaches to development. The use of a trained facilitator is often vital with large programmes. It is important that staff with different technical approaches can outline their view. It may be that the idea is rejected in favour of an alternative idea discussed at the meeting. The key is fair process and that the team are deciding the direction. This is the essence of agile allowing the client to work with the development team collaboratively to decide capabilities they which to develop and prioritizing the functionality and user stories, or in the case of research the investigations. (Collins, Graham, UCL Research and Consulting projects (2003–2013), 2013)*

It is self-evident that if one member of a team suggests a technical solution, another may disagree and point out an alternative and why in certain circumstances it is a better resolution.

> *Developing real-time modelling agile approaches with colleagues had a significant impact on communication and indirectly in one project resolving political issues by the clear focus on the technical problem rather than individuals. During a consulting assignment my colleagues in a small consulting group were asked to outline our object and business modelling approach. We had been asked for a solution and found on arrival at the client site already strained working relationships with a development project that had been*

ongoing for over a year. It was agreed to use our agile modelling approach to clarify the goals of the project and be clear on the direction. It was clear things were not going well, the project manager complained that he didn't know what kind of project he was working on, as it wasn't properly defined by the programme director. He didn't know whether it was a business transformation project or a business improvement project. The client really wanted the project management problems to be resolved and not upset the director of the consulting firm under contract. It was a mess and an expensive mess with technical teams having worked for a considerable time. Without a real-time modelling solution and experience in forming a unified team this impasse would not have been resolved.

Getting buy-in wasn't just a problem of clarifying the goals but getting support by under-standing each of the stakeholder's goals, communicating the direction over a short iteration with a clear product and defined time and engaging all stakeholders already working on the project.

The leader must be seen by others not to be gaining personally. In the case of the programme that had to be put back on track it was imperative to listen to the other parties and support each of their objectives so that the programme could move forward. Self-interest other than wishing for a successful outcome was not in the cards. Likewise the leaders in agile project management must lead by trusting their team and allow their team to deliver the project in a self-determined way. Self-organising teams and allowing them to report on progress are areas that the leader must embrace in agile project management. Much of what has been written on agile has been on what the teams do and how they track their progress. The burndown chart, keeping progress visible, and keeping tasks visible on the wall are for the team and the team leader. Keeping key tasks and communications visible, this is the 'whitebox' of progress reporting. It is the leader in agile project management who needs to understand this iterative process and be a resource provider, to remove all impediments to the team, who must trust his or her team to deliver in the technical approach they consider best. It is this beyond anything that defines the change to a leadership culture in agile.

The challenge in agile project management is not prescriptive plans and practices to follow but to populate the project planning process with appropriate patterns that are effective and add real value. For the time being the challenge must be to balance the planning so that you can achieve the flexibility to deliver increasingly complex projects and rapidly add new developments to enterprises and research establishments.

Bibliography

Bahsoon Rami, Emmerich Wolfgang: *Evaluating Architectural Stability with Real Options Theory*, 2004, Proceedings of the 20th IEEE International Conference on Software Maintenance (ICSM'04).

Bass Len, Clements Paul, Kazman Rick: *Software Architecture in Practice Second Edition*, 2003, SEI series in software engineering, Addison-Wesley.

Bosch Jan: *Keynote abstract Saturn Conference San Francisco*, May 2011, SEI. 16–20.

Brown Nanette, Nord Robert, Ozkaya Ipek: *Enabling Agility Through Architecture CrossTalk*, Nov/Dec 2010, .

CMMI® for Development: Version 1.3 CMMI-DEV, *V1.3, SEI*, November 2010.

Cockburn Alistair: *Agile Software Development*, 2002, Addison-Wesley.

Cohn Mike: *User Stories Applied*, 2004, Addison-Wesley.

Cohn Mike, Agile Estimating and Planning, Addison-Wesley, 2006, .

Collins Graham: *Experience in Developing Metrics for Agile Projects Compatible with CMMI Best Practice*, June 2006, SEI SEPG Conference Amsterdam. 12–15.

Collins Graham: Post-graduate course GZ07, academic year 2010-11, *Department of Computer Science, Faculty of Engineering Sciences, UCL*, 2011a.

Collins Graham: Developing Agile Software Architecture using Real-Option Analysis and Value Engineering, *SEI SEPG Conference Dublin*, June 2011. 2011b.

Collins Graham: Experience as lead consultant on commercial consulting project 1999-2000 included in GZ07 post-graduate teaching for GZ07 course, *Department of Computer Science, Faculty of Engineering Sciences, UCL*, 2013.

Elssamadisy Amr: *Agile Adoption Patterns: A Roadmap to Organisational Success*, 2009, Addison-Wesley.

Highsmith Jim: *Agile Software Development Ecosystems*, 2002, Addison-Wesley.

Highsmith Jim: *Agile Project Management*, second edition, 2010, Addison-Wesley.

Larman Craig, Development Agile & Iterative: *A Manager's Guide*, 2004, Addison-Wesley.

Leffingwell Dean: *Scaling Software Agility: Best Practices for Large Enterprises*, 2007, Addison-Wesley. (chapter 21 with Ken Schwaber).

Royce Walker: Measuring Agility and architectural Integrity, *Int. J. Software Informatics* vol. 5(3):415–433, 2011.

Rule P: *Grant, 'What do we mean by 'Lean'?*. Guest lecture for Professional Practice series, Department of Computer Science, Faculty of Engineering Sciences, UCL, 3rd, February 2011.

Schwaber Ken: *Agile Project Management, Microsoft*, 2004.

Shalloway Alan, Beaver Guy, Trott James R: *Lean-Agile Software Development: Achieveing Enterprise Agility*, 2010, Addison-Wesley.

Sutherland, Jeff, Future of Scrum: Support for Parallel Pipelining of Sprints in Complex Projects, Denver, CO: Agile 2005 Conference.

Abbreviations and Acronyms Used in Project Management

Abbreviation	Meaning	Usage
ACC	Annual Capital Charge	Finance
ACWP	Actual Cost of Work Performed	EVA
ADR	Alternative Dispute Resolution	Construct
ANB	Adjudicator Nominating Body	Construct
AoA	Activity on Arrow	CPA
AoN	Activity on Node	CPA
APM	Association for Project Management	PM
ARM	Availability, Reliability, Maintainability	MOD
BC	Business Case	PM
BCWP	Budgeted Cost of Work Performed	EVA
BCWS	Budgeted Cost of Work Scheduled	EVA
BoK	Body of Knowledge	PM
BOQ	Bill of Quantities	Construct
BS	British Standard	General
BSI	British Standards Institution	General
CAD	Computer Aided Design	General
CAM	Computer Aided Manufacture	General
CAR	Contractor's All Risk	Construct
CBS	Cost Breakdown Structure	PM
CDM	Construction, Design and Management	Construction
CEN	Comité Europeen de Normalization	General
CIF	Carriage, Insurance, Freight	Procurement
CM	Configuration Management	PM
CPA	Critical Path Analysis	PM
CPA	Contract Price Adjustment	Procurement
CPD	Continuing Professional Development	General
CPI	Cost Performance Index	EVA
CPM	Critical Path Methods	CPA
CSCS	Cost & Schedule Control System	EVA
CV	Cost Variance	EVA
CV	Curriculum Vitae	General

Abbreviation	Meaning	Usage
DCF	Discounted Cash Flow	Finance
DDP	Delivery Duty Paid	Procurement
DIN	Deutsche Industrie Normen	General
EAC	Estimated Cost at Completion	EVA
ECC	Estimated Cost to Complete	EVA
EV	Earned Value	EVA
EVA	Earned Value Analysis	PM
EVMS	Earned Value Management System	EVA
FCC	Forecast Cost to Complete	EVA
FF	Free Float	CPA
FLAC	Four Letter Acronym	General
FMEA	Failure Mode & Effect Analysis	MOD
FOB	Free on Board	Procurement
FOR	Free on Rail	Procurement
HAZOP	Hazard and Operability	General
HR	Human Resources	General
HSE	Health and Safety Executive	General
H&S	Health & Safety	General
IA	Investment Appraisal	Finance
IPMA	International Project Management Association	PM
IPMT	Integrated Project Management Team	PM
IPR	Intellectual Property Rights	General
IRR	Internal Rate of Return	Finance
IS	Information Systems	General
ISEB	Information Systems Examination Board	General
ISO	International Organization for Standardization	General
IT	Information Technology	General
ITT	Invitation to Tender	Procurement
JIT	Just In Time	General
KPI	Key Performance Indicator	PM
LAD	Liquidated Ascertainable Damages	Construct
LCC	Life Cycle Costing	PM
LD	Liquidated Damages	Construct
LOB	Line of Balance	Construct
LRM	Liner Responsibility Matrix	PM
MOD	Ministry of Defence	General
MTO	Material Take-off	Construct
NDT	Non-Destructive Testing	Construct
NOSCOS	Needs, Objectives, Strategy & Organizations Control System	MOD

Abbreviation	Meaning	Usage
NPV	Net Present Value	Finance
OBS	Organization Breakdown Structure	PM
OD	Original Duration	EVA
OGC	Office of Government Trading	General
ORC	Optimal Replacement Chart	Finance
ORM	Optimal Replacement Method	Finance
PBS	Product Breakdown Structure	PM
PDM	Precedence Diagram Method	CPA
PEP	Project Execution Plan	PM
PID	Project Initiation Document	PM
PERT	Program Evaluation & Review Technique	CPA
PFI	Private Finance Initiative	Finance
PM	Project Management	PM
PM	Project Manager	PM
PMI	Project Management Institute	PM
PMP	Project Management Plan	PM
PPE	Post Project Evaluation	PM
PPP	Public–Private Partnership	Finance
PRD	Project Definition	PM
QA	Quality Assurance	General
QC	Quality Control	General
QMS	Quality Management System	General
QP	Quality Plan	General
R&D	Research and Development	General
RFQ	Request for Quotation	Procurement
RR	Rate of Return	Finance
SFR	Sinking Fund Return	Finance
SMART	Specific, Measurable, Achievable, Realistic, Timebound	MOD
SOR	Schedule of Rates	Construct
SOW	Statement of Work	PM
SPI	Schedule Performance Index	EVA
SRD	Sponsor's Requirement Definition	PM
SV	Schedule Variance	EVA
TCP	Time, Cost & Performance	PM
TF	Total Float	CPA
TQM	Total Quality Management	General
TOR	Terms of Reference	General
VA	Value Analysis	General
VE	Value Engineering	General

Abbreviation	Meaning	Usage
VM	Value Management	General
WBS	Work Breakdown Structure	PM

See also list of acronyms.

Acronyms Used in Project Management

ARM	Availability, Reliability, Maintainability
BIM	Building Information Modeling
CAD/CAM	Computer Aided Design/Computer Aided Manufacture
CADMID	Concept, Assessment, Demonstration, Monitoring, In-Service, Disposal
CFIOT	Concept, Feasibility, In-Service, Operation, Termination
CS^2 (CSCS)	Cost & Schedule Control System
EMAC	Engineering Manhours And Cost
FLAC	Four Letter Acronym
HASAWA	Health And Safety At Work Act
IPMA	International Project Management Association
NAPNOC	No Agreed Price, No Contract
NEDO	National Economic Development Office
NIMBY	Not In My Back Yard
NOSCOS	Needs, Objectives, Strategy & Organization Control System
NOSOCS&R	Needs, Objectives, Strategy, Organization Control, System & Risk
PAYE	Pay As You Earn
PERT	Program, Evaluation & Review Technique
PESTLE	Political, Economic, Sociological, Technological, Legal, Environmental
PRAM	Project Risk Analysis & Management
PRINCE	Projects in a Controlled Environment
RAMP	Risk Analysis and Management for Projects
RIDDOR	Reporting of Injuries, Diseases & Dangerous Occurrences Regulations
RIRO	Rubbish In–Rubbish Out
SAPETICO	Safety, Performance, Time, Cost
SMAC	Site Manhours And Cost
SMART	Specific, Measurable, Achievable, Realistic & Time bound
SOW	Statement Of Work
SWOT	Strengths, Weaknesses, Opportunities & Threats

Glossary

Activity   An operation on a network which takes time (or other resources) and is indicated by an arrow.

Actual cost of work performed (ACWP)   Cumulative actual cost (in money or manhours) of work booked in a specific period.

Actual hours   The manhours actually expended on an activity or contract over a defined period.

Adjudication   Procedure for resolving a dispute by appointing an independent adjudicator.

Advance payment bond   Bond given in return for advanced payment by client.

Analytical estimates   Accurate estimate based on build up of all material and labour requirements of the project.

AoN   Activity on Node.

AoA   Activity on Arrow.

Arbitration   Dispute resolution by asking an arbitrator to make a decision.

Arithmetical analysis   A method for calculating floats arithmetically.

Arrow   A symbol on a network to represent an activity or dummy.

Arrow diagram   A diagram showing the interrelationships of activities.

Back end   The fabrication, construction, and commissioning stage of a project.

Backward pass   A process for subtracting durations from previous events, working backwards from the last event.

Banding   The subdivision of a network into horizontal and vertical sections or bands to aid identification of activities and responsibilities.

Bar chart   *See* **Gantt chart.**

Belbin type   One of nine characteristics of a project team member as identified by Meredith Belbin's research programme.

Beta (b) distribution   Standard distribution giving the expected time $te = (a + 4m + b)/6$.

Bid Bond   Bond required with quotation to discourage withdrawal of bid.

Bond   Guarantee given (for a premium), by a bank or building society as a surety.

Budget   Quantified resources to achieve an objective, task or project by a set time.

Budgeted cost of work performed (BCWP)   *See* Earned Value.

Budgeted cost of work scheduled (BCWS)   Quantified cost (in money or manhours) of work scheduled (planned) in a set time.

Budget hours   The hours allocated to an activity or contract at the estimate or proposal stage.

Business case   The document setting out the information and financial plan to enable decision makers to approve and authorize the project.

Calendar   Time scale of programme using dates.

Capital cost   The project cost as shown in the balance sheet.

Cash flow   Inward and outward movement of money of a contract or company.

Change control   The process of recording, evaluating, and authorizing project changes.

Change management   The management of project variations (changes) in time, cost, and scope.

Circle and link method   *See* **Precedence diagram.**

Close out procedure   The actions implemented and documents produced at the end of a project.

Close-out report   The report prepared by the project manager after project close-out.

Comparative estimates   Estimates based on similar past project costs.

Computer analysis   The method for calculating floats, etc., using a computer.

Conciliation   The first stage of dispute resolution using a conciliator to improve communications and understanding.

Configuration management   The management of the creation, maintenance, and distribution of documents and standards.

Conflict management   Management of disputes and disagreements using a number of accepted procedures.

Contingency plan   Alternative action plan to be implemented when a perceived risk materializes.

Counter-trade   Payment made of goods or services with materials or products that can be sold to pay for the items supplied.

Cost/benefit analysis   Analysis of the relationship between the cost and anticipated benefit of a task or project.

Cost breakdown structure (CBS)   The hierarchical breakdown of costs when allocated to the work packages of a WBS.

Cost code   Identity code given to a work element for cost control purposes.

Cost control   The ability to monitor, compare, and adjust expenditures and costs at regular and sufficiently frequent intervals to keep the costs within budget.

Cost performance index   The ratio of the earned value (useful) cost and the actual cost.

Cost reporting   The act of recording and reporting commitments and costs on a regular basis.

Cost variance   The arithmetical difference between the earned value cost and the actual cost. This could be positive or negative.

CPA   Critical path analysis. The technique for finding the critical path and hence the minimum project duration.

CPM   Critical path method. *See* **CPA**

CPS   Critical path scheduling. *See* **CPA.**

Critical activity   An activity on the critical path which has zero float.

Critical path   A chain of critical activities, i.e., the longest path of a project.

Dangle   An activity that has a beginning node but is not connected at its end to a node that is part of the network.

Deliverable   The end product of a project or defined stage.

Dependency   The restriction on an activity by one or more preceding activities.

Direct cost   The measurable cost directly attributed to the project.

Discounted Cash Flow (DCF)   Technique for comparing future cash flows by discounting by a specific rate.

Distribution schedule   A tabular record showing by whom and to whom the documents of a project are distributed.

Dummy activity   A timeless activity used as a logical link or restraint between real activities in a network.

Duration   The time taken by an activity.

Earliest finish   The earliest time at which an activity can be finished.

Earliest start   The earliest time at which an activity can be started.

Earned value hours   *See* **Value hours.**

End event   The last event of a project.

Estimating   Assessment of costs of a project.

EVA   Earned Value Analysis.

Event   The beginning and end node of an activity, forming the intersection point with other activities.

Expediting   Action taken to ensure ordered goods are delivered on time. Also known a progress chasing.

Feasibility study   Analysis of one or more courses of action to establish their feasibility or viability.

Feedback   The flow of information to a planner for updating the network.

Float   The period by which a non-critical activity can be delayed.

Free float   The time by which an activity can be delayed without affecting a following activity.

Forward pass   A process for adding durations to previous event times starting at the beginning of a project.

Front end   The design and procurement stage of a project. This may or may not include the manufacturing period of equipment.

Functional organization   Management structure of specialist groups carrying out specific functions or services.

Gantt chart   A programming technique in which activities are represented by bars drawn to a time scale and against a time base.

Graphics   Computer generated diagrams.

Graphical analysis   A method for calculating the critical path and floats using a linked bar chart technique.

Grid   Lines drawn on a network sheet to act as coordinates of the nodes.

Hammock An activity covering a number of activities between its starting and end node.

Hardware The name given to a computer and its accessories.

Herzberg's theory The hygiene factors and motivators that drive human beings.

Histogram A series of vertical columns whose height is proportional to a particular resource or number of resources in any time period.

Incoterms International trade terms for shipping and insurance of freight.

Independent float The difference between free float and the slack of a beginning event.

Indirect cost Cost attributable to a project, but not directly related to an activity or group within the project.

Input The information and data fed into a computer.

Interface The meeting point of two or more networks or strings.

Interfering float The difference between the total float and the free float. Also the slack of the end event.

Internal Rate of Return (IRR) The discount rate at which the Net Present Value is zero.

Investment appraisal Procedure for analysing the viability of an investment.

Key performance indicators (KPI) Major criteria against which the project performance is measured.

Ladder A string of activities which repeat themselves in a number of stages.

Lag The delay period between the end of one activity and the start of another.

Latest finish The latest time at which an activity can be finished without affecting subsequent activities.

Latest start The latest time at which an activity can be started without delaying the project.

Lead The time between the start of one activity and the start of another.

Leadership The ability to inspire and motivate others to follow a course of action.

Lester diagram Network diagram that combines the advantages of arrow and precedence diagrams.

Letter of intent Document expressing intention by client to place an order.

Line of balance Planning technique used for repetitive projects, subprojects, or operations.

Litigation Act of taking a dispute to a court of law for a hearing before a judge.

Logic The realistic interrelationship of the activities on a network.

Logic links The link line connecting the activities of a precedence diagram.

Loop A cycle of activities that returns to its origin.

Manual analysis The method for calculating floats and the critical path without the use of a computer.

Maslow's hierarchy of needs The five stages of needs of an individual.

Master network Coordinating network of subnetworks.

Matrix The table of activities, durations, and floats used in arithmetical analysis.

Matrix organization Management structure where functional departments allocate selected resources to a project.

Mediation Attempt to settle a dispute by joint discussions with a mediator.

Menu Screen listing of software functions.

Method statement Narrative or graphical description of the methods envisaged to construct or carry out selected operations.

Milestones Key event in a project which takes zero time.

Milestone slip chart Graph showing and predicting the slippage of milestones over the project period.

Negative float The time by which an activity is late in relation to its required time for meeting the programme.

Negotiation Attempt to reach a result by discussion which is acceptable to all sides.

Net Present Value (NPV) Aggregate of discounted future cash flows.

Network A diagram showing the logical interrelationships of activities.

Network analysis The method used for calculating the floats and critical path of a network.

Network logic The interrelationship of activities of a planning network.

Node The intersection point of activities. An event.

Organization breakdown structure (OBS) Diagrammatic representation of the hierarchical breakdown of management levels for a project.

Organogram Family tree of an organization showing levels of management.

Output The information and data produced by a computer.

P3 Primavera Project Planner.

Parametric estimates   Estimates based on empirical formulae or ratios from historical data.

Pareto's law   Doctrine which shows that approx. 20% of causes create 80% of problems. Also known as 80/20 rule.

Path   The unbroken sequence of activities of a network.

Performance bond   Bond that can be called by client if contractor fails to perform.

PERT   Programme Evaluation and Review Technique. Another name for **CPA.**

PESTEL   Political, Economic, Sociological, Technical, Legal, Environmental.

Phase   A division of the project life cycle.

Portfolio management   Management of a group of projects not necessarily related.

Post project review   History and analysis of successes and failures of project.

Planned cost   The estimated (anticipated) cost of a project.

Precedence network   A method of network programming in which the activities are written in the node boxes and connected by lines to show their interrelationship.

Preceding event   The beginning event of an activity.

Printout   *See* **Output.**

Procurement   Operation covering tender preparation, bidder selection, purchasing, expediting, inspection, shipping, and storage of goods.

Product Breakdown Structure (PBS)   Hierarchical decomposition of a project into various levels of products.

Program   The set of instructions given to a computer.

Programme   A group of related projects.

Programme management   Management of a group of related projects.

Programme manager   Manager of a group of related projects.

Progress report   A report that shows the time and cost status of a project, giving explanations for any deviations from the programme or cost plan.

Project   A unique set of co-ordinated and controlled activities to introduce change within defined time, cost, and quality/performance parameters.

Project context   *See* project environment.

Project environment   The internal and external influences of a project.

Project close-out   The shutting down of project operations after completion.

Project life cycle   All the processes and phases between the conception and termination of a project.

Project management   The planning, monitoring and controlling of all aspects of a project.

Project management plan (PMP)   A document which summarizes of all the main features encapsulating the Why, What, When, How, Where, and Who of a project.

Project manager   The individual who has the authority, responsibility, and accountability to achieve the project objectives.

Project organization   Organization structure in which the project manager has full authority and responsibility of the project team.

Project task force   *See* Task force.

Quality assurance   Systematic actions required to provide confidence of quality being met.

Quality audit   Periodic check that quality procedures have been carried out.

Quality control   Actions to control and measure the quality requirements.

Quality management   The management of all aspects of quality criteria, control, documentation, and assurance.

Quality manual   Document containing all the procedures and quality requirements.

Quality plan   A plan that sets out the quality standards and criteria of the various tasks of a project.

Quality policy   Quality intentions and directions set out by top management.

Quality programme   Project-specific document that defines the requirements and procedures for the various stages.

Quality review   Periodic review of standards and procedures to ensure applicability.

Quality systems   Procedures and processes and resources required to implement quality management.

Random numbering   The numbering method used to identify events (or nodes) in which the numbers follow no set sequence.

Requirements management   Capture and collation of the client's or stakeholders' perceived requirements.

Resource   The physical means necessary to carry out an activity.

Resource levelling   *See* **Resource smoothing.**

Resource smoothing   The act of spreading the resources over a project to use the minimum resources at any one time and yet not delay the project.

Responsibility code   Computer coding for sorting data by department.

Responsibility matrix   A tabular presentation showing who or which department is responsible for set work items or packages.

Retention bond   Bond given in return for early payment of retention monies.

Retentions   Moneys held by employer for period of maintenance (guarantee) period.

Return on capital employed   Profit (before interest and tax) divided by the capital employed given as a %.

Return on Investment (ROI)   Average return over a specified period divided by the investment given as a %.

Risk   The combination of the consequences and likelihood of occurrence of an adverse event or threat.

Risk analysis   The systematic procedures used to determine the consequences or assess the likelihood of occurrence of an adverse event or threat.

Risk identification   Process for finding and determining what could pose a risk.

Risk management   Structured application of policies, procedures, and practices for evaluating, monitoring, and mitigating risks.

Risk management plan   Document setting out strategic requirements for risk assessment and procedures.

Risk register   Table showing the all identified risks, their owners, degree of P/I, and mitigation strategy.

Schedule   *See* **Programme.**

Schedule Performance Index   The ratio of earned value cost (or time) and the planned cost (or time).

Schedule variance   The arithmetical difference between the earned value cost (or time) and the planned cost (or time).

Sequential numbering   The numbering method in which the numbers follow a pattern to assist in identifying the activities.

Situational leadership   Adaptation of management style to suit the actual situation the leader finds him/herself in.

Slack   The period between the earliest and latest times of an event.

Slip chart   See milestone slip chart.

SMAC   Site manhour and cost. The name of the computer program developed by Foster Wheeler Power Products Limited for controlling manhours in the field.

Software   The programs used by a computer.

Sponsor   The individual or body who has primary responsibility for the project and is the primary risk taker.

Stakeholder   Person or organization who has a vested interest in the project. This interest can be positive or negative.

Statement of Work (SOW)   Description of a work package that defines the project performance criteria and resources.

Start event   The first event of a project or activity.

Subcontract   Contract between a main contractor and specialist subsidiary contractor (subcontractor).

Subjective estimates   Approximate estimates based on 'feel' or 'hunch'.

Subnetwork   A small network that shows a part of the activities of a main network in greater detail.

Succeeding event   The end event of an activity.

Task   The smallest work unit shown on a network programme (see also Activity).

Task data   The attributes of a task such as duration, start and end date, and resource requirement.

Task force   Project organization consisting of a project team that includes all the disciplines and support services under the direction of a project manager.

Teamwork   The act of working harmoniously together in a team to produce a desired result.

Time estimate   The time or duration of an activity.

Toolbar   The list of function icons on a computer screen.

Topological numbering   A numbering system where the beginning event of an activity must always have a higher number than the events of any activity preceding it.

Total float   The spare time between the earliest and latest times of an activity.

Total quality management (TQM)   Company-wide approach to quality beyond prescriptive requirements.

Updating   The process of changing a network or programme to take into account progress and logic variations.

Value engineering   The systems used to ensure the functional requirements of value management are met.

Value hours   The useful work hours spent on an activity. This figure is the product of the budget hours and the percentage complete of an activity or the whole contract.

Value management   Structured means aimed at maximizing the performance of an organization.

Variance   Amount by which a parameter varies from its specified value.

Weightings   The percentage of an activity in terms of manhours or cost of an activity in relation to the contract as a whole, based on the budget values.

Work breakdown structure (WBS)   Hierarchical decomposition of a project into various levels of management and work packages.

Work package   Group of activities within a specified level of a work breakdown structure.

Examination Questions 1: Questions

The following sheets show 50 typical questions taken at random that may appear in the APMP written paper. Each question is followed by a bracketed number, which relates to the number of the topic as set out in the latest (fifth edition) version of the APM Body of Knowledge, which can be obtained from the Association for Project Management. The answers to these 50 questions are given in bullet, point format in the subsequent pages, against the same number as the questions.

(Note: At time of going to press, APM has not yet updated the Syllabus and Learning Objectives topic numbers to coincide with the latest BoK topic numbers. This small discrepancy does of course not make any difference to the subject matter.)

Candidates wishing to sit the American PMI multiple choice examination are advised to consult the latest issue of the *PMI Guide to the Project Management Body of Knowledge,* also known as the PMBOK Guide.

Guided Solutions to the questions can be found at
http://books.elsevier.com/companions/075066956X.

1. List 12 items (subjects) which should be set out in a PMP (2.4)
2. Explain the purpose and structure of a WBS (3.1)
3. Describe the most usual risk identification techniques (2.5b)
4. Explain the risk management procedure (2.5)
5. Set out the risks associated with travelling from Bath to London by road.
   Draw a risk register (log) and populate it with at least four perceived risks (2.5)
6. Describe a change management procedure. Draw up two forms relating to change
   management (3.5)

7. Draw a bar chart for the following activities:

Activity	Duration (days)	Preceding activity
A	5	–
B	7	A
C	9	A
D	7	B
E	2	C
F	6	C
G	2	E & D
H	3	F
J	4	G & H

What is the end date?

What is the effect of B slipping by 3 days?   (3.2)

8. Explain the difference between project management and programme management   (1.2)

9. Explain the purpose of stakeholder management and describe the difference
   (with examples of positive and negative stakeholders)   (2.2)

10. Explain what is meant by configuration management   (4.7)

11. Describe a risk management plan and give its contents   (2.5a)

12. State four risk mitigation strategies excluding contingencies   (2.5)

13. Explain the main tools used in quality management   (2.6a)

14. Explain the main topics of a quality management system   (2.6)

15. Describe the purpose of milestones and draw a milestone slip chart showing how
    slippage is recorded   (3.2a)

16. Explain the advantages of EVA over other forms of progress monitoring   (3.6)

17. Explain the purpose of a project life cycle and draw a typical life cycle diagram.   (6.1)
    Explain what is meant by product life cycle and expanded life cycle

18. Describe what documents are produced at the various stages of a life cycle   (6.1a)

19. Explain the difference between the three main types of project organization   (6.7)

20. Explain what is meant by communication management
    Give eight barriers to good communication and explain how to overcome them   (7.1)

21. Explain the advantages of a project team; list six features and give four barriers to
    team building   (7.2a)

22. Describe what is meant by conflict management and list five techniques   (7.4)

23. Describe the purpose of Belbin test and explain the characteristics of the eight
    Belbin types   (7.2c)

24. Explain Herzberg's motivation theory and Maslow's theory of needs   (7.2e)

25. Explain the main constituents of a business case. Who owns the business case?   (5.1)

26. Explain what are the qualities which make a leader   (7.3)

27. Describe the main stages of a negotiation process   (7.5)

28. Explain what is meant by cash flow. Draw the format of a cash flow chart   (3.4)

29. Describe a close-out procedure. List six documents that must be prepared and handed over to the client on close-out (6.5)
30. Describe six topics to be considered as part of a procurement strategy. List four types of contract (5.4)
31. Describe the selection process for employing contractors or subcontractors (5.4)
32. List and explain the phases of a project as suggested by Tuckman (7.2d)
33. Describe what is meant by the project environment (1.4)
34. List 10 reasons why a project may fail and suggest ways to rectify these failures. (2.1)
35. Explain the role of a programme manager and show the advantages to the organization of such a position (1.2)
36. Describe three methods for estimating the cost of a project and give the approx. % accuracy of each method (4.3)
37. Describe the principal reasons for an investment appraisal and give the constituents of such an appraisal (5.1a)
38. What are the four main types of estimating techniques and what is their approximate degree of accuracy? (4.3)
39. What is meant by resource levelling and resource smoothing? (3.3)
40. Describe the roles of the client, sponsor, project manager, and a supplier (6.8)
41. List at least six documents that have to be handed over at the end of a project (6.5)
42. Describe six common generic causes of accidents in industry (2.7)
42. Describe three main types of contract used in construction (5.4)
43. Draw a work breakdown structure for the manufacture of a bicycle. Limit the size to four levels of detail (3.1)
44. What is meant by internal rate of return (IRR)? Show how this can be obtained graphically (5.1a)
45. Describe the functions carried out by a project office (1.6)
46. What is the purpose of a post project review? (6.6)
47. Describe three methods of conflict resolution when mediation has failed (7.4)
48. State four main characteristics of a good project manager (1.1)
49. Describe two pros and cons of an AoA network and an AoN network (3.2)
50. Describe two pros and cons of DCF and payback (5.1a)

Adair, J. (2010). *Decision Making and Problem Solving Strategies*. Kogan Page.

Adams, J., & Adams, J. R. (1997). *Principles of Project Management*. PMI.

Alvesson, M. (2002). *Understanding Organisational Culture*. Sage.

Andersen, E. (2008). *Rethinking Project Management*. Pearson.

Andrews, N. (2011). *Contract Law*. Cambridge University Press.

APM. (2011). *APM Code of Professional Conduct*. APM.

APM. (2007). *APM Introduction to Programme Management*. APM.

APM. (2009). *Benefits Management*. APM.

APM. (2010). *Benefits Realisation*. APM.

APM. (2007). *Co-directing Change*. APM.

APM. (1998). *Contract Strategy for Successful Project Management*. APM.

APM. (2011). *Directing Change*. APM.

APM. (2002). *Earned Value Management: APM Guidelines*. APM.

APM. (2010). *From Physical Change to Benefits in the Built Environment*. APM.

APM. (2008). *Interfacing Risk and Earned Value Management*. APM.

APM. (2008). *Introduction to Project Planning*. APM.

APM. (2004). *Project Risk Analysis and Management (PRAM) Guide* (2nd ed.). APM.

APM. (2009). *Sponsoring Change*. APM.

APM. (1998). *Standard Terms for the Appointment of a Project Manager*. APM.

APM. (2010). *The Earned Value Management Compass*. APM.

APM. (2010). *The Scheduling Maturity Model*. APM.

Bagueley, P. (2010). *Improve Your Project Management: Teach Yourself*. Hodder & Stoughton.

Bahsoon, R., & Emmerich, W. (2004). Evaluating Architectural Stability with Real Options Theory. *Proceedings of the 20th IEEE International Confrernce on Software Maintenance (ICSM'04)*.

Balogun, J., et al. (2008). *Exploring Strategic Change*. Prentice-Hall.

Barkley. (2006). *Integrated Project Management*. McGraw-Hill.

Barkley. (2004). *Project Risk Management*. McGraw-Hill.

Bartlett, J. (2010). *Managing Programmes of Business Change*. Project Manager Today.

Bartlett, J. (2005). *Right First and Every Time*. Project Manager Today.

Basu, R. (2011). *Managing Project Supply Chains*. Gower.

Bass, L., Clements, P., & Kazman, R. (2003). *Software Architecture in Practice Second Edition*. SEI series in software engineering, Addison-Wesley.

Benko, C., & McFarlane, W. (2003). *Connecting the Dots*. Harvard Business School.

Boddy, D., Buchannan, D., Take the Lead, Prentice-hall (1002)

Boddy, D. (2002). *Managing Projects*. Prentice-Hall.

Bosch, J. (2011). *Keynote abstract Saturn Conference San Francisco*. SEI, 16–20 May.

Boundy, C. (2010). *Business Contracts Handbook*. Gower.

Bourne, L. (2009). *Stakeholder Relationship Management*. Gower.

Bradley, G. (2010). *Benefit Realization Management*. Gower.

Bradley, G. (2010). *Fundamentals of Benefit Realization*. The Stationary Office.

Broome, J. C. (2002). *Procurement Routes for Partnering: A Practical Guide*. Thomas Telford.

Broome, J. C. (2002). *Procurement Routes for Partnering*. Thomas Telford.

Brown, N., Nord, R., & Ozkaya, I. (Nov/Dec 2010). *Enabling Agility Through Architecture CrossTalk*.

Brulin, G., & Svensson, L. (2012). *Managing Sustainable Development Programmes*. Gower.

BSI, BS 6079-1:2010. (2010). *Guide to Project Management*. BSI.

BSI, BS EN ISO 9000:2000. (2000). *Quality Management Systems, Fundamentals and Vocabulary*. BSI.

BSI, BS EN ISO 9000:2000. (2000). *Quality Management Systems, Guidelines for Performance Improvement*. BSI.

BSI, BS EN ISO 9000:2003. (2003). *Quality Management Systems, Guidelines for Quality Management in Projects*. BSI.

BSI, BS EN ISO 900012008. (2008). *Quality Management Systems, Requirements*. BSI.

BSI, BS31100:2011. (2011). *Risk Management*. BSI.

BSI, PAS 2001:2001. (2001). *Knowledge Management*. BSI.

BSI, PD 7501:2003. (2003). *Managing Culture and Knowledge*. BSI.

BSI, PD 7502:2003. (2003). *Guide to Measurements in Knowledge Management*. BSI.

BSI, PD 7506:2005. (2005). *Linking Knowledge Management with other organizational Functions and Disciplines*. BSI.

Burke, R. (2011). *Advanced Project Management*. Burke Publishing.

Burke, R., & Barron, S. (2007). *Project Management Leadership*. Burke Publishing.

Camilleri, E. (2011). *Project Success*. Gower.

Cappels, T. (2003). *Financially Focused Project Management*. J. Ross.

Carroll, T. (2006). *Project Delivery in Business-as-Usual Organizations*. Gower.

Cavanagh, M. (2012). *Second Order Project Management*. Gower.

Chapman, C. B., & Ward, S. C. (2003). *Project Risk Management* (2nd ed.). Wiley.

Cialdini, R. B. (2008). *Influence, Science and Practice* (5th ed.). Pearson.

CIOB. (2009). *Code of Practice for Project Management for Construction and Development* (4th ed.). Wiley.

Cleden, D. (2009). *Managing Project Uncertainty*. Gower.

Cleden, D. (2011). *Bid Writing for Project Managers*. Gower.

Cleland, D. I. (2006). *Global Project Management Handbook*. McGraw-Hill.

Cleland, D. I. (2007). *Project Management: Strategic Design and Implementation*. McGraw-Hill.

CMMI. (November 2010). *CMMI® for Development, Version 1.3 CMMI-DEV, V1.3*. SEI.

Cockburn. (2002). *Alistair, Agile Software Development*. Addison-Wesley.

Cohn, M. (2006). *Agile Estimating and Planning*. Addison-Wesley.

Collett, P. (2003). *The Book of Tells*. Doubleday.

Collins, G. (2011). *Developing Agile Software Architecture Using Real-Option Analysis and Value Engineering*. SEI SEPG Conference, Dublin, June.

Collins, G. (2006). *Experience in Developing Metrics for Agile Projects Compatible with CMMI Best Practice*. SEI SEPG Conference, Amsterdam, 12–15 June.

Costin, A. A. (2008). *Managing Difficult Projects*. Butterworth-Heinemann.

Covey, S. (2004). *7 Habits of Highly Effective People*. Simon & Schuster.

Crane, A., & Matten, D. (2010). *Business Ethics* (3rd ed.). Oxford University Press.

Davies, R. H., & Davies, A. J. (2011). *Value Management*. Gower.

Davis, T., & Pharro, R. (2003). *The Relationship Manager*. Gower.

De Mascia, S. (2012). *Project Psychology*. Gower.

De Vito, J. A. (2011). *Human Communications* (12th ed.). Allyn & Bacon.

De Vito, J. A. (2012). *The Interpersonal Communications Book*. Pearson.

Dent, F. E., & Brent, M. (2006). *Influencing Skills and Techniques for Business Success*. Palgrave Macmillan.

El-Reedy, M. A. (2011). *Construction Management for Industrial Projects*. Wiley.

Elssamadisy, A. (2009). *Agile Adoption Patterns: A Roadmap to Organisational Success*. Addison-Wesley.

European Committee for Standardisation. (2004). *CWA 14924-4:2004, European Guide to Good Practice in Knowledge Management*. Guidelines for Measuring KM, CEN.

European Committee for Standardisation. (2004). *CWA 14924-5:2004, European Guide to Good Practice in Knowledge Management*. KM Terminology, CEN.

European Committee for Standardisation. (2012). *FprEN 16271:2012 (E), Value Management*. CEN.

Field, M., & Keller, L. (1998). *Project Management*. Cengage Learning EMEA.

Fisher, R., & Shapiro, D. (2007). *Building Agreement*. Random House.

Fisher, R., Ury, W., & Patton, B. (2003). *Getting to Yes*. Random House.

Forsberg, K., Mooz, H., & Cotterman, H. (2000). *Visualising Project Management*. Wiley.

Frigenti, E., & Comninos, D. (2002). *The Practice of Project Management*. Kogan Page.

Gambles, I. (2009). *Making the Business Case*. Gower.

Gardiner, P. D. (2005). *Project Management, a Strategic Planning Approach*. Palgrave Macmillan.

Garlick, A. (2007). *Estimating Risk*. Gower.

Gatti, S. (2007). *Project Finance in Theory and Practice*. Academic Press.

Goldsmith, L. (2005). *Project Management Accounting*. Wiley.

Goleman, D., Boyatzis, R., & McKee, A. (2002). *Primal Leadership*. Harvard Business School Press.

Goleman, D. (2007). *Social Intelligence*. Hutchinson.

Gordon, J., & Lockyer, K. (2005). *Project Management and Project Network Techniques* (7th ed.). Prentice-Hall.

Graham, N. (2010). *Project Management for Dummies*. Wiley.

Grimsey, D., & Lewis, M. K. (2007). *Public Private Partnerships*. Edward Elgar.

Hancock, D. (2010). *Tame, Messy and Wicked Risk Leadership*. Gower.

Harrison, F., & Lock, D. (2004). *Advanced Project Management*. Gower.

Haugan, G. T. (2002). *Effective Work Breakdown Structure*. Kogan Page.

Hersey, P. H., & Blanchard, K. H. (2012). *Management of Organizational Behaviour*. Prentice-Hall.

Highsmith, J. (2002). *Agile Software Development Ecosystems*. Addison-Wesley.

Highsmith, J. (2010). *Agile Project Management* (2nd ed.). Addison-Wesley.

Hillson, D. A. (2003). *Effective Opportunity Management for Projects*. Marcel Dekker.

Hillson, D. A., & Murray-Webster, R. (2005). *Understanding and Managing Risk Attitude*. Gower.

Hillson, D. (2009). *Managing Risk in Projects*. Gower.

Hopkinson, M. (2010). *The Project Risk Maturity Model*. Gower.

Hulett, D. (2009). *Practical Schedule Risk Analysis*. Gower.

Hulett, D. (2011). *Integrated Cost-Schedule Risk Analysis*. Gower.

ISO, ISO/FDIS 21500. (2012). *Guidance on Project Management*. ISO.

ISO, ISO 31000:2009. (2009). *Risk Management – Principles and Guidelines*. ISO.

Johnson, G., & Scholes, K. (2004). *Exploring Corporate Strategy*. Prentice-Hall.

Katzenbach, J. R., & Smith, D. K. (2005). *The Wisdom of Teams*. Harper Business.

Kemp. (2004). *Project Management Demystified*. McGraw-Hill.

Kerzner, H. (2004). *Advanced Project Management*. Wiley.

Kerzner, H. (2009). *Project Management*. Wiley.

Khan, F., & Parra, R. (2003). *Financing Large Projects*. Pearson Education Asia.

KOR, R., & Wijnen, G. (2007). *59 Checklists for Project and Programme Managers*. Gower.

Larman, C. (2004). *Agile and Iterative Development: A Manager's Guide*. Addison-Wesley.

Laudon, K. C. (2008). *Management Information Systems* (11th ed.). Prentice-Hall.

Leach, L. P. (2005). *Critical Chain Management*. Artech House.

Leffingwell, D. (2007). *Scaling Software Agility: Best Practices for Large Enterprises*. Addison-Wesley. (Chapter 21 with K. Schwaber).

Lewis, H. (2005). *Bids, Tenders and Proposals*. Kogan Page.

Lewis, J. P. (2003). *Project Leadership*. McGraw-Hill.

Lewis, J. P. (2005). *Project Planning, Scheduling and Control*. McGraw-Hill.

Linstead, S., Fulop, l, & Lilley, S. (2009). *Management and Organization*. Palgrave Macmillan.

Lock, D. (2007). *Project Management* (9th ed.). Gower.

Longdin, I. (2009). *Legal Aspects of Purchasing and Supply Chain Management* (3rd ed.). Liverpool Academic.

Mantel, S. J., et al. (2011). *Project Management in Practice* (4th ed.). Wiley.

Marchewka, J. T. (2012). *Information Technology Project Management with CD-ROM* (4th ed.). Wiley.

Margerison, C., & McKann, D. (2000). *Team Management: Practical New Approach*. Management Books.

Martin, N. (2010). *Project Politics*. Gower.

Maylor, H. (2005). *Project Management*. Prentice-Hall.

Meredith, J. R., & Mantel, S. J. (2010). *Project Management: A Managerial Approach*. Wiley.

Minter, M., & Szczepanek, T. (2009). *Images of Projects*. Gower.

Morris, P., Pinto, J. K., & Soderlund, J. (2011). *The Oxford Handbook on Project Management*. Oxford University Press.

Muller, R., & Turner, R. (2010). *Project-Oriented Leadership*. Gower.

Muller, R. (2009). *Project Governance*. Gower.

Neil, J., & Harpham, A. (2012). *Spirituality and Project Management*. Gower.

Newton, R. (2009). *The Project Manager*. Pearson.

Nickson, D. (2008). *The Bid Manager's Handbook*. Gower.

Nieto-Rodriguez, A. (2012). *The Focused Organization*. Gower.

Nokes, S. (2007). *The Definitive Guide to Project Management* (2nd ed.). Pearson.

O'Connell, F. (2010). *What You Need to Know about Project Management*. Wiley.

Oakes, G. (2008). *Project Reviews*. Assurance and Governance, Gower.

Obeng, E. (2002). *Perfect Projects*. Pentacle Works.

OGC. (2010). *An Executive Guide to Portfolio Management*. The Stationary Office.

OGC. (2009). *Directing Successful Projects with PRINCE 2*. The Stationary Office.

OGC. (2011). *Management of Portfolios*. The Stationary Office.

OGC. (2007). *Managing Successful Programmes*. The Stationary Office.

OGC. (2009). *Managing Successful Projects with PRINCE 2*. The Stationary Office.

OGC. (2008). *Portfolio, Programme and Project Offices*. The Stationary Office.

Pennypacker, J. S., & Dye, L. D. (2002). *Managing Multiple Projects*. Marcel Dekker.

Pennypecker, J. S., & Retna, S. (2009). *Project Portfolio Management*. Wiley.

Pidd, M. (2004). *Systems Modelling: Theory and Practice*. Wiley.

Pinto, J. (2010). *Project Management* (2nd ed.). Pearson.

PMI. (2008). *A Guide to the Project Management Body of Knowledge*. PMI.

PMI. (2006). *Government Extension to the PMBOK Guide*. PMI.

PMI. (2008). *Organizational Project Management Maturity Model (OPM3): Knowledge Foundation*. PMI.

PMI. (2003). *Organizational Project Management Maturity Model (OPM3): Overview*. PMI.

PMI. (2001). *PMI Practice Standard for Work Breakdown Structures*. PMI.

PMI. (2004). *Practice Standard for Configuration Management*. PMI.

PMI. (2009). *Practice Standard for Project Risk Management*. PMI.

PMI. (2007). *Project Manager Competency Development Framework*. PMI.

PMI. (2008). *The Standard for Portfolio Management*. PMI.

PMI. (2008). *The Standard for Program Management*. PMI.

Rad, P. F. (2001). *Project Estimating and Cost Management*, VA: Management Concepts.

Reiss, G., et al. (2006). *Gower Handbook of Programme Management*. Gower.

Reiss, G., et al. (2006). *The Gower Book of Programme Management*. Gower.

Reiss, G. (2006). *The Gower Handbook of Programme Management*. Gower.

Remington, K., & Pollack, J. (2008). *Tools for Complex Projects*. Gower.

Remington, K., & Pollack, J. (2007). *Tools for Complex Projects*. Gower.

Remington, K. (2011). *Leading Complex Projects*. Gower.

Robertson, S., & Robertson, J. (2006). *Mastering the Requirements Process*. Addison-Wesley.

Rodriguez, A. (2012). *Earned Value Management for Projects*. Gower.

Rogers, M. (2001). *Engineering Project Appraisal*. Blackwell Science.

Rose, K. (2005). *Project Quality Management*. J. Ross.

Royce, W. (2011). Measuring Agility and Architectural Integrity. *International Journal of Software Informatics*, *5*(3), 415–433.

Sant, T. (2004). *Persuasive Business Proposals*. Amacom.

Sanwal, A. (2007). *Optimising Corporate Portfolio Management*. Wiley.

Schwaber, K. (2004). *Agile Project Management*. Microsoft.

Schwindt, C. (2005). *Resource Allocation in Project Management*. Springer.

Senaratne, S., & Sexton, M. (2011). *Managing Change in Construction Projects*. Wiley.

Shalloway, A., Beaver, G., & Trott, J. R. (2010). *Lean-Agile Software Development: Achieveing Enterprise Agility*. Addison-Wesley.

Shermon, D. (Ed.), (2009). *Systems Cost Engineering*. Gower.

Shore, J. (2007). *The Art of Agile Development*. O'Reilly.

Shtub, A., Bard, J., & Globerson, S. (2004). *Project Management*. Pearson.

Sleeper. (2006). *Design for Six Sigma Statistics*. McGraw-Hill.

Smith, C. (2007). *Making Sense of Project Realities*. Gower.

Sutherland, J. (2005). *Future of Scrum: Support for Parallel Pipelining of Sprints in Complex Projects*. Denver, CO: Agile 2005 Conference.

Spencer, L. M., Spencer, S.M., Competence at Work, Wiley (1003)

Srevens, R., et al. (1998). *Systems Engineering*. Addison-Wesley.

Stenzel, C., & Stenzel, J. (2002). *Essentials of Cost Management*. Wiley.

Stutzke, R. (2005). *Software Project Estimation*. Addison-Wesley.

Sutt, J. (2011). *Manual of Construction Project Management*. Wiley.

Taylor, J. C. (2005). *Project Cost Estimating Tools, Techniques and Perspectives*. St. Lucie Press.

Taylor, P. (2011). *Leading Successful PMOs*. Gower.

Thacker, N. (2012). *Winning Your Bid*. Gower.

Thiry, M. (2010). *Programme Management*. Gower.

Thiry, M. (2012). *Project Based Organizations*. Gower.

Trevino, L., & Nelsom, K. (2010). *Managing Business Ethics* (5th ed.). Wiley.

Turner, J. R. (2008). *The Gower Handbook of Project Management* (4th ed.). Gower.

Turner, R. (2003). *Contracting for Project Management*. Gower.

Turner, R. (Ed.), (2008). *Gower Handbook of Project Management*. Gower.

Turner, R., & Wright, D. (2011). *The Commercial Management of Projects*. Ashgate.

Turner, J. R. (2003). *People in Project Management*. Gower.

Ursiny, T. (2003). *The Coward's Guide to Conflict*. Sourcebooks.

Venning, C. (2007). *Managing Portfolios of Change with MSP for Programmes and PRINCE2 for Projects*. The Stationary Office.

Ward, G. (2008). *The Project Manager's Guide to Purchasing*. Gower.

Ward, S., & Chapmen, C. (2011). *How to Manage Project Opportunity and Risk*. Wiley.

Wearne, S. H. (1993). *Principles of Engineering Organizations*. Thomas Telford Publications.

Weaver, R. G., & Farrell, J. D. (1997). *Managers as Facilitators*. Berret-Koehler.

Webb, A. (2003). *The Project Manager's Guide to Handling Risk*. Gower.

Wenger, E., McDermott, R., & Snyder, W. (2002). *Cultivating Communities of Practice*. Harvard Business School Press.

West, D. (2010). *Project Sponsorship*. Gower.

Williams, D., & Parr, T. (2006). *Enterprise Programme Management*. Palgrave Macmillan.

Winch, G. M. (2010). *Managing Construction Projects*. Blackwell.

Wright, D. (2004). *Law for Project Managers*. Gower.

Yang (2005). *Design for Six Sigma in Service*. McGraw-Hill.

Yescombe, E. (2002). *Principles of Project Finance*. Academic Press.

Yescombe, E. (2007). *Public-Private Partnerships*. Butterworth-Heinemann.

Young, T. L. (2003). *Handbook of Project Management*. Kogan Page.

Young, T. L. (2001). *Successful Project Management*. Kogan Page.

Words of Wisdom

Cash flow	More businesses go bust because of poor cash flow than low profitability
Cash flow	A bird in the hand is worth two in the bush
Claims	You need three things for a successful claim:
	(1) Good backup documentation
	(2) Good backup documentation
	(3) Good backup documentation
Communication	Listen carefully before talking; you have two ears and one mouth
Communication	Read twice, write once (for examinations)
Contract	A verbal contract isn't worth the paper it's (not) written on
Control	The wheel that squeals gets the grease
Delegation	Don't keep a dog and bark yourself
General	If it looks wrong, it probably is wrong
General	If it looks too good to be true, it probably is
General	A wise man learns from experience, a fool doesn't
Planning	The shortest distance between two points is a straight line
	The longest distance between two points is a shortcut
Planning	Forewarned is forearmed
Planning	It's later than you think
Procurement	If you don't inspect it arrives wrong
	If you don't expedite it arrives late
Quality	Quality is remembered long after the price is forgotten
Quality	A good product goes out
	A bad product comes back
Risk	Nothing ventured, nothing gained
Safety	It is better to be old than bold
Safety	Look before you leap

Index